W9-BGV-336

07/24
STRAND PRICE
$ 5 00

Man and Earth's Ecosystems

MAN AND EARTH'S ECOSYSTEMS

AN INTRODUCTION TO THE GEOGRAPHY OF HUMAN MODIFICATION OF THE EARTH

Charles F. Bennett, Jr.

University of California, Los Angeles

Cartography by
VINCENT G. MAZZUCCHELLI
California State University, Los Angeles

JOHN WILEY AND SONS, INC., PUBLISHERS
NEW YORK   LONDON   SYDNEY   TORONTO

Cover Photos:

Bunlap—Kal Muller; Woodfin Camp & Assoc.

El Paso Gas—Adam Woolfitt; Woodfin Camp & Assoc.

Dr. Eugene Joubert, Senior Research Officer, Ecology, Nature Conservation and Tourism Division, South West Africa Administration, Windhoek, S.W.A.

Copyright © 1975, by John Wiley & Sons, Inc.

All rights reserved. Published simultaneously in Canada.

No part of this book may be reproduced by any means, nor transmitted, nor translated into a machine language without the written permission of the publisher.

Library of Congress Cataloging in Publication Data

Bennett, Charles F.
  Man and Earth's ecosystems.

  Bibliography: p.
  Includes index.
  1. Man—Influence on nature. 2. Human ecology.
I. Title.
GF75. B46 1976 301.31 75-22330
ISBN 0-471-06638-9

Printed in the United States of America

10 9 8 7 6 5 4 3 2

To Carole

Preface

Man and Earth's Ecosystems: An Introduction to the Geography of Human Modification of the Earth is an attempt to combine history, physical geography, biogeography, cultural geography, and ecology into an introductory study of man's impact on the earth's ecosystems. This book is directed, primarily, to students who wish to gain a broad geographical and historical perspective on the environmental crises of our day.

I believe that success in understanding the nature of our human-induced environmental crises can be achieved only if we have a grasp of the world geography and history of human-caused environmental changes. It was this conviction that motivated me to design and introduce a course on "man and the earth's ecosystems" at the University of California, Los Angeles in 1966. The student response to the course was immediate, positive and sustained. Students are drawn from almost every traditional major on campus. This broad and sustained interest indicates that there are many students who wish to gain a geographical, ecological and historical understanding of man's influences on nature.

This book is a direct outgrowth of that course. The text not only reflects my interests but also the interests expressed by many of the thousands of students who have taken the course which is now the basic introduction to a new undergraduate major at UCLA called "Ecosystems Analysis and Conservation" and is offered by the geography department in addition to the regular geography major.

This book will have its greatest positive response among the professors and students whose environmental interests and concerns are directed toward achieving a broad understanding of geography, history and ecology as they relate to the genesis of contemporary human-induced environmental crises. The themes of this book form a counterpoise to some of the traditional approaches to understanding and solving human-induced environmental problems. I assert that such problems are basically *human* problems and that the human dimension must be studied and understood before satisfactory long-lasting changes will be achieved in the ways humans sometimes destructively use the world's ecosystems.

Comparatively few students will become professionals in the fields of environmental sciences and engineering. *All* students, however, throughout their lives will be required to make informed judgements on issues that affect their local, national and international environments. It is to this group that this book is principally directed. It is equally important, however, that present and prospective environmental scientists and engineers understand the themes presented in this book and incorporate them into their professional work.

The book is organized in the following way: Chapter 1 contains a resume of concepts and terms used in the main body of the text. Chapters 2 through 9 are devoted mainly to the geography and history of human-induced environmental changes and major emphasis is not, therefore, given to ecological problems of only recent origin. Chapters 2 through 7 are organized by major world regions and each region is presented in the order in which the genus *Homo* appears in the known fossil or cultural record. Thus, Chapter 2 is about subsaharan Africa and Chapter 7 is about Latin America. Chapter 8 is about islands and Chapter 9 contains discussions about the oceans.

At the end of the book is a list of suggested readings. These were chosen not only for their quality but with a view to the varied library facilities available to students.

Following Chapter 9 is a glossary of terms employed in the text. I have tried to include all the terms that ought to be de-fined. However, some readers may find it useful to consult more specialized sources, some of which are included in the lists of suggested readings.

Chapter 1 contains some maps showing physical and cultural aspects of geography and these maps should be referred to as the chapters are read. Other chapters have a limited number of maps showing materials that will assist the reader to understand some of the distributional aspects of topics discussed in the text. I have tried to provide the basic set of maps required for the book, but readers may wish to consult an atlas from time to time.

Charles F. Bennett

Los Angeles, California, 1975

Contents

Man and Earth's Ecosystems

Chapter One

Ecology and Human Geography: Some Fundamentals

1.1 Ecology

The word *ecology* was coined in 1869 by Ernst Haeckel. He employed two Greek words—*oikos,* meaning "a dwelling place," and *logos,* meaning "the study of." Thus, ecology is the study of how plants, animals, and their biological and physical environments interact and how they influence one another. Although ecological phenomena were observed long before Haeckel's time, the *scientific study* of organism-environmental relationships dates from the nineteenth century.

It has been chiefly since the publication of Darwin's monumental books that a scientific study of ecology has developed. No nation can be said to have a monopoly on such study, but much of the important pioneer work was done by scientists in Western Europe and the United States. Today, sophisticated ecological studies are being conducted in virtually all nations, and ecologists constitute an international body of scholars.

Among the concerns of ecologists since Darwin was the organization of the living and nonliving units under investigation. In this century, the concept of the *biological community* was advanced as the principal ecological unit. A *biological community* is a distinctive combination of plants and animals occupying an area. Usually attention was focused on the plants rather than the animals, and it became commonplace (and remains so) to speak of plant communities without concern for the animals present.

The community concept, though useful, tends to be over-focused on a *description* of the organisms occupying a given area rather than on the functioning of the living and nonliving units. The importance of understanding how biological communities *function* came to be one of the dominating concerns of ecologists. Thus, biological communities had to be characterized so that the focus of research would be clear from the terminology used to designate the objects of study.

The term that accomplished this is *ecosys-*

tem. This term was first used in the 1930s but did not assume its present importance until the 1960s. Today, the ecosystem concept is the principal organizing concept of ecology. Before we proceed to a definition and discussion of the functional aspects of ecosystems, it will be useful to indicate the *kinds* of ecosystems that it is possible to recognize.

Theoretically, there is almost no limit to the kinds of ecosystems that can be recognized. For example, we might consider the space under a grain of sand in a freshwater stream as an ecosystem. A drop of water taken from a rain barrel and placed under a microscope for observation is an ecosystem. Other examples of diminutive ecosystems include a fallen tree in a forest, a tree hole, water caught and retained externally by plants, an abandoned rodent burrow, and the area beneath a rock lying on a soil surface. In each case, the living and nonliving elements function together as a system. At the other end of the size scale is the entire planet Earth.

It can be seen that a list that attempted to include every kind of ecosystem in the world would be unwieldy. Thus, ecologists restrict themselves to a limited number of ecosystems when talking about them in a general way. The ecosystems most often employed when discussing large units of nonmarine earth space are the biomes. A *biome* is the largest terrestrial ecosystem (community) that is convenient to recognize. Although there is not yet a universally recognized classification of the world's biomes, the following list agrees fairly well with the varied terminologies in use.

Tropical Rainforest
Tropical Cloud Forest
Tropical Deciduous Forest
Tropical Scrub Forest
Tropical Savanna
Mangrove Forest
Desert
Mediterranean Woodland and Shrub

Middle Latitude Grassland
Middle Latitude Marshlands
Middle Latitude Deciduous Forest
Coniferous Forest
Tundra

As the list indicates, almost all the biomes are named for the principal physi-

Map 1.1

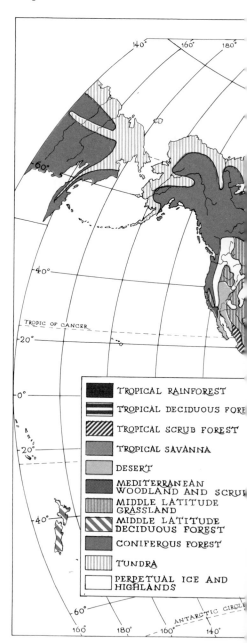

cal aspect of the vegetation (the *phytophysiognomic* aspect). All the biomes except the tropical cloud forest, mangrove forest, and middle latitude marshland, can be shown on maps of small scale, but size is not an accurate guide to a biome's importance (see Map 1.1, below).

The discussions in later chapters are focused chiefly on biomes, but other ecosystems will be introduced at appropriate locations. Here only two more units will be mentioned—the biotope and the ecotone. A *biotope* is the smallest ecosystem that is convenient to recognize. Examples of biotopes include a rotting log, a pool in a stream, a mudflat at low tide, and a hole in a tree. An *ecotone* is where two or more ecosystems come together and, frequently,

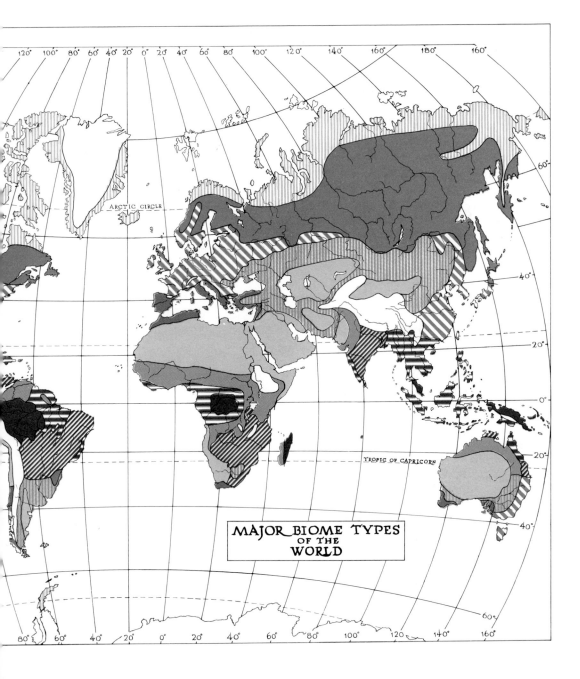

MAJOR BIOME TYPES
OF THE
WORLD

Figure 1.1 *Forest-grassland* ecotone *in the Amazon basin.*

intermingle with each other. One example is where a grassland biome intermingles with a forested biome (Fig. 1.1). Ecotones may involve small ecosystems and be local in distribution or large biomes and extend over hundreds of kilometers.

TROPICAL RAINFOREST

The tropical rainforest biome is considered by many scientists to be the most complex terrestrial ecosystem. This biome is located in the low latitudes, that is, within the tropics. Its greatest spread is within the Amazon Basin of South America and the Congo Basin in Africa, and it has a fairly major expression in tropical Asia.

Vegetation. The tropical rainforest varies from place to place in terms of species composition, but its principal physical aspect is similar wherever the biome occurs. The vegetation is dominated by a three-layer forest composed chiefly of numerous species of hardwood evergreen broadleaf trees. The tallest layer, or stratum, may reach almost a hundred meters. The trees in this layer do not form a continuous cover, that is, their crowns do not form a closed canopy. The next tallest layer may vary in height from thirty to fifty

meters; typically, their crowns meet and form a closed canopy. The third layer is more varied in height and may range from less than two meters to thirty meters. In this layer, there may be plants other than hardwood tree species such as palms and,

Figure 1.2 *Tropical rainforest at the eastern edge of the Andes (infrared photograph).*

when conditions are suitable, certain soft-wood tree species (Fig. 1.2).

The ground surface is seldom covered with low growth and is usually open except for the tree and palm trunks. In sharp contrast, the upper limbs of the trees are often covered on the upper surfaces with plants such as orchids and bromeliads, particularly in the American tropics. Neither orchids nor bromeliads are parasitical; they use the tree limbs only for support. They and other plants of similar habit are called *epiphytes* (Fig. 1.3).

Animals. In a tropical rainforest, most of the animals—particularly the vertebrate animals—live in the trees, probably because the ground, away from stream sides, offers only limited food sources. In addition to tree-dwelling (*arboreal*) vertebrates, there is always a vast host of invertebrates, of which many forms also occur at or near ground level.

Climate. The fundamental climatic aspects of the tropical rainforest biome are

Figure 1.3 *Epiphytes on a tree in the American tropics.*

no freezing temperatures and no drought season. There are seasonal phenomena in the tropical rainforest but they tend to be very subtle, especially when compared with middle and high latitude regions.

Soils. The soils within this biome tend to vary more than the physical aspects of the vegetation and the climate. However, some generalizations are possible. Because of the large amounts of rainfall, the more soluble elements in the soils tend to be leached away from the upper levels; this, in turn, tends to concentrate the relatively insoluble elements in the upper levels. Specifically, calcium tends to be leached away and iron, silica, and aluminum are concentrated in the upper levels. The oxides of iron impart a reddish tone to many of these soils. In addition to the leaching, the finer earth particles tend to be transported (*elluviated*) to greater depths, leaving behind a collection of relatively coarse earth particles.

Other features common to the red soils of this biome is that they usually lack humus and have low quantities of nitrogen and other chemical elements important to plant (particularly crop plant) growth.

Not all soils in this biome are as just described. Areas immediately adjacent to rivers, especially where the rivers enter the ocean, often have deep mineral-rich soils. These soils which are derived principally from water-deposited materials, are called *alluvial soils*. In addition, soils underlain by limestone may be comparatively rich in calcium and other minerals from the weathering limestone. And finally, some of the soils that have developed from volcanic ash or lava deposits are high in mineral content and possess great agricultural potential.

TROPICAL CLOUD FOREST

The tropical cloud forest ecosystem has a very limited area on the surface of the earth. Nevertheless it is very complex and has several unusual characteristics. The most outstanding of these characteristics is suggested by the biome's name, for this is an ecosystem that is enveloped by clouds and mists most of the time. Cloud forests are located at moderately high elevations in the tropics and in such positions where there is an almost constant supply of very moist air.

Cloud forests have not yet been the object of much study, but they are known to possess plants and animals that are adapted specifically to constant high atmospheric humidity. Inside such a forest, one constantly hears the drip of water, and underfoot, the ground is soft and yields easily to footsteps. Vegetation is abundant at the ground level and the trees are heavily burdened with epiphytes.

The world distribution of this ecosystem may best be thought of as a number of tropical islands on the land, because each unit is generally small in area and disjunctly distributed.

Figure 1.4 *Tropical deciduous forest in the wet season.*

TROPICAL DECIDUOUS FOREST

There are vast areas in the tropics that are covered by forest resembling the tropical rainforest, but which experience a period each year where there is a marked diminution of rainfall. During this dry season, many of the trees shed their leaves and some, in addition, bloom. This ecosystem may be conveniently termed tropical deciduous forest.

On the moister margins, this ecosystem may appear similar to the tropical rainforest but, as one examines the areas that have annual dry periods, the differences are easily seen. Chief among these differences is the *seasonal* loss of leaves, but another distinctive feature is the presence at ground level of often dense stands of plants. Also, this forest tends to be not so high as a tropical rainforest and may have only two well-defined tree strata. Often

there is a large quantity of epiphytes present (especially in the American tropics) and palms may be abundant. Moreover, there is a tendency toward unusual roots on the palms and some of the trees. Most of the tree species are hardwood, and species diversity is great (Fig. 1.4).

Animals. In general, there are more ground-level vertebrates than in a tropical rainforest, but there is often a large number and variety of arboreal vertebrates as well. Invertebrates also tend to be extremely numerous and play important ecological roles.

Climate. As mentioned above, this ecosystem has an annual period of reduced available water because of reduced precipitation. However, for any given place, the amount of the reduction varies from year to year. Contrary to widely stated opinion, annual and seasonal rainfall variability is often great and is an important ecological factor. During some years, a given location may *not* have a dry season at all; but in another year the "normal" dry season may be three times as long, and the wet season, when it arrives, may be much less wet than usual. Total annual precipitation for this ecosystem is highly variable, but usually it is at least 2500 mm a year. During the driest month(s) the rainfall may be less than 20 mm, but during the wettest month it may be 600 mm.

Soils. Even though there is a dry season during most years there is such a large quantity of precipitation that the soils tend to be as leached as the soils in the tropical rainforest. There is also a similarly wide range of major soil types, with some of them possessing large amounts of the chemical elements required for agriculture.

TROPICAL SCRUB FOREST

Vast areas of the low latitudes are covered with a scrubby forest consisting of low hardwood tree species, short shrubs with many thorns, some grass, and other types of low vegetation. This ecosystem is known by many vernacular names, and many subtypes are also recognized. The ecosystem occurs over a broad latitudinal range extending from relatively near the equator to at least 25 degrees poleward from the equator. In the lower latitudinal parts, many of the trees and shrubs are evergreen; but with increasing latitude, there are more deciduous species. One of the best examples of this is in northwestern Mexico. Here the forest becomes an expanse of leafless sticks during the long and very hot, dry season. But within a few days after the rainy season begins, the forest of sticks metamorphoses into a rankly verdant region.

Animals. Arboreal vertebrates are almost absent from this ecosystem, but there is usually a fairly rich assemblage of birds

and mammals that do not require trees for living space. Insect life is almost always abundant and extremely diverse. In those areas where there is a marked seasonality of rainfall, the maximum populations of active arthropod species coincides with the periods of rain.

Climate. As would be expected from the broad latitudinal distribution of this ecosystem, there is a broad range of temperature and precipitation. However, in all cases there is a pronounced dry season followed by a wet season during which precipitation is insufficient to maintain a multistrata forest. In the lower latitudes, freezing temperatures are unknown, but frosts are sometimes recorded for the more poleward limits. Rainfall variability is great throughout but tends to be greatest in the areas farthest from the equator. Air temperature is much more variable daily and seasonally than in the ecosystems described above. Again, the variability tends to be greatest in the areas farthest from the equator.

Soils. There is great variation in the soils. In some areas, they tend to be relatively high in calcium and nitrogen and also fairly deep. But there are great expanses of red soils that are low in plant nutrients and of soils that are thin and very stony.

TROPICAL SAVANNA

The designation *tropical savanna* is applied to a wide range of biological and physical complexes in the tropics. Originally the term was *sabana*, a word of Arawak Indian origin, and was applied to tropical areas in which grass was the dominant feature. With the passage of years, it came to refer to almost any low latitude situation that was not predominantly desert, on the one hand, or forest (or thick woodland), on the other. Today the term *tropical savanna*

Figure 1.5(a) *Grass savanna in Tanzania, Africa.*

Figure 1.5(b) *Acacia-grass savanna in Tanzania, Africa.*

Figure 1.5(c) *Palm-grass savanna in Africa.*

Figure 1.5(d) *Shrub-grass savanna in Africa.*

refers to an area in which grass is present either as a dominant plant (sometimes to the virtual exclusion of other seed plants) or as a conspicuous element, although there may be many shrubs, trees and/or palms. Thus, one encounters "thorn scrub savannas," "palm savannas," "short tree savannas," and "pure grass savannas," to name some of the possibilities (Fig. 1.5).

Animals. There is no close agreement between the major tropical savanna areas insofar as the presence and diversity of large vertebrates are concerned. In the African savannas, there is a richness of large animals not found anywhere else in the world, whereas in South America, savannas tend to be only sparsely populated by large wild animals. The differences are due principally to widely differing biological and geological histories. Invertebrate life is abundant, particularly during the times of the year when moisture and plant growth is greatest.

Climate. Although many textbooks have long included a category called "tropical savanna climate," such a climate really does not exist. That is, one cannot account for the vegetation of the savanna by direct reference to the climate, because the types of vegetation included under this biome category occur in areas with a total annual rainfall as low as 800 mm as well as in areas where the annual precipitation normally exceeds 2500 mm. It is true that most areas designated as tropical savannas experience an annual period of drought, but in the low latitudes there are areas of almost pure grass where in terms of precipitation alone there should be a stand of tropical deciduous forest. In short, this biome does not have a geographical distribution that coincides with precipitation or atmospheric temperature. There is often a certain degree of seasonal change in maximum and minimum daily air temperatures, but the range in a given locale is modest when compared with middle and high latitude weather (data collection) stations. Some places that the climatologists have classified as having a "tropical savanna climate" occasionally experience winter temperatures below 0°C.

Soils. Soils are so varied within this biome as to make a generalization difficult. However, the soils do tend to be heavy clays and contain only limited quantities of calcium and nitrogen. They also tend to be low in humus. During the dry seasons, the soils develop cracks; and during the period(s) of rain, they often tend to become waterlogged. In general, these soils can be farmed only with difficulty.

MANGROVE FOREST

One of the most singular biomes is the mangrove forest (often called mangrove). This tropical ecosystem is semiaquatic in that it is located in intertidal zones. The biome is very difficult to map because it almost always occurs as a thin band that may measure only a few meters in width. Thus the distribution of the biome as shown on Map 1.1 is merely a general guide to its latitudinal range.

As will be explained in greater detail, one of the generalizations of ecologists is that mature ecosystems are always diverse in their species composition. This generalization is usually well supported by the evidence, but the mangrove is an important exception. To illustrate we might imagine a strip of mangrove along a coast in southeast Asia. On the adjacent mainland there may be a mature and hence species diverse tropical deciduous forest. That is, within one square kilometer there might be over one hundred species of plants. But in the mangrove, there may be no more than a dozen plant species in one square kilometer.

The mangrove is dominated by tree species, most of which have hard, dense wood. Their roots are always in a wet soil that is repeatedly covered and exposed by the tides. The very position of a particular mangrove is subject to change, because this biome is a land builder—trapping sediments around the root systems and thus extending the area of land seaward. However, the process is slow and, in many instances, there is a rough equilibrium between the building of land by the mangrove forest and the taking away of that land by the restless tides (Fig. 1.6).

Animals. Vertebrate animal life tends to vary greatly in species diversity and total numbers from one part of the world to another. The mangroves of tropical Asia appear to be comparatively rich in bird and reptile species and invertebrates are abundant. In contrast, *some* of the mangrove forests of the American tropics appear almost devoid of vertebrate life (except during nesting season for some aquatic birds). Again, some mangroves in the American tropics have few insects and particularly few mosquitoes but other mangroves have mosquito populations so great as to make it almost impossible for a person to remain more than a short time within the biome.

Climate. There is no climate that is typical of this ecosystem. Mangroves occur along coasts that are backed by deserts and along coasts that are backed by rainforests. They require comparatively high atmospheric temperatures and do not appear to occur poleward of where freezing air temperatures occur. In general, the tallest mangrove forests occur in the lower latitudes and the shortest occur at the poleward limits. These phenomena probably relate chiefly to climatic conditions.

Soils. The soils on which mangrove forests grow are waterlogged and saline. They generally have little or no vertical structure and are in a more or less continuous state of being added to or reduced, according to the whims of tides. The phenomenon for which the ecosystem is best known is that of land building. In some parts of the world, mangrove forests have been responsible for adding thousands of square kilometers to the total area of dry land.

DESERT

The desert biome occurs in a wide latitudinal range. Some deserts are located very near the equator, such as along the west coast of South America; other deserts are found well within the middle latitudes.

Figure 1.6 *Mangrove forest at low tide.*

One problem with describing the distribution of this ecosystem is that there are varying opinions as to the climatic criteria that correctly define a desert. It is almost impossible to assign a single value such as average annual precipitation to define a desert. In fact, using the arithmetic mean (average) to describe desert precipitation is highly misleading because of the high degree of *variability* of annual amounts of precipitation in deserts. Thus, the following is the best description insofar as moisture is concerned: a desert is a place where the *potential* evapotranspiration exceeds precipitation. *Evapotranspiration* is the general evaporation plus the water given up by plants (transpiration). The result is an environment in which there is usually a scarcity of water except very locally or very intermittently such as just after a shower.

Desert vegetation reflects the limitations imposed by the lack of water. Many desert plants are adapted to drought; they are referred to collectively as *xerophytes*. Many desert plants escape the rigors of drought by remaining in seed form except for the very brief periods when there is soil moisture. Then they grow, flower, and seed within a matter of days. Usually there is an incomplete vegetation cover and much bare land visible (Fig. 1.7).

Animals. Animal life in deserts tends to fall into two large groups—animals that are *physiologically* adapted to the environment and those that are *behaviorally* adapted. Of course, many of the animals belong to both groups. An example of physiological adaptation is seen in the kangaroo rat (*Dipodomys*), which is found in deserts in North America. This rodent can live a healthy life without ever taking a drink of water. Moreover, this animal is also behaviorally adapted to the desert because it is strictly nocturnal insofar as above-ground activity is concerned. Thus it avoids the heat of the day.

Another behavioral adaptation, aside from being nocturnal, is being crepuscular (active at dusk and/or dawn). Many mammals and some birds fall within both of these categories.

It is often thought that deserts support few animals. On the contrary, mammals, birds, reptiles, and arthropods can be abundant—one must simply be in the right place(s) at the right time(s) to observe them.

Climate. Most deserts are very hot during some parts of the year; that is, daytime shade temperatures may exceed 50°C on some occasions. Typically, there is a marked change in the air temperature at

Figure 1.7 *Desert vegetation in the southwestern United States.*

night, even during the hottest parts of the year, with the lowest temperature being below 18°C. There are a few deserts along parts of the west coast of South America and the coast of Southwest Africa that are extremely foggy much of the time.

Because soil is usually an excellent insulating mechanism, the air temperatures of burrows tend to be nearly isothermal and significantly lower in the daytime than the prevailing air temperatures above ground.

Soils. Although highly varied in some respects, desert soils tend to have some common features—thin development, large stone content, high mineral content, and low organic content. Locally, however, the soils may be very sandy without any stones in evidence. And some desert soils possess excellent attributes for agriculture and can be so used if adequate water is available.

MEDITERRANEAN WOODLAND AND SHRUB

This Mediterranean biome is located disjunctly in six major units: the "rim" lands of the Mediterranean Sea; the extreme south part of Africa; extreme southwest and extreme southeast Australia; central Chile; central and southern California.

Although there is little taxonomic (species) similarity between one unit and another the principal physical aspects of the vegetation are strikingly similar in all the units. The woody vegetation often possesses adaptations for long periods of summer drought and some of the shrubs are adapted to fire. In these cases, the seeds of the plants appear to require the passage of fire (and the heat thereby generated) in order to break dormancy. Moreover, some plants are adapted to make rapid growth recovery after a fire has passed. Indeed, it appears certain that fire—from one source or another—plays a fundamental role in the ecology of this biome.

Most of the trees and shrubs are broadleaf evergreen species, although this is sometimes difficult to determine near the end of the long, hot summer. In the Mediterranean Sea and California units, oaks (*Quercus*) are abundant, but they are lacking in the Southern Hemisphere units because oaks are not native there. However, one can sometimes observe trees that *look* like oaks although they may not even be distantly related to true oaks—further evidence of the physical similarity of the units.

A substantial part of the shrubs are flowering. Some of these and other plants are sources of certain oils used by the perfume and other industries.

Grass may be abundant in parts of this biome, especially in the first few years following a burn of the woody vegetation (Fig. 1.8).

Animals. The animal life of this biome is not particularly distinctive. Most of the species found in the biome are also found in adjacent biomes. On the other hand, mammals, birds, and reptiles may often be abundant, both in number and kinds.

Climate. Perhaps the most distinctive feature of this biome is the climate. Typically, the biome receives most of its precipitation in the form of rain, concentrated during the winter months. The summer

Figure 1.8 *Mediterranean shrub vegetation, Spain.*

period, which may exceed six months, may be one of complete drought, or there may be short-lived but violent periods of intermittent rainfall. Reliability of precipitation is only a little greater than for desert biomes. An examination of the precipitation record for almost any weather recording station shows that there are wide variances between the amounts of precipitation that occur during the driest and wettest years.

Soils. Hillsides tend to have thin soils, often with large quantities of stones. Soils in valley bottoms or other relatively level areas are frequently deep and excellent for agriculture—the chief limiting factor often being the lack of water for irrigation. However, on most slopes, there are great expanses of heavy clay soils that are poorly drained during the periods of rainfall and that tend to harden and crack during the summer. Such soils are often extremely difficult to cultivate. In all areas, salinity may be high and it is usual for soils to test higher than 7.0 on the pH scale.

MIDDLE LATITUDE GRASSLAND

Extensive areas of grasslands are found in the middle latitude areas of both Northern and Southern Hemispheres. The species composition of these areas are varied and are distinctive in that one does not find, for example, native North American grasses in the grasslands of Australia—unless introduced by humans.

There is disagreement respecting the origin of these grasslands. Although the conventional belief is that they exist in more or less direct response to climatic and soil factors, there is evidence to show that their *total areal extent* may reflect the actions of humans over varying periods of time in the past (Fig. 1.9).

Animals. Most of the areas in this biome have (or had) a fauna that appears to have been long adapted to the special

vegetation conditions. In North America, the fauna included a great host of animals ranging from small rodents to vast herds of bison (buffalo). Birds, reptiles, and amphibians were also abundant. Arthropods were often seasonally abundant, and there were even seasonal population explosions of some arthropods, such as grasshoppers (Insects: order Orthoptera).

This biome was also well developed in Argentina and Uruguay. In those areas, too, there was a rich and abundant fauna, although the large mammals present in North America were lacking. In Eurasia, the biome is not known for faunal richness—at least not on the scale of North America. But this may be because of major human-induced changes in times long past. In Australia, the biome is populated by a wide range of animals, of which the marsupial mammals are perhaps the most distinctive.

Climate. Although long accepted that there is a grassland climate, more recent investigation has shown that such a climate may not exist. There are certainly wide variations in the amounts of rain and snow, and trees can and do grow unaided in some parts of the biome.

Precipitation, typically, is distributed through most of the year. Snow may occur (abundantly) in the winter (but not in all areas of the biome), and rain, in the other seasons. Rainfall reliability is often low and drought is sometimes a significant ecological factor—especially for humans attempting to farm. It is within this biome that most of the major wheatgrowing areas are presently located. However, some parts of the biome almost never experience a serious drought and it is here that crops such as maize, which require copious and reliable quantities of rainfall, are found.

Soils. In general, soils in this biome tend to have considerable depth, well-developed vertical structures (soil horizons), and large quantities of mineral and

Figure 1.9(a) *Pampa of Argentina.*

Figure 1.9(b) *Prairie remnant in the Middlewest.*

Figure 1.9(c) *Steppes of the U.S.S.R.*

Figure 1.9(d) *Pastures in the south of Australia.*

organic matter; they lend themselves admirably to plow agriculture. Indeed, some of the finest soils—insofar as agriculture is concerned—occur within this biome.

MIDDLE LATITUDE MARSHLANDS

Although frequently too diminutive to map at other than topographic (large) scale, this biome is considered to be among the most productive of the world's ecosystems. As its name indicates, the vegetation is adapted to either perpetual flooding of the roots or, in the case of some plants, to periodically high water tables (Fig. 1.10).

Animals. Marshlands offer habitat to an impressive array of animals. Many species of birds and mammals are found in this biome; it is one of the most important areas for nesting birds. Reptiles, especially certain snakes, may be abundant, and freshwater marshes are often the optimal habitat for many amphibians. In both saltwater and freshwater marshes, the arthropod life can be exceedingly abundant—especially during the summer months.

Figure 1.10 *Freshwater marsh.*

Marshes are often of critical importance to the life cycles of certain fish and crustaceans. To some crustaceans, salt marshes are critical; their loss can have a profound negative effect on, for example, shrimp populations in adjacent marine ecosystems.

Climate. There is no high correlation between climate and marshes. In general, however, freshwater marshes tend to be most numerous in areas of abundant precipitation and least numerous in arid regions.

Soils. The chief characteristic of the soil(s) is that it is almost permanently waterlogged. Thus, the kinds of plants that can grow are greatly restricted.

MIDDLE LATITUDE DECIDUOUS FOREST

An outstanding aspect of this biome is its comparatively large number of broadleaf deciduous tree species. Many of the trees are familiar, for example, oaks, *Quercus;* maple, *Acer;* beech, *Fagus;* chestnut, *Castanea;* tulip tree, *Liriodendron.* There are also many under-story shrub species, most of which are also deciduous (Fig. 1.11). In some areas one encounters needle-leaf trees, especially pines, *Pinus* (Map 1.1).

This biome is best developed in the Northern Hemisphere, where it once covered vast areas. Today, its distribution in the Northern Hemisphere is much less than formerly, due to human actions.

Animals. In general, this biome supports a rich assemblage of birds and mammals and, in some areas, of reptiles and amphibians as well. Because hard winters are frequently a feature of these areas, the animals adapt by seasonal migration, hibernation, or physical attributes that permit their being active throughout the year.

Arthropod life is usually abundant but is

Figure 1.11 *Middle latitude deciduous forest in the summer.*

active mostly during the warmer months. Indeed, summer in this biome may appear to rival the tropics in terms of the abundance and variety of insects and other arthropods present.

Climate. There are marked seasonal aspects to the climate in this biome. The rule is that there are cold winters and warm to hot summers. As expected, the climatically milder parts of the biome are near the equatorward limits and the colder parts near the poleward limits. Precipitation is usually distributed throughout the year, with winter rain and snow and summer rain. In the summer months, atmospheric humidity is often high and may make one feel that he is near the equator.

Soils. In general, the soils in this biome are acidic and contain large amounts of organic material, especially humus. The soils tend to erode easily if the tree cover is removed, and if great care is not taken, the soil is washed away. Although varying

greatly from place to place, the soils of this biome are attractive for agriculture.

CONIFEROUS FOREST

As the name of this biome indicates, the chief vegetation is coniferous (cone-bearing) trees. The biome has its best development in the Northern Hemisphere, particularly between 45° and 70°N. However, extensive areas also occur southward, such as the coniferous forests of the southeastern United States, Mexico, parts of Central America, and the Caribbean (Fig. 1.12).

Although there are many kinds of conifers, two genera tend to dominate many of the regions occupied by this biome, namely, pines, *Pinus,* and firs, *Abies.* Both of these genera are confined to the Northern Hemisphere, unless moved elsewhere through human agency. (To be exact, *Pinus* does just get south of the equator in tropical Asia, but this is so limited as

Figure 1.12 *Pine forest in the southeastern United States.*

to be only a curiosity in plant geography.)

In the Southern Hemisphere, the biome is poorly developed and has few species. One of the most important genera is *Podocarpus,* which occurs in South America, Africa, New Zealand, and many other places, including middle latitude Asia.

Animals. Because this biome is distributed over such an enormous area and has such a broad latitudinal development, the animal life is extraordinarily varied. In the boreal (northern) parts of the biome, many mammals possessing economically useful furs and large mammals such as deer (*Cervus*) and moose (*Alces*) occur. Bird life may be abundant during the summer nesting season, and arthropods may also be extremely abundant during the summer months.

Climate. The climate ranges from short warm summers and long, cold winters in the northern part of the Northern Hemisphere to essentially tropical rainforest conditions in Central America. One of the more interesting aspects of this biome is the broad range of climates in which it occurs.

Soils. Soils range from poorly drained, acid soils to excessively drained, calcic soils. Soil humus may be abundant in the higher latitudes and sparse in the low latitudes. The soils may be fairly deep or they may be thin and extremely stony.

TUNDRA

With the tundra, we arrive at the poleward limits of terrestrial vegetation. The essential vegetational aspect is a lack of trees— even shrubs may be completely absent. Sedges and rushes tend to predominate and grasses are often present, although they are seldom, if ever, really abundant. Very little tundra occurs in the Southern Hemisphere (see Map 1.1). Some authorities speak of tundra occurring in the high elevations of tropical mountains, but these distributions are too limited to be mapped at the scale of the map presented in this book (Fig. 1.13).

Animals. In the boreal tundra, animal life, vertebrate and invertebrate, can be abundant. This is particularly true during the short summer period when great flocks of migratory waterbirds are nesting

Figure 1.13 *Tundra on Alaska's north slope.*

and when there is a population explosion of insects (particularly flies and mosquitoes). But even in the winter, there is no lack of animals. Caribou (*Rangifer*), muskox (*Ovibos*), wolf (*Canis*), fox (*Alopex*), and snowshoe hare (*Lepus*) are among the resident mammals that remain active throughout the winters.

Climate. Short, cool summers followed by long, extremely cold winters typify the climate of this biome. Contrary to much popular opinion, precipitation is generally modest and the atmosphere is usually comparatively dry because of the low waterholding capacity of extremely cold air. Any precipitation that occurs is usually snow.

Soils. The soils are usually thin and poorly drained. Sometimes there is a frozen subsoil, the so-called permafrost phenomenon. With only a few exceptions, tundra soils are not adapted to agriculture.

THE FUNCTIONING OF ECOSYSTEMS

A major revolution has recently occurred in ecology. Ecosystems (communities) ceased being studied from a natural history point of view and began to be analyzed in terms of energy flow and nutrient cycling. Moreover, great emphasis has come to be placed on the measurement of biological productivity. Thus, a functional and integrative approach has largely supplanted the descriptive approach.

Thus far, we have been using the term *ecosystem* without presenting a detailed definition of it. In the definition given here the key elements are indicated by numbers in parentheses and are discussed in greater detail below.

"An ecosystem, or ecological system, is considered to be a unit of biological organization made up of all the organisms in a given area (that is,

community) interacting with the physical environment so that a flow of energy leads to characteristic trophic structure and material cycles within the system" (E. P. Odum, Science, *Vol. 164, p. 262, 1969*).

The term *ecosystem* was first proposed by the botanist A. G. Tansley in 1935, although the *concept* that living and nonliving entities in a given area interact as systems is older than Tansley's statement. However, it was not until after World War II that the ecosystem concept gained general acceptance among professional biologists and other scientists. The major reasons for this lag must include the lack of computors, the lack of general acceptance of information theory (cybernetics), the feeling that the ecosystem concept was too difficult in practical terms to permit the ecosystem to be the principal object of scientific investigation, and a general unwillingness to abandon the traditional descriptive approach to the study of ecology.

The term *community* is synonymous with ecosystem. "Ecosystem" is a way of saying "community" in *functional terms.*

Physical environment is employed here to designate the nonliving elements of an ecosystem. These elements are numerous but generally may be reduced to energy and nutrients.

Energy is the capacity for performing work and in the earth's ecosystems, originates almost entirely from the sun. Energy cannot be created nor destroyed although

Figure 1.14 *Trophic levels in a shallow experimental pond.*

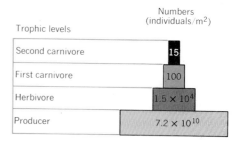

Trophic levels	Numbers (individuals/m²)
Second carnivore	15
First carnivore	100
Herbivore	1.5×10^4
Producer	7.2×10^{10}

it can be transformed. The amount of energy in the universe at any given moment is constant (the First Law of Thermodynamics). Energy, as it moves through a system, becomes increasingly diffuse with more and more being lost to heat, that is, entropy (the Second Law of Thermodynamics). Thus there must be a regular solar energy input into the earth's ecosystems in order to counteract the constant tendency toward entropy.

The term used to refer to the way an ecosystem is organized to maximize the use of an always limited supply of energy is *trophic structure* (Fig. 1.14).

The trophic structure is largely dependent on the organization of energy flow (movement through the ecosystem) but energy alone will not maintain the living components of an ecosystem. There is also the need for *materials* to provide the added food required by plants and animals. In major contrast with energy, a given unit of material, such as an atom of calcium, may (theoretically) move or cycle round about

an ecosystem for as long as there is some form of life present that is capable of taking calcium as a part of its life processes. A given unit of energy (e.g., one calorie) may *persist* in a given ecosystem for a long time and may even become fossilized, as in the case of coal. But that energy unit has only been detained in its journey toward entropy and cannot be repeatedly reused or recycled as can the atom of calcium. Thus, materials are permanent residents of an ecosystem (until or unless transported into other systems), whereas a unit of energy has a residence time limited to how and when it is consumed and ultimately will be dissipated in the form of heat (Fig. 1.15).

The basic tendency of an ecosystem is to move toward becoming so well organized that the use of available materials and energy is accomplished in the most efficient manner possible. To achieve this end, the living organisms in a given place tend to pass progressively through a series of stages in which each succeeding stage is more stable than any of its predecessors.

Figure 1.15 *A natural, balanced ecosystem.*

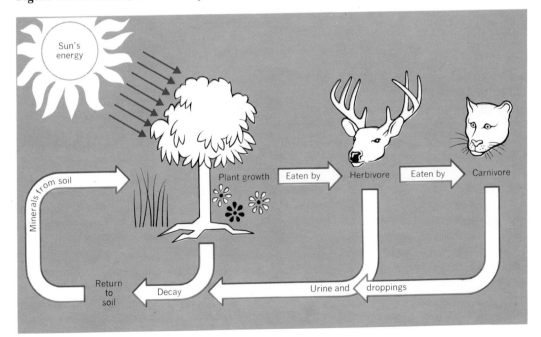

This orderly process is termed *ecological succession,* and the end product of the succession is termed a *mature ecosystem* (or a *climax ecosystem*). Each recognizable part of a given ecological succession is called a *seral stage* and the entire succession is termed a *sere.* Generally, two major categories of ecological succession are recognized—primary and secondary succession.

Primary ecological succession is not often observed, because it occurs where a community has not previously existed or where there is a substitution of a terrestrial succession for a previously existing aquatic ecosystem or *vice versa.* One of the more dramatic examples of primary succession that has been observed occurred on the slopes of the volcano Krakatoa, on whose slopes all (or almost all) life was destroyed during a major eruption in the nineteenth century. The comparative rapidity with which succession took place surprised many observers.

Secondary ecological succession refers to the community and environmental changes that occur when an existing ecosystem is altered. This may be the result of such human disturbances as cutting a forest for timber or plowing a prairie for crop production or such natural disturbances as the destruction of a forest by high winds or burns or the shifting of along-shore marine currents that alter temperature and/or substrate conditions. On any given site—for example, a farm, secondary ecological succession may be repeated hundreds of times. Although there are many factors other than human that induce secondary succession, those caused by humans are of great significance and tend to dominate over much of the earth. It is one of the growing concerns of modern ecologists that such human-induced successions be well understood (Fig. 1.16).

The efficiency of an ecosystem's organization of energy flow and materials cycling can be measured—in fact this is increasingly the basic focus of research in ecosystems analysis. The variables most often measured and examined are *gross primary productivity, net primary productivity,* and *net community productivity.*

The common feature of these variables is *productivity.* Primary productivity is the *key* element in the functioning of an ecosystem. Its meaning is simple but we do not yet understand the chemical and physical complexities involved. Primary production is *photosynthesis*—the process whereby solar

Figure 1.16 *Secondary ecological succession on a farm. Cultivated field in foreground, middle area returning to brush, forest (probably secondary) in background.*

energy is fixed and stored by plants in a form that is then available for food for the plant and other organisms. This food consists of energy and materials.

Gross primary productivity refers to the total, or gross, product of photosynthesis in a given ecosystem. However, not all this production is stored, because the plants require part of it for their own life processes (respiration). Therefore, *net primary productivity* is that part of the production that remains stored in plants after the plant's needs are met, and *net community productivity* is the part of the production that remains after the other ecosystem members, herbivores principally, have consumed their required part(s) of the net primary production. Ecosystems, then, are populated by producers (chiefly green plants) and consumers, includ-

ing the decomposers (e.g., bacteria).

Consumers in an ecosystem are organized in response to the first and second laws of thermodynamics. Thus, utilization of primary production is *structured;* the structures are commonly referred to as *trophic levels.* The first trophic level is that of the *autotrophs* (self-nourished), which are mostly green plants. The rest of the levels are occupied by *heterotrophs* (other-nourished). They are ordered as follows: the second level is occupied by herbivores (both vertebrate and invertebrate); the third level, by primary carnivores; and the fourth level, by secondary carnivores (those that feed on other carnivores). One sometimes reads or hears of *decomposers* as a separate element in an ecosystem. Actually it is more accurate to view decomposition as a part of the functions of all the liv-

Figure 1.17 *Simplified representation of the food web of a small meadow pond. Arrows point in the direction of energy flow.*

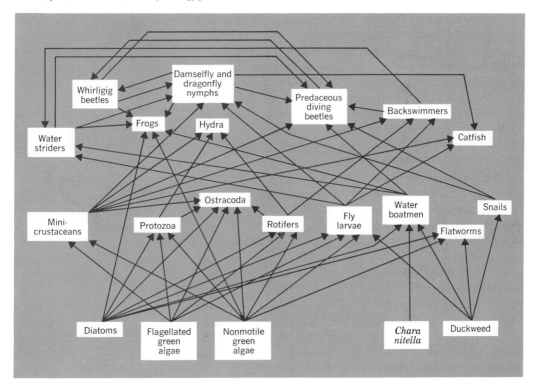

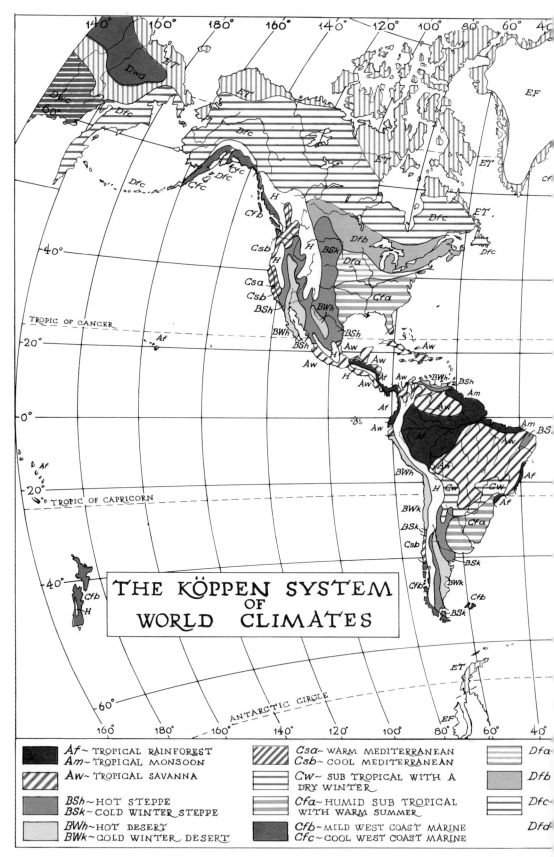

THE KÖPPEN SYSTEM OF WORLD CLIMATES

Af ~ TROPICAL RAINFOREST	*Csa* ~ WARM MEDITERRANEAN	*Dfa*
Am ~ TROPICAL MONSOON	*Csb* ~ COOL MEDITERRANEAN	
Aw ~ TROPICAL SAVANNA	*Cw* ~ SUB TROPICAL WITH A DRY WINTER	*Dfb*
BSh ~ HOT STEPPE	*Cfa* ~ HUMID SUB TROPICAL WITH WARM SUMMER	*Dfc*
BSk ~ COLD WINTER STEPPE		
BWh ~ HOT DESERT	*Cfb* ~ MILD WEST COAST MARINE	*Dfd*
BWk ~ COLD WINTER DESERT	*Cfc* ~ COOL WEST COAST MARINE	

Map 1.2

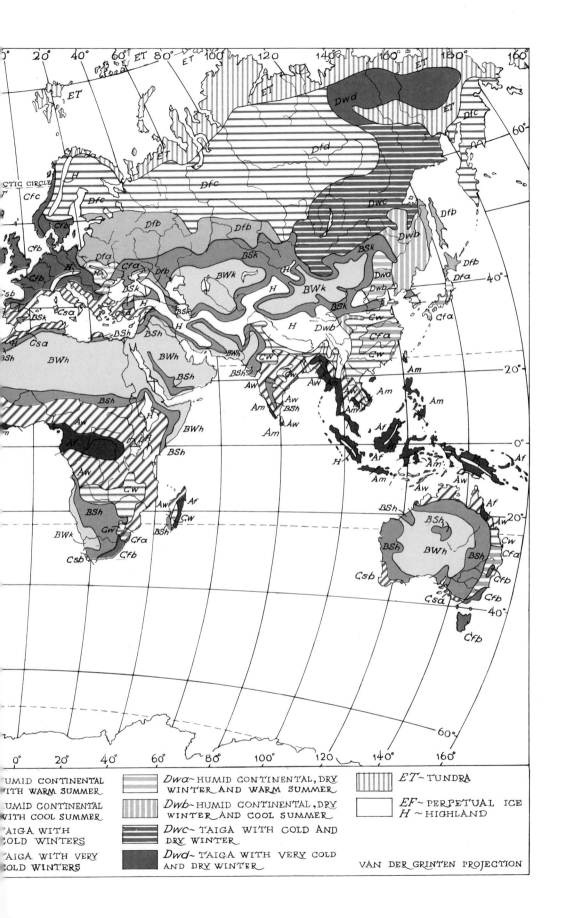

UMID CONTINENTAL
ITH WARM SUMMER

UMID CONTINENTAL
ITH COOL SUMMER

AIGA WITH
OLD WINTERS

AIGA WITH VERY
OLD WINTERS

Dwa~ HUMID CONTINENTAL, DRY WINTER AND WARM SUMMER

Dwb~ HUMID CONTINENTAL, DRY WINTER AND COOL SUMMER

Dwc~ TAIGA WITH COLD AND DRY WINTER

Dwd~ TAIGA WITH VERY COLD AND DRY WINTER

ET~ TUNDRA

EF~ PERPETUAL ICE
H ~ HIGHLAND

VAN DER GRINTEN PROJECTION

ing and some of the nonliving elements of any ecosystem.

There is an upper limit (probably five) to the number of trophic levels possible in any ecosystem—a limit imposed by the second law of thermodynamics. (It must be remembered that as energy is passed from one level to another, efficiency decreases because of the dissipation of heat, that is, entropy.)

The movement of food in an ecosystem, particularly in a mature ecosystem, is complex. If charted or mapped, its movement within and between trophic levels in all but the simplest of immature ecosystems forms a weblike pattern; thus it is customary to describe the movement of food as the *food web*. Limited but linked parts of the food web of an ecosystem are called *food chains* (Fig. 1.17).

Another attribute of ecosystems is *relative diversity*, or, as it is often stated, *diversity*. Diversity, when applied to ecological systems, usually refers to the total number of plant and animal *taxa* (species and subspecies) present in a given ecosystem at a given time.

In general, the taxonomic diversity in any given ecosystem tends to increase as the ecosystem succession moves toward maturity, with maximum diversity occurring when maturity is achieved and maintained. This is not to say that the diversity of an individual group of animals, such as birds, is always greatest under mature conditions. Sometimes maximum diversity may occur when the ecosystem is in a state of major disturbance. But the overall diversity tends to be greatest when the ecosystem is mature.

The tendency toward increased diversity as an ecosystem matures appears to be closely related to what is commonly called the balance of nature. Diversity tends to protect an ecosystem against the shock effects of environmental disruptions and thus is a major factor in maintaining the equilib-

rium that is a prime characteristic of a mature ecosystem.

Other measures of diversity aside from taxonomic such as biochemistry, have been suggested. However, for the time being taxonomic diversity appears to be the most useful measure available.

Thus far we have described the major terrestrial ecosystems (biomes) and some of the structures and functions of an ecosystem. However, to understand the basics of how ecosystems function, we must understand macroclimates and bioclimates, photoperiod, plant geography, and animal geography.

Macroclimates and Bioclimates. The term *climate* refers to the atmospheric conditions (often given in terms of averages) that prevail in an area over comparatively long periods of time. The climate of any large area is called a *macroclimate*. There is no useful way to accurately designate the smallest area that would be considered a macroclimate. In general, when people talk about climate they are talking about macroclimates and have in mind two variables—air temperature and atmospheric precipitation.

Figure 1.18 *Relation of sun's rays to earth.*

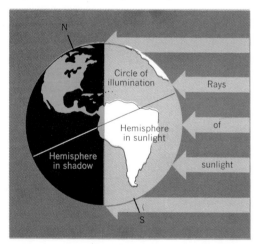

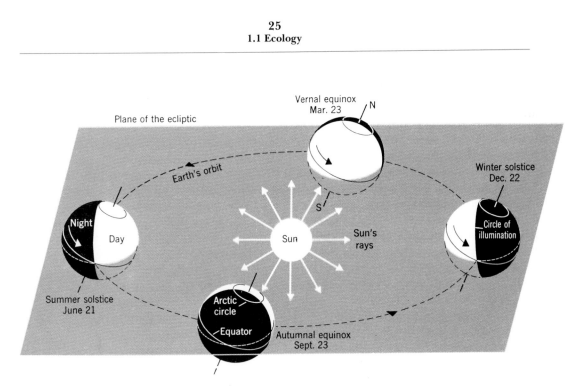

Figure 1.19 *The seasons; equinoxes and solstices.*

It is also useful to recognize that portion of the atmosphere that is in direct contact with living organisms. This is called the *bioclimate* (or *microclimate*). There is an almost infinite number of bioclimates because each living nonaquatic organism has one of its own. However, it is possible to make generalizations about bioclimates, just as one can about macroclimates.

Macroclimates have been described for the entire planet and there are several climatic classifications in use. However, all these classifications suffer from various shortcomings, chiefly the result of generalizing from limited weather data. Nevertheless, a general classification of macroclimates is of considerable value for it permits one to visualize global climate patterns. One must not try, however, to discover close fits between the distribution of one or more macroclimates and the geographical distribution of plants and animals.

One climatic classification that has proved useful was devised by Vladimir Köppen in the early part of this century. The Köppen Climatic Classification was originally an attempt to link climate to the major world distribution patterns of vegetation. However, only if one maintains a very open mind can such relationships be seen from the Köppen map (Map 1.2 on pages 22–23). The map contains a brief key to the climates, the main elements of which are air temperature and atmospheric precipitation. These elements are given in mean (average) values. However, plant and animal distribution are not controlled by such values. The principal use of the Köppen classification and others that rely on mean values to describe the behavior of climatic parameters is that they offer a general idea about the world geographic distribution of air temperature and atmospheric precipitation over lengthy periods of time.

Photoperiod. A major attribute of climate is seasonality. No latitude is entirely

free from seasonal changes in climate, although such changes are least at 0° latitude
and greatest at 90° latitude. The phenomenon of seasonality is caused by the following earth-sun relationships: (1) the earth's
vertical axis is inclined 23.5° from the per-
pendicular of the plane of the earth's orbit
around the sun; (2) the earth rotates on its
axis producing alternate periods of daylight and darkness; and (3) the earth
moves in an orbit about the sun and
requires approximately 365.25 days to
complete each orbital trip (Fig. 1.18). Incli-
nation of the earth's axis, rotation, and orbital motion produce seasonality on this
planet. Thus, there are seasons when atmospheric precipitation occurs and seasons when it does not or is much reduced.
And there are seasonal differences in the
quantity of solar energy received on any
place on earth, with the least change occurring in the lower latitudes and the greatest
change occurring in the higher latitudes.
Thus, there is also a seasonal aspect to the
reception of light. Light seasonality is almost nonexistent at the equator but becomes more and more evident as one proceeds poleward in either the Northern or
Southern Hemispheres.

Light on earth comes mostly from the
sun and is important to the functioning of
an ecosystem. Much of the importance of
light relates to primary production, that is,
to photosynthetic activity. This being the
case, all living elements in an ecosystem
must be adjusted to the seasonal behavior
of light. The term applied to the length of
daylight is *photoperiod*. At 0° latitude, the
photoperiod is essentially constant the year
round, with twelve hours of light and
twelve hours of darkness. However, as one
moves away from the equator, the seasonal
behavior of the photoperiod becomes
quite noticeable. As you can see from Fig.
1.19, the most dramatic aspects of photoperiod occur at the poles. In the summer, the sun does not set—the so-called

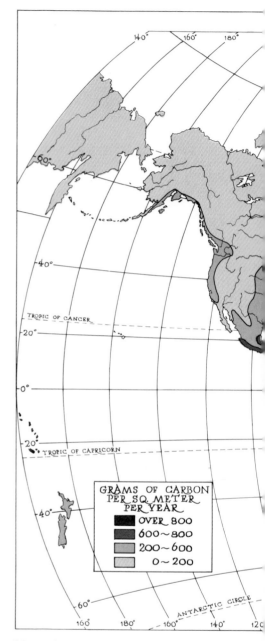

Map 1.3

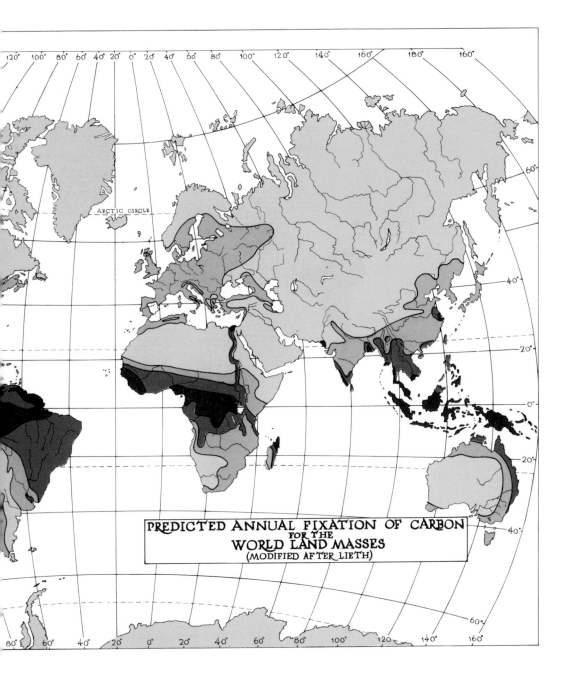

PREDICTED ANNUAL FIXATION OF CARBON
FOR THE
WORLD LAND MASSES
(MODIFIED AFTER LIETH)

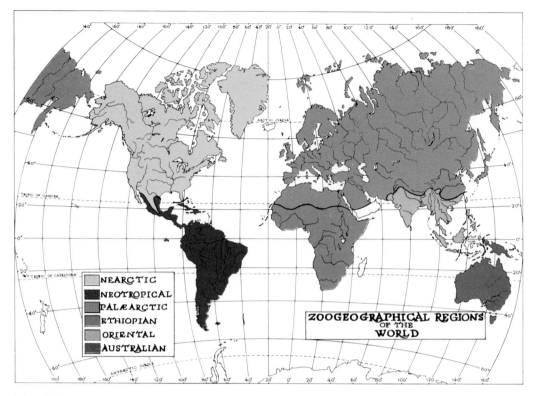

Map 1.4

midnight-sun effect—and in the winter, the sun scarcely makes its appearance at all. Photoperiod is a particularly influential environmental element because it is a stimulant to seasonal changes in the behavior of plants and animals, such as leaf fall, flowering, mating, and migration.

At present, research is underway that aims to formulate concepts relating world climate to primary productivity in ecosystems. An important step in this direction is exemplified by the work of Dr. Helmut Lieth. Dr. Lieth has produced a map showing the *predicted* annual fixation of carbon (Map 1.3 on page 26). If you compare Lieth's map with that of Köppen, you will note that there are some general similarities insofar as the major regional units are concerned.

Plant Geography. The geographical distribution of vegetation must reflect to some degree the geographical distribution of the ecological requirements (energy, food) of the plants. In addition, the distribution of vegetation reflects the dispersal by wind, water, animals, man, continental drift, and complex evolutionary histories.

The geography of plants may be examined from a number of viewpoints but only two will be mentioned here: the floristic distributions (species, genera, families, and so on) and the phytophysiognomic distributions (the *physical* attributes of vegetation, such as grassland and shrub land). In general, it is easier to relate the phytophysiognomic distributions of plants than the floristic distributions to macroclimate patterns. Map 1.1 (on page 3) is in many respects also a map of the world's major phytophysiognomic units.

Animal Geography. As with plants, the geographical distribution of animals is sensitive to the geographical distribution of food and energy. Animal distributions also reflect the role of dispersals over the long and short run of time. And because animals are directly or indirectly dependent on plants for food (and often for shelter as well), animal distributions are related to plant distributions.

It has long been the custom, when taking a global view of animal geography, to direct attention mainly to the distribution of terrestrial vertebrates and only secondarily to that of aquatic vertebrates. This is because the first of the modern zoogeographers were chiefly interested in the terrestrial vertebrates. As one can see from Map 1.4 on page 28, there are six major zoogeographical regions in the world and two lesser ones, Madagascar and New Zealand. These two have traditionally been included with the Ethiopian and Australian regions, respectively, but they are distinct entities and should be set apart.

1.2 Human Geography

The biological origins of *Homo sapiens* are coming to be better known each year, in part because of the discovery of major fossil beds in Africa—in Olduvai Gorge, Tanzania, and Lake Rudolph, Kenya. It appears that humans underwent most of their early evolution in Africa, although there is some disagreement among physical anthropologists as to when *Homo* actually appeared there. Some hold that a species of man was present in East Africa almost 2.7 million years ago. Other authorities, however, believe that the fossil material in question represents not *Homo* but another genus of primates. Part of the argument hinges on the interpretation given to certain stones found with the primate bones. Some authorities believe that the stones represent *worked* stone and that this is *prima facie* evidence that humans fashioned them. Others are not certain that the stones were worked and, even if

they were, that this would be sufficient to characterize the primate fossil material as being human.

In the distant past, *Homo* drifted out of Africa into Eurasia, probably moving through relatively warm and moist low latitude areas. The timing and routing of the earliest dispersals are unknown, but some of the oldest dates we have for *Homo* in Eurasia are for material (human and animal fossils) found in cave deposits near Peking, China. The general consensus of opinion is that this material dates from at least 400,000 years BP (before the present). The fossil human at Peking was long classified as *Sinanthropus pekinensis,* but modern scientists have designated Peking Man as *Homo erectus,* a species that appears to have been widespread in parts of eastern Asia including Java, where a fossil man was long classified as *Pithecanthropus erectus*—the so-called Java Ape Man. Today, the

Java man is viewed as also belonging to *Homo erectus,* because the Javan materials have been dated as of similar antiquity as those from Peking. This suggests that by that time, *Homo* had adapted itself to an extremely broad spectrum of ecological conditions, ranging from tropical low latitude ecosystems to ecosystems in the higher middle latitudes, where long and cold winters were probably the rule.

Homo erectus finally arrived at the evolutionary stage we recognize as modern man, *Homo sapiens,* perhaps 50,000 to 60,000 years ago. (One puzzle still not adequately solved is how such a geographically far-flung primate could have evolved in such a way as to result in only one species.)

We have seen that *Homo* dispersed from Africa into Eurasia at some as yet undetermined date but one that must be at least 400,000 years BP. Humans appear to have been confined to Africa and continental Eurasia for a long time, because they do not appear in the archeological records of Australia or the New World until recently. *Homo sapiens* somehow crossed the water gaps between the Asian mainland and Australia by about 30,000 years BP.

The date of the entry of *Homo sapiens* into the New World is still the object of intense study and some disagreement. As new archeological sites are discovered and older data reevaluated, the date keeps being pushed back. There is fairly strong agreement that man was in the New World by 23,000 to 25,000 years ago and one group of scholars argues that the date was 50,000 years ago. New archeological finds and new interpretations of the available data will surely alter the current prevailing opinions.

A major means of dating the campsites and living sites of humans in the prehistoric period is radioactive carbon dating techniques. Often referred to simply as carbon dating, the method is based on the fact that a radioactive isotope of carbon, carbon 14 (^{14}C), which is produced continuously in the atmosphere, is taken up by all living organisms. Upon death, the carbon 14 atoms contained in the tissues of an organism decay at a known rate. Thus it is possible to measure the rate of carbon 14 emission from a piece of charcoal or other organic sample and determine the time when that material ceased to live. Thus, because all living things take up carbon, any preserved tissue, plant or animal, may be tested by this method. The technique, first developed by Willard Libby, has undergone many refinements. There are several problems associated with this dating method, but the most important for our discussion is that with the increasing age of a specimen, there is an increase in the probable error respecting the date derived from it. For most purposes, material older than 60,000 years cannot be successfully dated by the ^{14}C techniques.

Thus, we saw that all the major land masses had been discovered by at least 23,000 to 25,000 years ago. Later, many more dispersals took place: people entered the Pacific Ocean region, beginning at least 1000 years ago; people bearing Islamic cultures went to North Africa, Iberia, and the islands of southeast Asia; Europeans dispersed to the New World, beginning in the late fifteenth century; Africans were dispersed to the New World, beginning in the sixteenth century and extending into the nineteenth century; and Europeans dispersed to certain Pacific Islands and Australia. These movements are major aspects of the human ecological story and will be noted again in the chapters to follow.

HUMAN ECONOMIES—A SURVEY

In this section, we shall survey the more important types of human economies, from the simplest to the most complex. However, because of space limitations, broad generalizations will be used. The main purpose is to present a general history of the types of economies that have occupied human groups from the earliest times to the present. *It is largely through understanding human economic activity that one can grasp the sources of most of the human activities that have led to modifications of the earth's ecosystems.*

Specialized and Unspecialized Gathering Economies. Most available evidence indicates that in its earliest phases human economy differed little from that of many of the other animals with which humans were in contact. No food was produced and a wide range of plant and animal tissue provided the required sustenance. The basic differences lay in the human use of handfashioned tools of wood or stone. The *use* of tools, per se, is not limited to humans and it might be argued that the *fashioning* of tools is not limited either, because it has been shown that other primates occasionally fashion tools. But in terms of ecological dependence on crafted tools,

humans have been set apart from other animals since early times. It is largely through the manufacture and use of tools according to a predetermined plan that humankind has forged ahead of other animal species in terms of world ecological dominance.

The basic primordial human economy can be variously characterized, but *unspecialized gathering* is probably as accurate a term as any now in use. The designation *unspecialized* does not means always simple and unsophisticated. Rather, it means that the unspecialized gatherers typically relied on a wide array of plant and animal foods. Some of the societies belonging to this economy possessed only limited tools and technologies, whereas some groups had many complex techniques of exploiting and preparing the wild foods available to them (Map 1.5, below).

During the early periods of this economy, the total world human population

Map 1.5

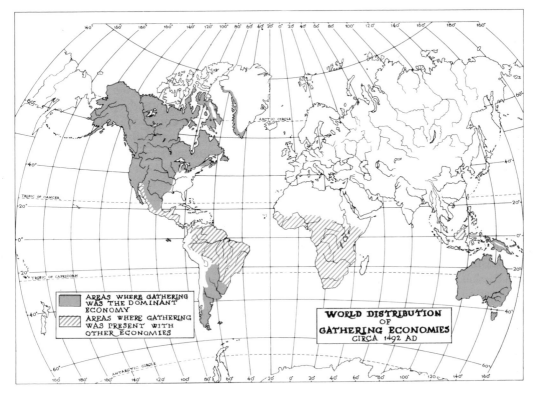

AREAS WHERE GATHERING WAS THE DOMINANT ECONOMY

AREAS WHERE GATHERING WAS PRESENT WITH OTHER ECONOMIES

WORLD DISTRIBUTION OF GATHERING ECONOMIES CIRCA 1492 AD

was small and confined to Africa and parts of Eurasia. The basic social unit may have consisted of the nuclear family (mother, father, children) and a grandparent or two or it may have consisted of two or three genetically related nuclear families. Social cooperation among families probably developed early, because it affords clear adaptive advantages for hunting and other gathering activity. These social units probably wandered over fairly definite territories, which they may have defended against other bands.

Lacking fire, all food had to be consumed raw. In areas where ecosystems were rich in larger mammals and other vertebrates, hunters may have been able to supply most of the protein needs of their people. But in some ecological situations, perhaps in the majority of them, returns from hunting were sometimes unreliable. Periods of meat abundance would be followed by periods of scarcity. The earliest human hunters may have positioned themselves at the edge of herds in order to kill a newborn animal or an old member too feeble to keep up with the herd. People may also have helped themselves to the kills of other predators or have been carrion feeders.

If data obtained from the few surviving societies of gatherers are indicative of past conditions, we may assume that although the men were the hunters, it was the women and children who played the major role in keeping food available. While the men may have been out chasing a large herbivore (not always with success), it was the women's and children's everyday grubbing beneath stones, prying apart of rotten wood for insect larvae, harvesting of shellfish on shorelines, and gathering of seeds that may have kept human life and culture alive over the first several hundred thousand years of human tenure on this earth.

As noted above, in the early days humans did not know how to make or control fire. *The acquisition of fire by humans rep-*

resents one of the most important technological milestones of history and one of the most significant ecological events in the past half million or so years. It is therefore extraordinary that we still do not know where and under what circumstances mankind first learned to use fire. A major reason for this lack of knowledge is that we cannot determine whether the charcoal we find is unquestionably the product of human-set fire. The mere presence of charcoal in association with human artifacts is not proof that the charcoal came from a fire kindled by a human. However, if the site where the artifacts are found were within a cave where a natural fire could not easily have entered, the case for a human origin would be quite strong. Although this still would not tell us whether humans kindled the fire, it would indicate that humans were able to obtain fire from a natural source and had learned how to transport it and use it in a controlled manner.

The oldest charcoal of presumed human causation thus far reported was found in some caves near Peking. If the dating of the associated archeological and paleontological materials is correct, the people living there approximately 400,000 years BP knew the use of fire. It might be noted here that Peking is at almost 40° north latitude. This suggests that in the winter heat may have been a requirement. However, we have not yet discovered similar old charcoal-laden sites elsewhere.

It should not be deduced from the Peking date that fire was in general use among humans by 400,000 BP. The use of fire may have been only local at that time and for a long time afterward.

This leads us to wonder why humans troubled with fire at all if they did not need it. It may be of significance that fire has long played a role in religious ritual. Sacred fires appear in a number of religions, and the idea of the "eternal flame" continues to evoke responses from some sophisticated persons today. Could it be the apparent mystical nature of fire that first caught the attention of humans?

Whatever the original reason for taming fire, eventually it was employed for heating, cooking, and hunting. Fire was used to drive animals so that a small number of hunters could provide the greatest meat yield. Some writers have suggested that the *fire drive* was a standard Paleolithic technique and that it provided, in conjunction with weapons, much of the meat. There is no question that the fire drive was widely distributed in ecosystems where the technique could be profitably employed, but great regional variations in the frequency of its use seems to have been the rule.

Whenever and wherever people first loosed fire into the ecosystems, those times and places marked, collectively, a great turning point in the ecology of this planet. From then on, no matter how small their population might be, people were to exert an ecological impact on the world probably unequalled by any other animal species during the past two million years.

Gathering economies differed greatly in respect to the degree to which the people had developed technologies and the degree to which there was specialization. Specialized gathering appears never to have been common, and some of the cultural units that have been so designated were actually highly dependent on the generalized gathering of women and children. One outstanding example of specialization is the Polar Eskimos who inhabited parts of the west coast of Greenland. In this difficult environment, the people fashioned an economy that depended overwhelmingly on the walrus, *Odobenus rossmarus*. In addition to that large marine mammal, they also hunted seals (*Phocidae*) of several species and small whales (order Cetacea) and took birds and eggs during the short arctic summer. The diet of these people was almost entirely meat, and most of the raw materials for garments and weapons also came from marine mammals, particularly the walrus.

It is not possible to make generalizations about the size of the populations of gathering societies. The population of a particular society varied, in part because of the ecological carrying capacity of the ecosystem(s) it lived in. Carrying capacity was determined largely by the availability of food and technology, and both were highly variable. Other variables were infant mortality, adult mortality, food taboos, and other cultural factors (including religion) that might have reduced the carrying capacity of a given habitat. Some writers believe that recurrent famine was a basic part of the history of gathering societies. Although there were many factors that prevented societies from growing too numerous for the productive capacity of their ecosystems, the unreliability and annual variability of air temperature and precipitation must have led occasionally to sharply lessened wild food production. And this may have led, in turn, to human deaths through starvation.

As recently as five hundred years ago gathering societies were still widely distributed in the world but today they have all but disappeared.

Plant and Animal Domestication. Before we can discuss agriculture, the next type of economy, we must review some of the basic aspects of plant and animal domestication, because these phenomena were mixed with agriculture.

One of humankind's most significant modifications of the earth was and is plant and animal domestication.

Many definitions have been offered for *domestication,* but most of them have serious shortcomings. One element often included in the definitions of animal domestication is that the animal must breed in captivity. However, such animals as the Indian elephant (*Elephas maximus*), which has long been used by human societies in Asia, breeds only rarely in captivity. With the growth of the science of genetics, there has been an increasing tendency to include genetic criteria in the definitions of domesticated plants and animals. Thus, one finds definitions to the effect that a domesticated plant or animal is one whose reproduction is under human control. The oldest definition and the one still most often encountered is that a domesticated plant or animal is one that man has brought under his control for some *economic* purpose such as for food or for raw material, e.g., wool. There is also a widespread idea, which is false, that plant and animal domestication began and ended in the dim past before records were kept.

Most of the definitions break down because they focus on biological (genetic) criteria rather than cultural criteria. The first and most important fact regarding plant or animal domestication is that the organism has been taken by a person or persons into some sort of human cultural context. And this occurs long before there can be any observable genetic changes in the organism. Domestication *may* and often does involve genetic changes, but this should not be the fundamental criterion for defining it. The definition offered here for plant or animal domestication is as follows: *A domesticated plant or animal is one that people have consciously taken into the cultural context of their living space for any purpose whatever.*

This definition clearly excludes those organisms that are found in association with people but that are not there as a result of human desire, such as *commensals* (e.g., house flies) and *parasites* (e.g., bedbugs, human body lice, and fleas).

We do not presently have any idea as to when man began to keep plants or animals. Pet keeping appears to be a very old culture trait, so this may be the oldest reason for keeping animals. We usually view plant domestication as utilitarian—to use plants for food—but there is reason to suspect that the earliest human interest in plants focused on magic, not on food. In some parts of the world, there appears to be an old set of traditions respecting plants kept for their flowers—not just for the color but because the flower form and/or color suggested some natural revered phenomenon, such as the sun. Part of the present-day flower culture of Mexico stems from this.

There has long been a question as to where the first plants used for food were brought under cultivation. The conventional view is that grasses were the first important crop plants—wheat and barley in the Old World and maize (corn) in the New World. However, another group of scholars argues that root and tuber crops may have been developed earlier in lowland low latitude ecosystems. There are some good arguments supporting this latter hypothesis, but there is little supporting archeological evidence. Plant materials do not preserve well in such ecosystems, whereas seeds do preserve well in many arid regions.

Similar questions surround the origins of animal domestication, particularly the domestication of the animals that currently play major economic roles in human societies—cattle, dogs, sheep, pigs, goats, horses, fowl.

Many persons assume that these animals were domesticated to provide the products or services we now derive from them. Thus, cattle are said to have been domesticated for meat and hides and plow pulling, sheep for wool, dogs for watch animals, chickens for eggs, and so on. Very likely there were many reasons why these and other animals were first domesticated. Some reasons may have been pet keeping, religious ritual (perhaps the single most important reason), companionship, and, occasionally, some economic good. Certainly in recent times, animals have been domesticated for a wide variety of reasons—fruit flies, for genetic studies; guppies, for aquarium fish; shell parakeets, for their color and personalities as pets; and

Figure 1.20(a) *Burning the plant debris on a newly cleared plot for slash and burn cultivation.*

Figure 1.20(b) *Slash and burn cultivators planting a crop in the Amazon basin.*

zoo animals, for observation. In fact, the present may be the greatest period of animal domestication yet attempted by man.

Once having acquired a certain number (from one to many) of domesticated plants and/or animals, humans were in a position to develop the next stage of their economic development—agriculture.

Agriculture. In this era of space travel, it can be easily forgotten that the most important technical achievement of mankind after the acquisition of fire was the inven-

tion of agriculture, that is, the invention of human-controlled food production. Our urban life styles tend to obscure the fact that we are still tied to agriculture for our food. Moreover, it is through the development and geographical dispersal of agriculture that man has caused and is causing many of the ecological changes we shall examine.

The term *agriculture* includes many activities and systems and several important subsystems. The systems discussed in this book are slash and burn cultivation (or shifting cultivation), sedentary hoe cultivation, plow cultivation, plantation agriculture, simple animal husbandry, and industrial animal husbandry.

The oldest organized system of agriculture that we know of is *slash and burn cultivation* (Fig. 1.20). The times and places this system was developed in the Old and New Worlds (assuming independent inventions in both hemispheres) are not known. Undoubtedly a long period of experimentation with planting seeds or roots and tubers preceded the emergence of an agricultural *system*. (This period is sometimes called *incipient agriculture*.) We do not know whether the original experimentation was performed to develop a system of producing food or as an aspect of ritual. In any case, a crop-growing system did emerge and was dispersed widely over the world. Although now slash and burn cultivation is much reduced in areal distribution, it is still one of the most important of the world's agricultural systems.

In this form of agriculture, the cultivated fields are moved at fairly short intervals (one to four years is common), with the previously cultivated areas *fallowed* (rested) for periods that may exceed decades. The houses of the farmers may also be shifted after brief intervals or may remain in place for protracted periods. The timing of the field shifts depends on a number of variables. It has long been held that soil exhaustion is the most important cause of the shifts, but, in many cases, other factors may be more important. For

example, in those instances where the farmer has no tool capable of rooting up grass, he often finds it impossible after two or three years of farming to keep grasses from invading his cultivated fields. Thus, the easiest solution is to move on. Another factor involves religion. It has been shown that entire villages of shifting cultivators can be induced to move by a shaman (medicine man) if he has seen some portent of danger.

The basic techniques and tools employed in slash and burn cultivation are relatively simple. A suitable tree- or brush-covered site is cleared by cutting the vegetation with a knife (once made of stone but now almost always of steel), and the debris is allowed to dry where it falls. When sufficiently dry, the trash is ignited and burned and then the field is planted. These operations are usually timed to take

Map 1.6

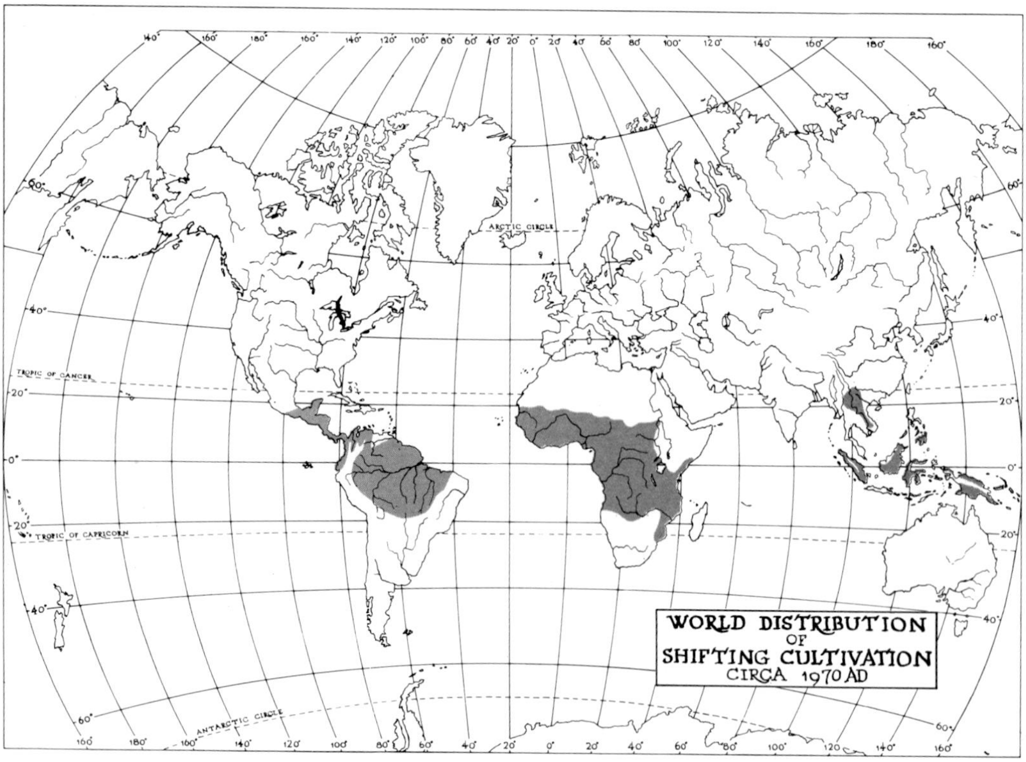

WORLD DISTRIBUTION
OF
SHIFTING CULTIVATION
CIRCA 1970 AD

advantage of rain at the beginning of the growing season. Of course the farmer is at the mercy of natural forces—late rains may require replanting or rains that come too early may make it difficult to burn the cut debris.

Planting is usually accomplished with the aid of only a heavy pole sharpened at one end called a *digging stick*. This is poked into the ground and seed is dropped in the hole, covered with soil, and tamped into place by the farmer's foot. If root crops are planted, more effort is required to make the holes large enough, but although demanding of hard work for a short time, the operation is simple. In some areas where slash and burn cultivation is practiced, a hoe or hoe-like tool is employed instead of a digging stick. This allows the farmer a greater degree of control over weeds and enables him to work the soil more than is possible with the digging stick. However, the use of a hoe does not indicate that the system is a part of sedentary hoe culture.

Under slash and burn cultivation, it is not usual to set the plants out in straight rows, but the plots can be remarkably productive. They have several natural advantages for societies possessing limited technical means, chief among which is that almost any slope, however steep, can be cultivated.

That this system of agriculture has produced and continues to produce profound ecological changes cannot be denied. However, it is an ecologically sound system, providing the human population does not grow too great for the carrying capacity of an area.

A number of attempts have been made to determine the "normal" population

Map 1.7

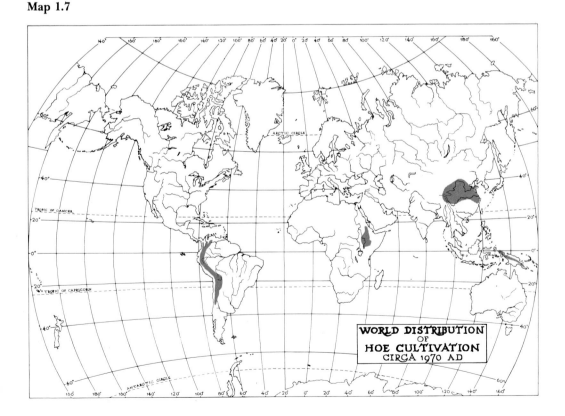

WORLD DISTRIBUTION
OF
HOE CULTIVATION
CIRCA 1970 AD

Figure 1.21(a) *Plow cultivation in the Nile valley.*

Figure 1.21(b) *Plow cultivation in Oaxaca, Mexico.*

densities for this agricultural system but none has been satisfactory because the range is too great. The system itself has many variants and it is practiced under many different ecological conditions, which means that population sizes also vary greatly. The most frequent error when trying to determine normal population densities is asserting a too-low carrying capacity for the system. Actually, under optimal conditions, a surprisingly large number of people can be supported.

The world geographical distribution of slash and burn cultivation, although reduced in recent centuries, is still geographically widespread, particularly in the low latitudes (Map 1.6 on page 36). So common is this system in the low latitudes that many

persons have regarded it as a tropical system of agriculture unsuited to the middle latitudes. This is not true. Slash and burn cultivation once had a very important development in the middle latitudes, particularly in parts of Eurasia.

The next clearly defined development in agriculture is *sedentary hoe cultivation*. As its name suggests, this system is notable for the use of a hoe, or some other digging and soilworking tool, and for being fixed in place. Actually, fields *are* rotated (cultivated and fallowed) but not in the somewhat random manner of slash and burn cultivation. Also, slash and burn cultivators tend to move their dwelling sites but in sedentary hoe cultivation areas, relatively large, fixed villages work and fallow the same pieces of land for decades. However,

Figure 1.21(c) *Plow cultivation in the Middle West.*

Figure 1.21(d) *Plow cultivation in Sri Lanka.*

there is a considerable degree of similarity between the two systems because large, fixed villages have been associated with shifting cultivation also. The chief differences are the intensity of land use and the techniques employed. In sedentary hoe cultivation, fertilizer is in common use, regular crop rotation may be followed, field sites are fixed (and there is some form of enduring land ownership), irrigation *may* be employed, and there is typically a high input of human labor with relatively high yields. However, animal power is never a significant part of the system and is frequently lacking altogether.

This system tends to develop "civilized" landscapes, because unwanted vegetation is kept in check and the field patterns, transport routes, village sites, and special tree plantings are highly visible.

This system is generally productive of high caloric yields per unit of land cultivated, which provides a much greater base for human population densities than even the maximal conditions encountered under shifting cultivation. As is the case with shifting cultivation, the area given over to this type of agriculture is on the wane (Map 1.7 on page 37).

Plow cultivation does not appear to have evolved from hoe cultivation. It seems to have first developed, rather locally, in the Nile Valley, which represents one of those technological leaps that have enormous impact on human affairs. However, this was not at all apparent where the plow and (presumably) plow cultivation was invented. On the contrary, the oldest plows appear to be so crude and inefficient as to have scarcely been a significant factor in food production. These early types of plows (hundreds of thousands are still in use) do little more than scratch the top of the soil. In fact, they are so inefficient that it has been suggested that the first plows were less agricultural tools than phallic symbols used in a ritualistic fertilizing of the earth.

However inefficient the prototype plows, the idea contained the ingredients that led to what is often characterized as *the agricultural revolution*. This came not in the Nile Valley but in parts of Western Europe, where plow cultivation underwent major technological developments. These developments had the effect, on the one hand, of greatly increasing food production and, on the other, of dramatically modifying the vegetated landscapes. Plow cultivation came to include not only a complex array of special types of plows but also many associated tools, such as discs and harrows, which are dragged over plowed land to reduce the larger soil clumps to smaller bits, and machines to plant, cultivate (weed) and harvest. First drawn by oxen, the equipment was later pulled by horses

that had been specially bred for the work. Mules also came to play an important role in those ecological situations where their attributes made them superior to horses. And finally, all the animals were replaced by machinery—an agricultural revolution still in progress in many parts of the world.

Plow cultivation has many subtypes—from the simple, inefficient systems still encountered in parts of the Mediterranean region and the Indian subcontinent to the heavily mechanized factories in the field one may encounter in the irrigated Central Valley of California and the more intensively farmed areas of the United States Middle West (Fig. 1.21).

Under the more primitive types of plow cultivation a relatively high degree of ecological diversity is retained. However, as the system becomes more efficient (measured by per acre yields), the landscapes tend toward ecological sameness and lowered diversity, and vast areas are frequently devoted to a single crop (Fig. 1.22).

Plow agriculture has made possible the increase of human population to a degree undreamed of a few thousand years ago. There can be no question that this system,

Figure 1.22 *Kansas wheat farm during the harvest.*

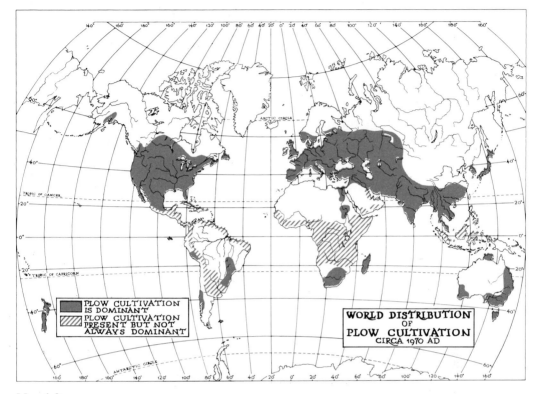

Map 1.8

in its more modern phases, can be made to yield enormous returns in relation to the input of labor and materials. Although the system is also at the bottom of some of our pressing ecological problems, it is presently producing most of the commercial crops of the industrialized nations of the world. On Map 1.8, above, which shows the present world distribution of plow cultivation, there is no attempt to differentiate primitive from modern plow cultivation. In general, the plow cultivation in low latitude areas is relatively primitive, although there are exceptions to that generalization just as there are to the generalization that advanced plow cultivation is the only subtype in the middle latitudes.

The term *plantation agriculture* is employed here in the following sense: an agricultural system based on the intensive production of a crop for export in which there is a relatively heavy dependence on low cost, local, unskilled labor. This type of ag-

riculture is *now* found mainly in low latitude areas. Examples of plantation crops includes bananas, rubber, coconuts, jute, abacá, coffee, and sugar (Fig. 1.23). Typically, the system imposes on the nations in which it occurs a *crop monoculture* such as coffee in Colombia, sugar in Cuba, rubber in Malaysia, jute in Bangladesh, and bananas in Honduras.

Simple animal husbandry may or may not be a part of each agricultural system described above. In some instances, certain animals are closely integrated into the agricultural system, for example, swine, ducks, and carp in Chinese sedentary hoe cultivation. Slash and burn cultivators usually do not have any well defined form of animal husbandry, but this can change if the opportunity arrives. For example, some Amerinds in the New World tropics who practice shifting cultivation have added pigs and chickens to their agricultural system since A.D. 1492.

Figure 1.23(a) *Banana plantation in Honduras.*

Figure 1.23(b) *Rubber plantation in Malaysia.*

Figure 1.23(c) *Coffee plantation in Colombia.*

Figure 1.23(d) *Cuban sugar cane plantation at harvest time.*

Animal husbandry is, of course, a basic part of premotorized plow cultivation. Not only are draft (pulling) animals important but there is typically a wide assortment of so-called barnyard creatures such as fowl, pigs, goats, and sheep. As plow cultivation becomes more motorized, the significance of simple animal husbandry is reduced, and modern factories in the field commonly have as few animals around (other than pets) as one might encounter in a suburban residential backyard.

Animal husbandry may stand alone and apart from one of the basic agricultural subtypes. Thus, *pastoral nomadism* is an ecological system built largely on livestock that is moved in a nomadic pattern. Such economies are becoming rare but can still be found in the Sahara and parts of East Africa and Eurasia. The apparent freedom of movement of the people in such economies has often captured the romantic fancy of more sedentary folk, but the system is not efficient in terms of food production. Those who live this way must almost always establish some kind of symbiotic relationship with cropgrowers in order to assure that they have plant food and other goods.

Between simple animal husbandry and *industrial animal husbandry* there are many subtypes, but because our purpose is to acquire a basic framework for understanding how the economic activities pursued by man relate to the earth's ecology, we shall pass to the "modern" aspect of animal husbandry.

The distinctive aspect of this type is that animals have ceased to be integral parts of a total farm system and have become specialized products grown under controlled conditions in the shortest amount of time. Thus we now have chickens that have been bred to be little more than egg-producing machines capable of laying over 300 eggs per year and fryer chickens, hogs, and cattle raised by automated techniques. In short, any animal used for food in large quantity is subject to these practices. The concentration of large numbers of animals in limited space leads to huge accumulations of feces and other waste products—with attendant ecological problems (Fig. 1.24).

Industry. Surprisingly, there is still no simple definition for the term *industry*. The meaning of the word is not to be sought in dictionaries but in the way people use it in everyday writing and speaking. It is customary, for example, to exclude agriculture from industry, although certain modern farming practices are essentially the same as one might encounter in a factory. For many developing nations, industry means the ability to produce metals, such as steel, and a wide variety of manufactured goods, including various types of machinery.

In this section we will be concerned chiefly with the development of industry and will confine our remarks to extractive industry and manufacturing. Extractive refers to such activities as mining, fishing, and forestry, and manufacturing refers to the processes by which the products of extractive industry are converted into marketed goods.

Extractive Industry: Fishing and Hunting. Although it might be correct to speak of subsistence hunting or fishing as being industries, the conventions of use do not so recognize them. Thus, industrial fishing and hunting refers to the commercialization of these activities.

We do not know how old commercial fishing is but we do know it predates the invention of writing. It may have begun, in a simple way, as far back as Paleolithic time, because fish may have been bartered for some desired item. By the time records were kept in the Mediterranean region of the Old World, commercial fishing was a well-developed activity and involved the collection of many fish, crustaceans, and molluscs. Later, when men began to venture short distances out on the Atlantic,

Figure 1.24(a) *Cattle feedlot in the Middle West.*

Figure 1.24(b) *White leghorn chickens in an egg factory.*

most of the voyages were in search of fishing grounds. By the eleventh century, a whale fishery (called a *fishery* even though the animal is a mammal) had been developed in the Bay of Biscay. The whales were killed a short distance from shore and towed back to land, where their fat was *tried* (cooked) for oil and some of the meat, particularly the tongue, was salt-cured. This meat was sold widely in Europe. Other fisheries were developed to supply the growing need for inexpensive animal protein occasioned by growing human populations. As market demand grew, there was a gradual improvement in boats and gear, and fishermen sailed increasing distances, particularly into the North Atlantic, in search of whales and codfish. By the time Columbus discovered America, commercial fishermen had discovered and were exploiting many of the major fishing grounds of the North Atlantic Ocean.

In the succeeding centuries not all nations moved at the same rate of commercially exploiting marine animal life. As might be expected, those nations that possessed direct access to the sea were most likely also to develop important commercial fisheries. Thus, many of the western European countries developed in this manner, and in the early years, the United States did too. Somewhat later, Japan and Russia entered the field. Today, thousands of vessels are engaged in this activity, and the reduction of marine animal resources has been increasing rapidly during the past few decades.

Japan presently leads the world in fishing but the northwest European countries and the U.S.S.R. also account for a large percentage of the annual take from the world's oceans. Thousands of fishing vessels are equipped with sophisticated electronic equipment to help locate fish, and aircraft are flown over likely fishing water in an effort to locate the quarry.

Excluding the whales, the larger part of the commercial fish catch is used for food by humans. However, a growing amount is being processed into other products, ranging from oil for printer's ink to feed for chickens. One of the most valuable fish caught by American fisheries is menhaden. Virtually none of the enormous menhaden catch is eaten. Rather, the fish is used for oil and as a source of fish meal fertilizer.

There are so many commercial fisheries, it would be almost impossible to show them all on a single map. Thus, only a selection of the principal commercial fisheries are depicted on Map 1.9 below. Note that most of the major commercial fisheries are in middle and high latitude waters. This is because warm waters generally do not support the concentrations of a few species

Map 1.9

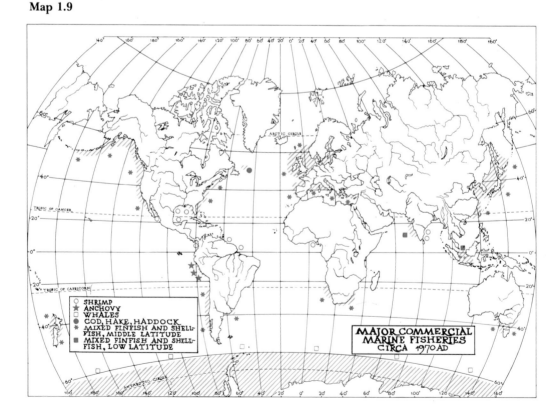

that permits a highly developed, specialized commercial fishery. The one exception shown on the map is along part of the west coast of South America. It is there because of the cool waters of the northward-flowing Humboldt Current.

Commercial hunting, including the trapping of furbearing mammals, is now a modest activity and has only a localized development (whaling being considered a fishery). Commercial hunting reached its apogee near the end of the nineteenth century and has been in a decline since. The decline is due chiefly to laws that have been passed in many parts of the world. Unlike many marine animals, land animals can seldom maintain themselves when subjected to commercial exploitation. Thus we do not generally find long histories of commercial hunting for a limited group of land animals as we do, for example, with the fishing for codfish in the Atlantic. Commercial hunting has frequently been associated with the expansion of a pioneer settlement fringe or with the preliminary exploration of a region prior to regular settlement. Once large numbers of people are present, the wildlife cannot withstand commercial exploitation without becoming extinct.

Mining. Mining must be among the oldest of human exploitative activities. It may be said to have begun when some person, in the long ago past, picked up a piece of stone in order to use it for something. The deliberate seeking out of stone and native metal laid the groundwork for one of the most significant human economic activities.

Passing over the period during which man had to be content with stone, we shall continue our brief account of mining with the use of metal. We have long been accustomed to reading about the Iron Age, the Bronze Age, and so on. Actually those and similar designations lose some of their usefulness when we look at the entire world. For example, in parts of Central America

and northern South America skilled metallurgists were working gold into beautiful objects long before the Europeans arrived. The gold had been mined from stream deposits by *placer mining*—perhaps the oldest organized mining technique. Placer mining involves the separation of the desired material from the undesired sands and gravels that accompany it. The technique ranges from hand-held pan to elaborate machine-operated dredges (Fig. 1.25).

Hard-rock or vein mining was practiced in the parts of the New World where copper occurred in an almost pure state. But there was no Iron Age in the Western Hemisphere—at least no evidence of worked iron for the pre-Columbian period has been accepted by the archeologists.

Turning to the Old World, we find a history of gold mining, copper mining, and, later, iron mining. Iron (its chemical symbol is Fe) has become indispensable to modern manufacturing industry, which is based to an extraordinary degree on steel. But only relatively recently did the manufacture of steel become so important and iron was in use long before. Iron mining now involves mining monumental quantities of ore that must be processed to become refined iron and then steel. Thus although iron is one of the most abundant elements in the earth's crust, only occasionally does it occur in such concentrations as to make its mining economically worthwhile. Presently known major deposits of iron ore are located chiefly in the Northern Hemisphere. To some extent this distribution represents the degree to which mineral exploration has been carried on, but it may indicate that concentrations of this valuable ore are comparatively scarce in much of the Southern Hemisphere.

Of course, in addition to iron, many metals and minerals have long been the object of mining activity. Mining directly and/or indirectly greatly modifies the environment, particularly when the products of the mines are smelted or, in the case of fossil fuels (coal and petroleum), burned for fuel.

Forestry. The scientific management and utilization of forests and woodlands is called *forestry*. The term is also frequently employed to characterize forest-woodland utilization, managed or not. Most technologically advanced nations have educational and governmental agencies to manage forest-woodlands. Unfortunately,

these agencies often started only because the forest-woodlands were partially or largely destroyed.

From the earliest periods of human cultural development to the present, wood has been important to man. Without wood and products, we would have to reshape some of the physical characteristics of human existence—in both the most technologically sophisticated and the most technologically undeveloped nations. But although we recognize the Stone Age, Bronze Age, and Iron Age, little attention has been given to the fact that laid across all these ages has been the continuing and enduring "wood age."

Part of the reason for this cultural myopia is that in many archeological sites, wood does not preserve well, so we have come to focus on those objects, such as stone, ceramics, metals, that do preserve. This has skewed our understanding of the important roles that wood plays in human societies, and our current preoccupation with heavy industry further blinds us to the fact that the products of forests and woodlands are critical to human economies.

Woods suitable for construction and manufacturing are generally grouped into two broad categories—hardwoods and softwoods. Of the two, the softwoods are in

Figure 1.25(a)
Prospector panning for gold in Alaska.

Figure 1.25(b)
Mechanized gold dredge.

Figure 1.26 *Logging operation in British Columbia.*

greater demand worldwide, because they are used for a multitude of construction and manufacturing purposes and the trees that produce them grow relatively fast. Among the softwoods in greatest demand are the conifers, or cone-bearing trees. And of these, pines (*Pinus*) are the most valuable and most sought after (Fig. 1.26). It is one of those curiosities of biogeography that with but a minor exception true pines are confined to the Northern Hemisphere, except where transported and introduced by man into the Southern Hemisphere.

Many hardwoods are economically important. Of these, oak (*Quercus*), walnut (*Juglans*), maple (*Acer*), and ash (*Fraxinus*) are commonly used for furniture. Teak (*Tectona*), mahogany (*Swietenia*), and ebony (*Diospyros*), some of the hardwoods from low latitude forests, also appear in the marketplace. Hardwoods figure prominently in the manufacture of charcoal which, even in this era of fossil fuel consumption, remains important in many parts of the world. In addition, hardwoods once played many roles now filled by metal.

Land Transportation Routes. An essential feature of human activity is developing well-defined land transportation routes. Basically, this is also true of a number of other mammals. The notable difference is the degree to which man engages in such activity and the profound effects his activity produces in the ecosystems through which the routes pass. Although we do not know how early people came to establish and use relatively long-lasting trails, this was a well-defined trait by the time agriculture came into being.

An important impetus to this development was the growth of political states. Political leaders, realizing the necessity of tying claimed territory together in order to maintain political, military, and economic control, set up elaborate road systems. No better example of this can be found than the Roman roads, whose principal reason for being was the rapid movement of foot soldiers to any part of the Roman Empire; they were specially designed to effect that end. However, the roads also served the needs of commerce.

In the New World, the expansion plans of the Inca imperialists led them to construct foot roads that, although not equal to the Roman roads, were impressive achievements and served to unite the empire. These roads were superior to most of those constructed for several centuries after the European conquest of the Andean region.

In spite of the obvious utility and necessity for well-maintained land surface routes, the Roman roads marked the apogee in European road quality for centuries. Until the advent of the railroad in the nineteenth century, vehicular traffic moved with great difficulty except in a few parts of a few countries.

The railroad effected a major revolution in transportation and set off a series of further economic and ecological revolutions. Not the least of these was caused by the movement of large numbers of people into pioneer areas. The result was acceleration of the exploitation associated with mining and forestry and the spread of agriculture, with its attendant alteration of ecosystems.

The automobile appeared haltingly on the scene at the end of the nineteenth century and soon challenged the horse and mule. However, the automobile revolution is mainly a phenomenon of the twentieth century and is still continuing to spread into the farthest reaches of the world. The automobile has come to assume an extraordinary position, not only in human industrial economies, but also in the complex status systems of human societies the world over. It is naive to speak of the automobile simply as a means of transport because it is much more than that.

Few gods in the past have claimed the devotion accorded to the automobile and it appears that its "needs" are sacrosanct. Not only does this creature of human creation consume huge quantities of resources with fantastic speed, but through the construction of roads and highways, it is reshaping landscapes at an economic-ecological cost that leaves one in awe. Yet, most of this awe-inspiring behavior is restricted to the industrialized nations of the Northern Hemisphere. Enormous regions of South America and Africa are almost without any form of surface routes other than the most rudimentary. It does not require much imagination to picture the ecological outcome if highways and roads become as frequent in Africa and South America as they presently are in Western Europe.

The impact on the ecology of roads constructed in isolated regions has received almost no attention. These roads often have the effect of spreading rural poverty and causing the indiscriminate waste of resources. However, they do offer opportunities undreamed of before and broaden the lives of the people. Certainly, roads should receive more ecological and social attention than they have in the past.

HUMAN POPULATION

Of all the topics discussed in this chapter none is more central to the theme of the book than human population. However, this subject—often expressed as the *population explosion*—has been receiving so much attention in the public press that many persons are turning away from it in boredom. One is told with great frequency that the world is rapidly becoming overpopulated or that there is no such thing as (human) overpopulation, only underproduction of goods. Somewhere between these positions one hopes to find the truth.

As is now generally known, most of the time that *Homo sapiens* has been on earth it was relatively rare. Many attempts have been made to estimate the world human population in remote times but these are simply guesses. Most of us have seen graphs that plot the growth curve of human population from the earliest times to the present and are familiar with the nearly horizontal line that supposedly reflects the size of the world population for thousands of years (Fig. 1.27). If viewed

Years ago	Cultural stage	Area populated	Assumed density per square kilometer	Total population (millions)
1,000,000	Lower paleolithic	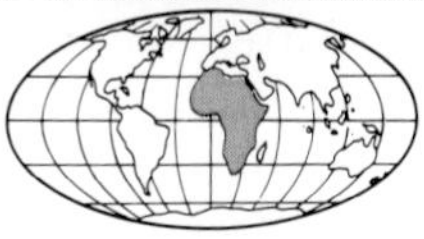	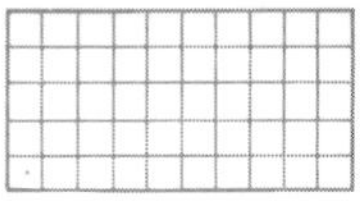 0.00425	0.125
300,000	Middle paleolithic	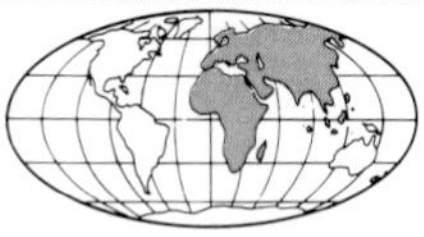	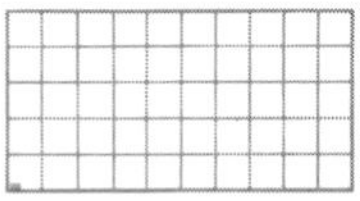 0.012	1
25,000	Upper paleolithic	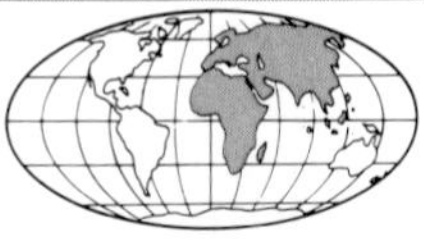	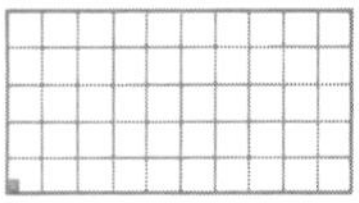 0.04	3.34
10,000	Mesolithic	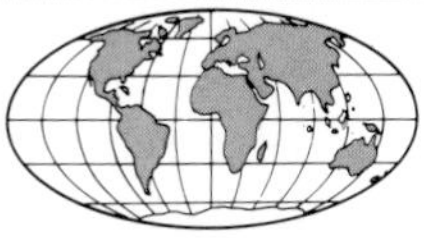	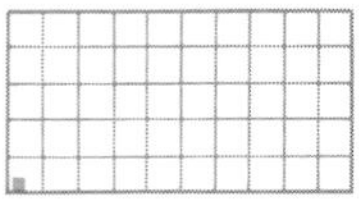 0.04	5.32
6,000	Village farming and early urban	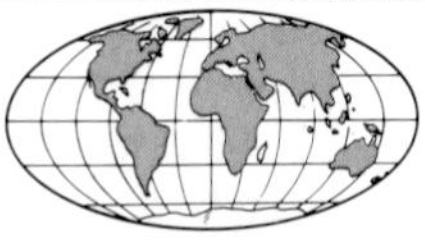	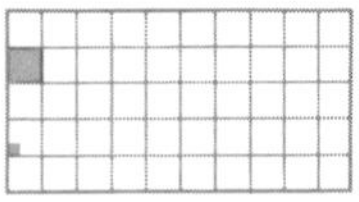 1.0 0.04	86.5
2,000	Village farming and urban	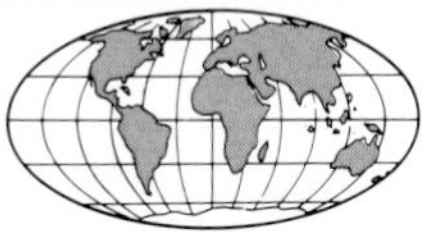	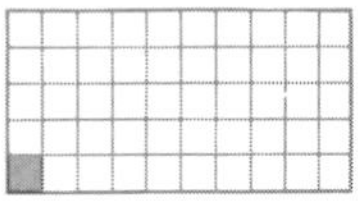 1.0	133
310	Farming and industrial	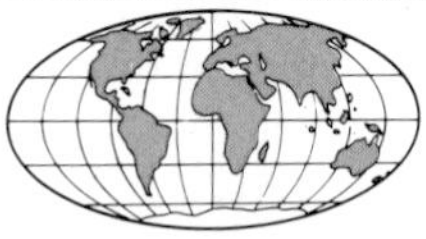	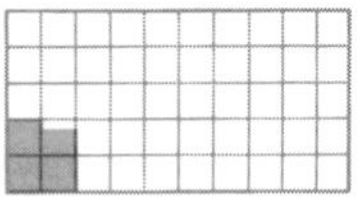 3.7	545
210	Farming and industrial	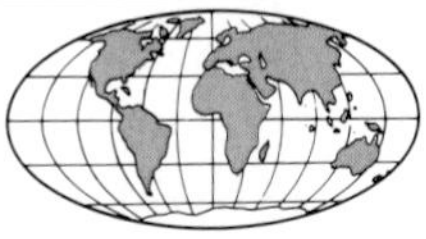	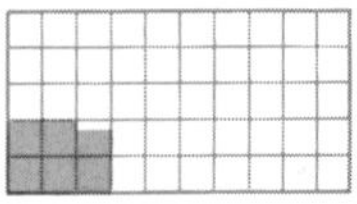 4.9	728
160	Farming and industrial	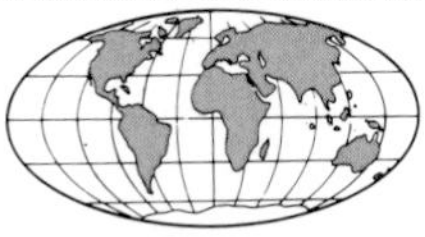	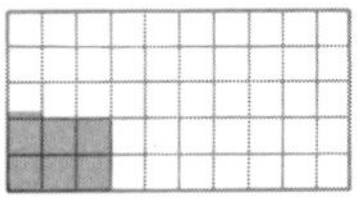 6.2	906
60	Farming and industrial	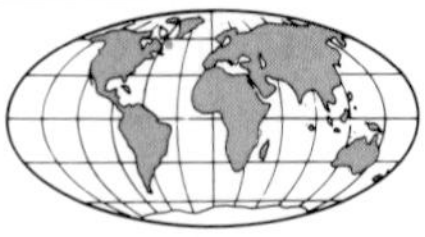	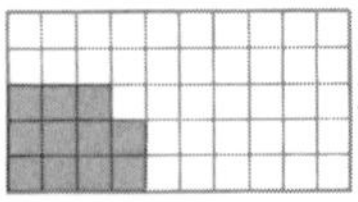 11.0	1,610
10	Farming and industrial	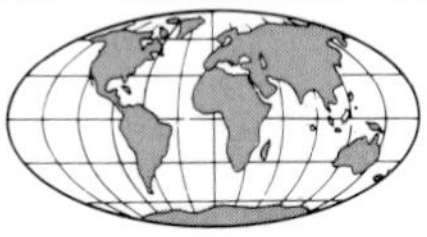	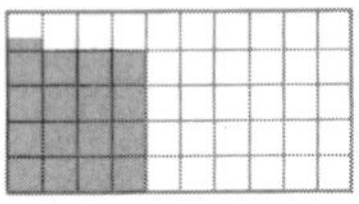 16.4	2,400
A.D. 2000	Farming and industrial	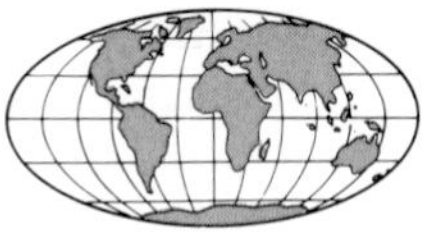	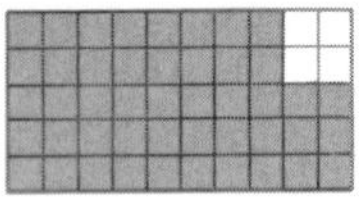 46.0	6,270

Figure 1.27 *(From "The Human Population" by Edward S. Deevey, Jr. Copyright ©*
September 1960 by Scientific American, Inc. All rights reserved.)

simply as a generalization based on almost nonexistent data, it probably presents a sound picture. That is, humans were not numerous and total world population grew slowly. We know essentially nothing about the early local and worldwide fluctuations in human populations and almost nothing about the dynamics influencing local and world population size. We are safe in asserting that with agriculture came an increase in our population and an increase in the rate of growth—but to what extent we are not sure. Agriculture built an ecological support under us that permitted more people to live in the world. The next important event for the increase in human population was industrialization. There is a positive correlation between industrial growth and human population growth—at least in the early phases of industrialization.

Perhaps the greatest cultural impact on population growth was caused by the development of "modern medicine." Only a few decades ago, disease was a major worldwide source of mortality among the young and thus was a significant factor in human demography. The twentieth century has witnessed a medical revolution that has made life far more pleasant but has also triggered a population growth that is without precedent in human experience. Infant mortality, long one of the saddest facts of human existence, has dropped dramatically in most of the world and is falling sharply in those regions where the impacts of medicine and sanitation are now beginning to be felt. Almost no society today can remember when infant mortality exceeded three hundred per one thousand live births.

Because of high infant and child mortality, we "protected" ourselves with high birth rates. This survival response is no longer necessary but continues strong, particularly among many people living in underdeveloped nations. For a wide variety of reasons, people continue to have many children even after the infant mortality rate has fallen sharply. Children are often seen as economic goods in societies lacking social security or as a means of looking after older persons who are no longer able to contribute to the labor force. There is also a tendency for men to measure their masculinity by the number of children they have sired (the *machismo* complex) and a high social value is sometimes placed on male children. This may induce a couple to have more children if they have had only girls. Another factor is the pervasive cult of motherhood, which holds that a woman can only find self-identity through reproduction. And finally, some religions encourage large families— or at least discourage most attempts to limit family size through the use of contraceptives.

The world human population has now reached a size that would have been undreamed of except by Malthus. The doubling time is now approximately thirty five years, and if current growth rates continue, we shall soon have six billion people on earth. This figure, by itself, has little usefulness, however. Overpopulation must relate numbers to environmental factors. This may appear simple to do but is actually so difficult as to be nearly beyond accomplishment at present. The problem is that there is *no one standard human ecology*, no single model to which we may refer. It has become common to denounce population growth and to insist that the human population cease to grow at all. However, one man's overpopulation is another man's great blessing. Therefore, we cannot argue effectively against the clear danger that unchecked growth will and must at some future time result in an ecological disaster for our species.

We must recognize that we place ourselves in positions difficult to defend when we make assertions as to how many persons have a right to be born. We can and should attempt to understand the ecological implications involved in population increase, but we should not assume that our personal concept of life and how it should be lived in respect to the world's resources is the only valid one.

Different life styles lead to different cultural landscapes, and the ecological impact of differing life styles vary greatly. The ecological meaning of 100,000 additional persons born in rural India is different from 100,000 additional persons born to the white middle-class sector of the United States. The length of life of the two groups would vary and their effects on natural resources would be markedly different. The per capita impact on natural resources of the United States persons would be much greater than that of the Indians. If all cultures accepted the Indian rural model as the ideal, the world could accommodate more persons than many of us may believe desirable. Were the United States middle-class model to be emulated worldwide, it would be impossible to achieve, given the present level of technology and availability of resources. The Indian rural model is too meager even for Indians but the United States model is too excessive. What then is a reasonable model? That is a fundamental question of our time but is only rarely raised. What we have is a race, by many of the world's peoples, for a goal approximating the United States model, which they cannot hope to attain. Little thought is presently given to the alternative life styles available and how these may better harmonize with environmental reality. Of all the world's nations, it may be the United States that has the greatest population problem—if the size of the population is weighed against the magnitude of resources used by its inhabitants.

Chapter Two

Human Influences on the Ecosystems of Subsaharan Africa

2.1 Cultural Background

The study of man's modification of the earth properly begins with Africa, for it is on this continent that the most complete record of early human evolution has been unearthed. Thus, man has had the greatest length of time in this region to influence the ecosystems.

Although there have been a few attempts to reconstruct the cultures of the earliest men in Africa, there is as yet too limited a data base to indicate other than that these people possessed no more than the rudiments of material culture and that tool making and use was of a low order of development. The archeological riches of Africa are only now beginning to be explored in a scientific manner and finds may be made of tools and other artifacts relatable to the earliest cultures.

It appears that agriculture was not known in this region until approximately 5000 years BP and even at that time had not spread widely. Thus, gathering economies dominated the African cultural scene, south of the Sahara, for most of the time man has been there.

Agriculture probably started as shifting cultivation. Some time after the spread of agriculture, herding societies arrived—perhaps by drifting southward through the Nile corridor. Other antecedents of present-day herders may have entered via routes over the Sahara.

The Sahara has played a significant role in the affairs of *Homo sapiens,* as it has in the ecological and biogeographical affairs of other living things in Africa. Its chief influence on human culture is that it was a major barrier to the dispersal of culture traits and complexes both out of and into subsaharan Africa. Thus, we find major cultural events occurring in the Mediterranean world that were unfelt and unknown in the region south of the Sahara. Similarly, events occurring in the south had little or no impact outside the region.

However, the isolation was not complete within the past 8000 to 10,000 years.

There must have been some movements of preagricultural peoples and ideas through the Nile corridor north and south and some overseas contacts, although they probably occurred more recently.

By about the fifth century B.C., relatively regular trade contacts had been established across the Sahara, linking the Sudan corridor with the Mediterranean region. In more remote parts of Africa, regular trade contact seems to have come later (although our knowledge is based on inadequate data), with evidences of Roman culture showing up in coastal East Africa by about A.D. 400.

Much better documented are the beginnings of the Arab presence in East Africa. This began about A.D. 1250 and lasted until the nineteenth century, although Arab coastal rule was first challenged in the fifteenth century. This challenge was by Portuguese nationals who were reaching out for a sea route to the East Indies and its spice trade. The first successful European navigation around Southern Africa was by the Portuguese mariner Vasco da Gama, in 1497 (to India in 1497 and return to Portugal in 1499). His voyage marked the apogee of Portuguese efforts to discover a sea route to India, a task begun with voyages part way down the west coast of Africa as early as 1444.

The Portuguese were followed closely by other European seafarers, attempting to obtain gold, ivory, and slaves. They traveled mainly along the west coast of Africa.

For a time the Europeans were content to maintain only local coastal stations, where trade could be carried on with Africans and where ships could be repaired and provisioned. In this, the Europeans were frequently aided and encouraged by African chiefs, who saw chances for personal gain in such trade. Most of these stations were established on the west coast, for it was closest to European ports and precious metal and slaves could be obtained with relative ease in that area.

Soon, the trade in slaves eclipsed almost all other enterprises combined. It had been a practice long before Europeans came on the scene, but the growth of plantations in the New World provided a demand for slaves probably never approached in earlier days. The European-dominated slave trade was underway before the end of the fifteenth century and continued until the early part of the nineteenth century. During this time hundreds of thousands of people were shipped away, but this period still has not received the attention it deserves from scholars.

Many European nations became engaged in the slave trade, and their most obvious legacy in Western Africa was their acquisition of territory that later became colonies. However, the founding of European colonies did not follow the slave trade immediately. It was not until the nineteenth century that European penetration and partitioning of subsaharan Africa became a major phenomenon. But once started, no time was lost; by the end of the nineteenth century, Africa was little more than the supply yard for European nations. This political condition was to persist, with only minor changes, until after World War II. As recently as 1940, only one political unit south of the Sahara could be said to be politically independent— Liberia, which was established in 1822 by former slaves from the United States.

England controlled a vast area of colonies and protectorates, which today are all independent states. The other European powers that had colonies were France, Belgium, Germany (until the end of World War I), Spain, and Portugal. Today, European-derived minorities are in control of two independent African nations, the Republic of South Africa and Rhodesia.

Arab dominance in Eastern Africa was challenged soon after the appearance of Europeans along the coast, and Arab influence waned slowly until it was eclipsed by the British in the nineteenth century. However, Arab cultural influences remain significant along parts of the East African coast, as illustrated by place names such as Dar Es Salaam.

The cultural influence of the Europeans was varied and continues to the present day. In West Africa, where a rich array of native languages are spoken, English and French became the general languages. In East Africa and parts of central Africa, Swahili (Kiswahili) became the general language. Swahili contains many borrowed words from Arabic and English, but it is a true African language and not a recent invention.

The Europeans were responsible for a monumental dislocation of the lives of millions of Africans during the years of the slave trade and the effects of this on recent African history have not yet been adequately studied. A host of foreign culture traits and complexes were introduced, including such diverse elements as the Parliamentary form of government, from England, and peanuts, from South America. In the pages ahead, we shall examine some of the specific cultural changes that relate to the modification of Africa's ecosystems. However, we still have much to learn about Africa's prehistory, and our assessments will surely be revised as we gain new information.

2.2 Influences on Vegetation

The vegetation maps in this chapter, like all maps, are compromises in that they show those features that are most relevant to the discussion and leave out other features because of limited space. Map 2.1*a* on page 58 shows the distribution of noncultivated vegetation as it might appear had there been no human disturbance. Map 2.1*b* on page 59 shows the general pattern of vegetation as it is at present.

Preagricultural Period (1,750,000 years BP to 5000 years BP). Before fire came into use in subsaharan Africa, man probably did not have much impact on the vegetation. Some authors assume that fire was a part of human cultures in Africa for hundreds of thousands of years. Although this may prove to be the case, at present there is almost no evidence to support such a view. Others have suggested 60,000 years BP as the approximate date of the beginning of fire-use, but, again, there is no unequivocal evidence to support this. As was pointed out in Chapter 1, it is extremely

difficult to discover charcoal that is unequivocally of human origin. Fires set by wandering nomads (Fig. 2.1) on the open

Figure 2.1 *Bushmen of the Kalahari gathered around a fire.*

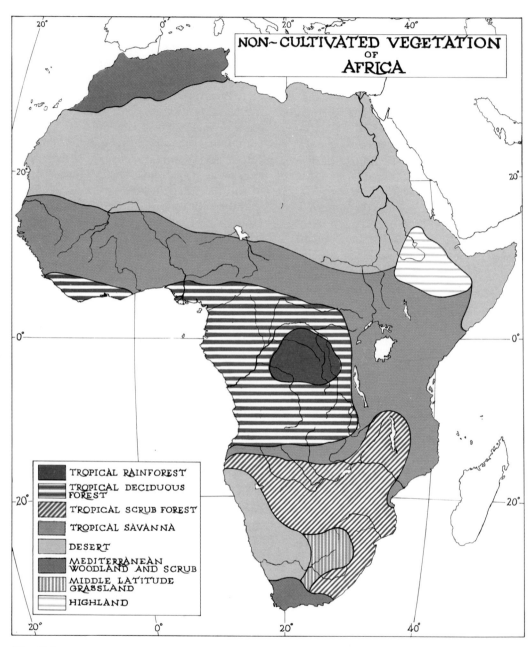

NON~CULTIVATED VEGETATION OF AFRICA

TROPICAL RAINFOREST
TROPICAL DECIDUOUS FOREST
TROPICAL SCRUB FOREST
TROPICAL SAVANNA
DESERT
MEDITERRANEAN WOODLAND AND SCRUB
MIDDLE LATITUDE GRASSLAND
HIGHLAND

Map 2.1a

ground leave charcoal that is indistin-guishable from charcoal left by a fire set by lightning.

It has been argued that fire-use must be ancient in this region, because it was so long inhabited by hunters and gatherers and "fire was a standard Paleolithic hunt-

ing tool." This argument is an extrapola-tion to prehistoric time of human behavior observed in historic time. Although fire may long have been a hunting tool, the hunters probably were careful to burn their hunting and gathering lands only in a limited manner. Otherwise, they might

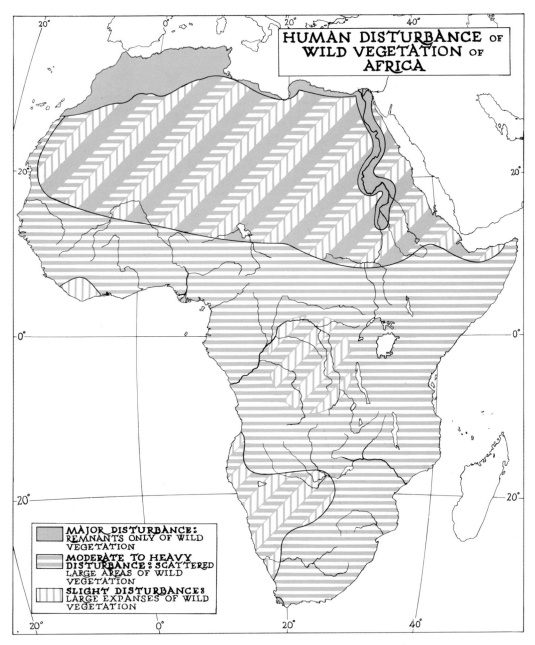

Map 2.1b

quickly destroy the ecological base for subsistence. Anyone who sees burned land in this area can appreciate how dead it is until rains bring forth new plant growth. Moreover, wild plant food probably represented the bulk of the gatherers' diet. Thus, they had to be careful not to destroy the vegetable food on which they depended.

Given the present paucity of information, we offer only the following generalizations:

1. Evidence for the human use of fire in subsaharan Africa is extremely

sparse for any period before the agriculture period, because the evidence has not yet been discovered.

2. Fire, when it was acquired by gatherers, must have been used with considerable caution to prevent unwanted destruction of the environment. However, fire is extremely difficult to control in these wild ecosystems, so escaped fires may have been moderately frequent.

3. The impact of preagricultural folk on the vegetation was probably local and dependent on the population size and how they used fire. The dryer areas were probably the most vulnerable, but we cannot yet document this hypothesis.

Slash and Burn Cultivation (5000 years BP (?) to present). When and where some form of organized food production appeared in subsaharan Africa are not known. The *idea* of plant cultivation may have been *autochthonous* (native) to the region or may have come in from the north; the answer to this is not yet known.

Various reasons have been advanced to account for the relatively late appearance of agriculture in this region, but none is satisfactory. It must be emphasized that the date given above for the start of agriculture is tentative. However, it is fairly certain that by approximately 3000 years BP, slash and burn cultivation had dispersed widely and had displaced to a major degree the previously dominant gathering economies. Moreover, prior to this there must have been a long period of experimentation with native plant species, because there is a relatively wide range of domesticated plants in this area.

The essential relationship between ecology and slash and burn cultivation is fire. Thus, the period of adaptation and spread of crop cultivation in subsaharan Africa must be viewed as the opening of a major chapter in the history of human modification of ecosystems. By at least 3000 years BP, fire, as a part of shifting cultivation, must have been exerting a marked effect on the vegetation cover of much of subsaharan Africa, with the possible exception of the moister parts of the tropical rainforest.

By approximately the same time, pastoral people—herders—were beginning to appear on the scene, at least in the Uganda region. (More archeological study may place them even farther south by that time.) There is also the possibility that herders were present in the Chad region of West Africa at an earlier date than is presently known for herders in East Africa.

We know almost nothing about the population sizes of these groups *circa* 3000 years BP. But fossil pollen records taken from the Lake Victoria area indicate a decrease in tree pollen and an increase in grass pollen about that time. The evidence is highly circumstantial but there is no clear evidence of climatic change, so significant forest removal by man appears to be a possible answer. There are similar indications for northern Angola, where, instead of changes in the pollen record, there is evidence of sharply increased soil erosion. This suggests, but does not prove, vegetation clearance by man.

The early herders were probably limited to small areas because much of Africa is inhabited by the tsetse fly (*Glossina*). This fly transmits a disease called nagana, which is fatal to almost all livestock animals and, occasionally, man (in man, the disease is known as sleeping sickness). Because the tsetse fly requires brush for shelter, many brushy areas are unsuitable for livestock. It is probable that herders were able to increase the area available for grazing only by setting the land afire. This technique also encourages new grass growth and is therefore used in many parts of the world. We do not know to what degree this use of fire influenced the vegetation cover, but there is a growing awareness that it may have been one of the more significant aspects of human ecology in East Africa and other parts of Africa as well.

The term slash and burn cultivation includes a wide variety of subsystems, each of which possesses its own ecological implications. One must remember that not just one system of slash and burn cultivation existed in Africa. Several have been described, and further research may bring to light further variations. Two varients are described below.

In Zambia, Rhodesia, and other parts of Southern Africa, a system called *Chitemene* is widely practiced. It differs from the more common types of slash and burn cultivation in that in addition to burning the vegetation cleared on the plot to be cultivated, a much larger area of woodland is cut. The debris from this cutting is also burned on the cultivated land in order to provide a greater quantity of nutrients for the crop or crops. The chief crop in the area is finger millet (*Eleusine coracana*), which is used chiefly for beer, but a wide variety of crops is planted under this system. The Chitemene system is highly destructive of the wild vegetation cover.

In West Africa, a somewhat different form of slash and burn cultivation is prac-

ticed. Here after the desired plot is cleared and burned, the soil is worked up into hills or mounds (Fig. 2.2). These are then planted with the desired crops, of which yams (*Dioscorea* spp.) are prominent. This system can result in the rapid reduction of woodland and forest to poor scrub and grass if the fallow period is not sufficiently long.

Some time ago, shifting cultivation was denounced in Africa and elsewhere in the tropics as being ecologically unsound. More recently, scholars and agricultural and forestry experts are tending toward a moderate defense of the system. They almost unanimously state that the real problem is the increased human pressure on the land—the direct result of the growth of the farm population.

The first problem we face in examining this thesis is the almost complete lack of reliable data on African populations for even as recently as 1900. Thus, statements about the pressure of a growing population rest largely on the informed guesses and intuitively based judgments of people who have some first-hand experience in Africa. But admitting these difficulties, even the fragmentary demographic data

Figure 2.2 *Yam cultivation in Ghana.*

of recent decades justifies the statement that African populations have been increasing during the past century, with an increasing increment in recent decades.

We can safely hypothesize that the population densities of farmers and herders in subsaharan Africa had reached an approximate ecological balance before interference from outsiders. We may also hypothesize that human population growth was slow in the period from the *widespread* establishment of slash and burn cultivation, *circa* 3000 years BP, to the less widespread establishment of herders at a somewhat later time. Warfare, disease, and other checks on population growth probably prevented serious ecological dislocations, except perhaps locally. The total area of forest and woodlands undoubtedly were reduced during this time, but the period of maximal ecological change to the vegetation probably accompanied the adoption and spread of agriculture and herding. The new population equilibria probably fluctuated around a slowly rising mean with an attendant slow altering of the ecosystems, until new influences from outside the region were felt.

It may be hypothesized that by 2000 years BP, the major configurations of the nondesert biomes of subsaharan Africa had assumed the characteristics encountered by European explorers in the fifteenth century. The region of chief human induced ecological stress during this time may have been West Africa, because this area had to support a higher density of human population than most of the other areas of Africa (as is still the case). The moist forest margin was probably being slowly pushed back, but the rate at any given place may have been modest and intermittent, depending on the demographic conditions.

As previously discussed, herders were restrained from rapid expansion by the presence of the tsetse fly, although they may have used fire to push back the brush and make new areas available (Fig. 2.3). Present evidence suggests that once the safe areas were discovered and occupied, the herding societies settled down and added area slowly by the annual burning, which encourages the growth of new grass.

There may have been a significant ecological upset when a few crops of Asian origin—bananas, plantains, and coco-yams (*Colocasia*)—were introduced. The time of these introductions has not been definitely established but it seems to have occurred shortly after A.D. 1. Although who brought these crops to the African mainland is also unknown, available evidence points toward the people who settled Madagascar, probably early in the Christian era if not somewhat before. These crops are better

Figure 2.3 *Masai herders and their cattle, Kenya.*

suited to moist forest conditions than to the drier savannas, so we can infer that previously unfarmed parts of the forest became more attractive to farmers. Another feature of these crops is that they continue to yield for several years after a single planting and thus would permit use of a given plot for a longer time than would previously have been the case with grains.

Of at least equal and perhaps greater ecological significance during this period (about 2000 years BP) was the spread of the use of iron. There is still no agreement as to when the Iron Age made its appearance in subsaharan Africa, but its use appears to have been widely established by 2000 years ago. Wood was required for smelting, but this use (charcoal) apparently was not of sufficient scale to result in vegetation changes over large areas.

Our attention has thus far focused on fire, but another important impact on the vegetation was caused by the need for wood for fuel and construction. Because of the almost complete absence of population data, we cannot estimate the quantity of wood used for these purposes, but it must have been great. Also the cutting of wood must have been selective because certain woody species would be preferred over others. Thus, this would tend to have an effect on the species composition of the woodlands.

The period of the European slave trade produced many effects on Africa. It was long believed that so many persons were removed that a serious demographic reduction took place. But the data purporting to show how many humans were exported are often suspect. Perhaps more relevant to our discussion is that the slave trade resulted in social turmoil and cultural instability, and raiding for human merchandise was frequent. The area most affected was West Africa, but the impact was felt all across the Sudan corridor and into East Africa, where the Arabs had long been engaged in the same trade. But whether there was widescale abandonment of farmed areas during this time awaits further study. Thus it is difficult to adopt a working hypothesis other than that there were important ecological changes taking place during this time in the areas most affected by slave raiders.

Turning from the slave trade and examining some other aspects of this same time period, we discover that as a result of European contacts with the west coast of Africa and tropical America, an exchange of plants resulted. This must be characterized as the second agricultural revolution in Africa. Many plants came from the New World and some of these were to reshape the crop complex of much of the shifting cultivation of Africa. Chief among these plants were maize (*Zea mays*) and manioc (*Manihot*), although the peanut (*Arachis*) may be a serious challenger for prime position among the crops of American origin. But maize and manioc quickly transformed the agricultural landscapes of much of Africa, pushing most of the autochthonous crops into positions of secondary importance.

The question of human population growth in Africa was raised earlier. It was pointed out that we can assume a series of demographic plateaus, particularly one that may have occurred about 2000 years BP. Several possible disturbances were discussed with the implication that these may have led to moderate increases in population. It was further suggested that over the several thousand years between the time that slash and burn cultivation became widely established and herding was established, there was a modest increase in population. The ecologically desirable areas were most affected by these economies, with resultant changes in the vegetation, but the new ecosystems partly created and maintained by human agency were stable to the extent that the human input was stable.

The establishment of the Europeans was to alter that. Leaving aside the effects of slavery on African ecosystems, the input of new plants provided a much more reliable source of calories than had previously been available. Much more food could be produced with no particular increase in labor. And because food reliability relates directly to human population, it must be inferred, though supportive numerical data are lacking, that this agricultural change brought an increase in population. Not a large increase perhaps, but a shift upward nevertheless.

The next and probably much more important influence was that of medicine and sanitation. The impact was probably felt most in West Africa because of its large urban clusters of people. These clusters made it relatively easy for foreign medical officers, medical missionaries, and others charged with health responsibilities to make ecological inputs that would have maximal demographic effect. At first, the inputs must have been mainly treating wounds and the ill. However, when the germ theory of disease became established in the last half of the nineteenth century (after the work of Pasteur, in France, and Robert Koch, in Germany), water supplies were scrutinized, attention was given to the safe disposal of human wastes, inocculation against certain communicable diseases was practiced, and simple ideas of hygiene were diffused, particularly among the children. A little preventive medicine and a little sanitation go a long way toward influencing upward the number of humans on this earth.

The diminution of tribal warfare has been cited by some writers as the chief cause for the sudden growth in African population in the past century. However, a large amount of the so-called warfare was ritualized symbolic aggressive behavior and did not result in mass slayings. During the expansion of agriculturalists thousands of years ago, there may have been loss of life when hunter-gatherers attempted to retain their territories, but this demographic loss was more than offset by the eco-

logical successes of the supplanting cultures.

With the increase of human numbers associated with death control came the latest and still unfolding chapter of the influence of slash and burn cultivators on the vegetation of Africa. It has been within this period that most of the outright destruction of ecosystems has occurred.

It has been observed that at earlier population levels, slash and burn cultivation was not an ecological evil in Africa. Most of the available evidence supports this view. When the ecologically permissible population densities are not exceeded, this system can go on and on. But let a people exceed those permissible densities, even for a short time, and the vegetation and most other aspects of the ecosystem will be degraded. This is what is now occurring in much of Africa.

One of our chief sources of verification of this is the accounts of travelers of the first half of the last century (and occasionally later). These books contain much information about how Africa looked then and, in a general way, what Africans were doing to the ecosystems. One is struck by the many and often vivid allusions to fires set by herdsmen or farmers (Fig. 2.4). Europeans, whose own slash and burn cultivation days were not far behind them, were often shocked and frightened by these activities. Of course, a veld fire in South Africa or a wall of flame racing across the Serengeti Plain in East Africa was probably enough to awe even the most blasé.

It would be easy to suggest that widespread destruction or alteration of vegetation was well underway by that time. However, we have no reason to believe that what was observed was not a normal condition, because if the observed use of fire were greater than required to maintain an ecological *status quo,* we could not account for the fact that herders and farmers lived in these areas for thousands of years without destroying their ecosystems.

Figure 2.4 *Fire burning uncontrolled in East Africa.*

The major degradation of the subsaharan Africa ecosystems started only recently. The evidence for this is the accounts of persons who saw the "old Africa" of the late nineteenth and early twentieth centuries, and then, returning after an absence, discovered that the landscapes they had come to love were no more or were in the process of being altered. In addition, there are numerous government reports that detail the destruction of the vegetation.

The destruction caused by slash and burn cultivators and herders is not confined to any particular region. However, there is undoubtedly a relationship between the growth of human population and the rate of vegetation changes in an individual region. For example, Uganda has had a fairly rapid increase in human population for at least a century and very little of the forests noted as recently as one hundred years ago are still standing. And the remaining African forests are in retreat before this human onslaught, which the ecosystems are no longer able to withstand. No major ecosystem is exempt from this change, and the result is a rapidly increasing spread of plant species that are resistent to fire and drought, the reduction of the area of forests, and the southward extension of the border of the Sahara. One authority has stated that only about one third of the potential forest area in Africa is presently forested.

Some persons do not believe that so much could happen in little more than a century, but we have ample evidence that only a few decades are required, given the population imbalances, to produce dramatic changes in the vegetation cover.

Much of the vegetation of Africa is shaped and modified by human agency. We are just now beginning to learn the scope of this influence and we must grasp this if man is to continue to live in African ecosystems. Slash and burn cultivation and nomadic herding have become ecological anachronisms in most of Africa. If left uncontrolled as human numbers increase, they can cause Africa to become a dying land.

European- and Arab-Dominated Land Use. As noted above, Europeans established themselves on the west coast of Africa in the fifteenth century and the Arabs were established on the east coast of Africa by the thirteenth century (although the Arab influence had been felt even earlier). Both groups introduced practices that modified the plant geography of Africa.

Unfortunately, we have only sparse data on the Arab influence on the vegetation. We do know that they were relatively uninterested in agriculture. There are crop plants in this region that could have been introduced by the Arabs, but confirmation of this is lacking at present. Late in the Arab period, cloves were introduced to Zanzibar and Pemba, both of which are islands but are considered here because they are part of Tanzania. The clove plantations were established in the first half of the nineteenth century, at which time coconut plantations also became important. It is not yet established when the coconut arrived in Africa, but its commercial production dates from this time and was first the occupation of persons under Arab control. In the second half of the nineteenth century, sugar cane plantations were also set out on Zanzibar, but sugar had probably been introduced here earlier. Although Arab control in this area gave way to European control late in the nineteenth century, the agricultural landscapes created by the Arabs dominate the islands today as they did a century ago. Most of the vegetation cover is therefore the result of human modification of the previously existing moist tropical forest (tropical deciduous forest).

European land use, insofar as it relates to vegetation, may be conveniently divided into the following categories: commercial agriculture with mixed crops; commercial agriculture with specialty crops; livestock; and forestry. With the founding of colonies, largely in the nineteenth century, a need arose to find ways by which the settlements could support the settlers and, if possible, provide a surplus for export. There is no single sequence by which this occurred over all of Africa. Therefore, we shall begin this discussion with the first area of permanent European settlement,

which was the extreme south of the continent.

After an initial period of settlement around Cape Town in the mid-seventeenth century, some of the Dutch settlers moved into the bush to lead a life uncontrolled by the Dutch authorities. These persons, known as Boers (farmers), developed, in the eighteenth century, a life based on cattle keeping. With their ox-drawn carts, they moved into territory previously controlled, at least nominally, by Africans. Although their cattle were of both African and European origin, they more-or-less adopted the range management practices they observed among the African peoples whose lands they took. That is, they used fire to remove unwanted dried grass and encourage the growth of new grass. Thus, ecologically speaking, the cultural turnover in this region, in its earlier phase, was not important. Changes in land use were to come later, however, when the nomadic *Voer Trekers* disappeared. Then it became clear that uncontrolled burning of the veld was unacceptable for sedentary ranchers. Although laws were passed early to control burning, they appear to have been ignored until the population increase and British victory in the Boer War gave the government a relatively secure control of the region. However, by that time, severe damage had been done to the woody vegetation of much of the area.

The next most important event in South Africa was the discovery of diamonds and other valuable minerals. This resulted in the exploitation of the remaining wood reserves for mine timbers and fuel. Thus, areas were stripped of their tree cover. Belatedly, the government sought to control this exploitation and, in the early twentieth century, began to look for trees that might be introduced to the region that would provide the needed wood. This effort resulted in the introduction of numerous tree species. The most important of these, in terms of the size of the area covered in South Africa, are pines—from Mexico and the Caribbean and Mediterranean regions. Australia yielded *Eucalyptus* species. These and other exotic trees now lend a degree of wildness to what would otherwise be deforested expanses in several parts of Southern Africa. During the last century, pasture grasses were also introduced, and some of these are well established on the better managed cattle farms.

The developments that took place in West and East Africa followed different courses and their effects on the vegetation cover were therefore different from those in South Africa. In West Africa, the European colonial powers followed the pattern of establishing plantation-type operations to provide exports and make use of the abundant, inexpensive African labor. Chief among the plantation crops was cacao (*Theobroma cacao*), which is produced from a tree of American tropical origin (Fig. 2.5). It was introduced near the end of the nineteenth century and became so well established that Ghana has for some time been a world leader in its production and export. Cacao (or cocoa, as it is known in Africa) is not tolerant of high wind and requires conditions of high atmospheric

Figure 2.5 *Young cacao plantation in Ghana.*

humidity most of the year. Therefore, large tracts of tropical deciduous forest have been either removed or reduced to accommodate this crop. Reactions of ecologists to this crop are mixed. Some feel that its spread has been destructive to the vegetation and others point out that the ecological changes are less severe than they would have been if the same areas had been given over to slash and burn cultivation. There is no question, however, that the cultivation of cacao has resulted in the waste of valuable hardwood species and a marked alteration of the physical character of the vegetation. The forests have been exploited for commercially valuable trees; and exotic trees, including rubber (*Hevea brasiliensis*), teak (*Tectona grandis*), and tropical cedar (*Cedrela mexicana*) have been introduced as plantation crops.

In East Africa various attempts were made by the colonial governments to establish crops for which there would be a market outside the territory. Many of these crops were exotic to the region. They include cinchona (*Cinchona*); coffee (*Coffea arabica*), which is autochthonous to Africa, probably Ethiopia; peanut (*Arachis*) and papaya (*Carica papaya*) from the New World tropics; pyrethrum (*Chrysanthemum cinerariaefolium*) probably a native of the Dalmation region; sisal (*Agave* sp.), a native of the Americas; sugar cane (*Sacharum*), a native of southeast Asia; tea (*Thea*), from Asia; and black wattle (*Acacia mollisima*), a native of Australia. The effects of these crops on the vegetative landscape have been relatively local because the crops are confined to those areas the Europeans chose as suitable for white settlement (Fig. 2.6). However, in some areas the lands have been transformed so much that they resemble parts of Western Europe and the United States.

Forested areas in much of East Africa had been greatly reduced prior to European entry, and most of the remaining forests were on mountains such as Mount

Figure 2.6 *Wheat fields of European farmers in East Africa.*

Kenya, Mount Kilimanjaro, and the Arusha Mountains. In recent decades, plantations of *Eucalyptus* and pines (all exotics) have been set out, and these, if successful, may cover fairly large areas within the next fifty years. At present, however, these plantings are local and do not greatly modify the prevailing ecological conditions.

Our review of the vegetation of subsaharan Africa has shown that the greatest source of human influence has been and continues to be African. Non-African cultural influences are found in those areas where European colonization was of a scale and duration that resulted in significant modification of the vegetation. Most important, however, is the fact that almost all the vegetation cover of subsaharan Africa is in some degree an artifact of human creation and maintenance. Man is and has long been a major factor in the plant ecology of this region.

2.3 Influences on Animal Life

Subsaharan Africa comprises one large zoogeographical unit—the Ethiopian region. Traditionally, Madagascar has been included in the region but there are good reasons to consider Madagascar a separate entity. It is so treated in this book.

Today, the Sahara forms a major physical barrier to the dispersal of many animals out of and into the regions south of its arid expanses. One must keep in mind, however, that the present climatic conditions did not always exist. Even in late prehistoric time (late paleolithic), much of the Sahara was wetter and supported a richer mammalian fauna than it now does. Afri-

can lions ranged as far north as Greece as recently as 2000 years ago. Thus, the effectiveness of the Saharan barrier to the movement of land animals, including *Homo,* has varied over time.

One of the prominent features of subsaharan Africa is the similarity of faunal conditions over much of the land. However, it is useful to recognize two major subunits based on the vegetation and the animals in each of these units. One subregion is the forest area in the center of the continent and the other is the woodland and grass areas. Although the vegetative cover is much more complex than this breakdown might suggest, our purpose is to discuss the impact of man on the animal life in a general fashion so a more detailed breakdown of the region is not required.

The Ethiopian region is the richest region in the world in terms of large vertebrates. Such richness may suggest that there is no human impact on animals but the opposite is true. Man has had and is having a profound influence on the animal life of this region.

Preagricultural Period. As we have seen, human population during the preagricultural period in Africa must have been numerically small. This suggests that although men, armed with fire, altered the vegetation cover to some degree, they were unable to have any significant influence on the animal life. In recent years, however, some scholars have suggested that hunters might have exterminated some animal species during the Pleistocene period, but at present, there is only limited evidence to support this hypothesis. Nevertheless, if *Homo* functioned as a predator, he exerted direct and indirect forces on the gene pools of the animal species he exploited—just like any other African predator such as hyenas, lions, and leopards. Unfortunately, the ethnographic literature does not give us enough information about this aspect of hunting to allow us to construct a reliable predation model. It might be suspected that when hunting was a random activity, its selective importance was slight. But as hunting tools and techniques were improved and a taste for certain mammals developed, there was probably a tendency to concentrate on fewer species. Hunting may also have been done to acquire animals with particular pelt colors. Thus, we may reasonably hypothesize that man played some role in the evolutionary trends of the animals he hunted.

Of possibly greater importance to animal life were the effects of the vegetation changes caused by human-set fire. It was suggested earlier that vegetation alteration by gatherers was modest and, in the absence of archeological evidence, of fairly recent vintage. But man was undoubtedly creating a new series of vegetation complexes, most of which would tend to favor nonforest herbivores over forest forms. At present we can only wonder if the species diversity among the mammals relates in any way to these conditions. But it is astonishing that in the Ethiopian region, where man the hunter was present longest, the most varied assortment of large mammals existed into historic times.

Agriculture—Herding. The adoption and spread of agriculture appears to have resulted in an increase in human numbers. This fact alone indicates that there must have been increased pressure on the wild animal resource for food, because animal domestication was minor and we have no evidence that African agriculturalists kept *any* animal as a principal source of animal protein. Thus, while slash and burn cultivation substituted the cultivated field for the wild forest and savanna as the major source of calories in the diet, hunting must have increased in importance to provide the protein that was not supplied by the crop plants.

Was this increased exploitation sufficient to result in the extermination of any of the animals taken for food? The available evidence suggests that if such extinctions did occur, they were rare, and that the spread of agriculture had a salutory effect on many of the mammal populations. This may have been because fire opened up and kept open large tracts of land that would otherwise have been unsuited to many of the grazing mammals. Moreover, the size of human populations was controlled by factors other than the meat supply. Plant food availability, disease, and other factors kept human populations well below the point at which they would have exerted devastating impact on the numbers of animals.

The relative unimportance of animal

domestication in this region has been noted often and puzzles persons interested in this aspect of culture history. Only three animals were domesticated—the guinea hen (*Numida*), forest hog (*Potamochoerus porcus*), and honey bee (*Apis*). The guinea hen is not an important source of food—except the wild forms, which are hunted in many areas. The forest hog is no longer kept as a domesticated animal and may never have been of any economic importance. And the honey bee domestication was apparently East African and of only local significance, although the honey from wild hives has long been one of the most sought-after woodland products.

Thus, until recently, domesticated animals were unimportant in the ecology of African farmers. They have acquired *exotic* domesticates, including chickens, dogs, pigs (*Sus*), cattle, goats, and, in some instances, horses. But even today we seldom see these animals integrated into the farming ecosystems.

The occupation of parts of grassy Africa by herders resulted in the vegetation modifications described earlier. This, in turn, exerted influences on the native grazing mammals not unlike the influences of the slash and burn cultivators. There was a major difference, however, which has increased in importance in recent years. The herders' domesticated animals were in direct competition for the available plant food in the grassland ecosystems. To some degree, the effects of this competition on the wild animals may have been offset by the fire practices of the people, for the fires probably ranged beyond the areas actually used. Thus, native grazing and browsing mammals benefited by the expansion of grassland and brush areas.

Herders were often faced with the problem of large predators, particularly lions. However, all available evidence shows that wild game was sufficiently plentiful and was preferred by the native predators—except when an individual predator may have acquired a taste for cattle (or man).

At least one herding group, the Masai, institutionalized predator control by requiring that a young man, in order to achieve manhood status, assist in the killing of a lion using nothing more than a hand held spear. Whether this resulted in any important diminution of the lion population may be doubted, however, for lions were numerous in the Masai area until very recently.

Before discussing more recent time, we should assess the influences of farmers and herders on African wildlife prior to the arrival of Arabs and Europeans. In general, nonforest mammals prospered and may have achieved and maintained population sizes much greater than they would have if man had not induced and maintained the vegetation changes. The accounts written by the first Europeans on the scene describe an almost incredible faunal richness, in terms of both species diversity and population sizes. Whether by accident or design, man's use of Africa's ecosystems before the arrival of the Arabs and Europeans resulted in one of the most extraordinary arrays of large mammals the world has ever witnessed.

Arab Period. As we saw earlier, Arab settlement on the east coast of Africa dates from around the thirteenth century and continued until the latter part of the nineteenth century, although by then direct Arab influence was greatly diminished. For most of their stay in East Africa, the Arabs were traders, interested particularly in elephant ivory and other animal products. Ivory retained much economic importance until, in recent decades, plastic displaced it to a major degree. For the most part, the Arabs obtained ivory from the stores that the Africans kept for domestic purposes. There was some direct ivory hunting by Africans, but it was of limited significance (the ivory was a by-product of hunting elephants for meat). At present, there is no evidence that Arab operations were of such a scale as to have exerted any measurable influence on the animal life in Africa.

European Period. The impact of the Europeans on the animal life follows well-defined cultural (national) and historical lines. It is by identifying these lines that we can best obtain insights into this next to most recent period in subsaharan Africa.

Until the Dutch settlement of the Cape Town area in the middle seventeenth century, Europeans had had only an ephemeral influence on the animals of Africa. True, the Europeans, like the Arabs, obtained ivory and other animal products but apparently never in such quantity as to result in seriously disturbing the ecosystems from which the materials were withdrawn. However, in southern Africa this period ended when the Dutch decided that farming and perhaps livestock raising should be encouraged in order to provision Cape Town, at that time a maritime rest stop. This led to the establishment of the Boers—the European farmer class mentioned earlier—who, later, chafing under what they felt were harsh governmental restrictions, moved farther and farther from the sources of government con-

trol. The Boers increasingly focused on pastoral activity, crossing cattle from Europe with native stock. They continued the African herder's practice of burning the grasslands to provide maximum feed for the cattle, but they shot an extraordinary number of the larger mammals native to the South African region. As best as can be determined from published accounts, much of this shooting was not for food or to eliminate wild competitors on the veld but simply for recreation (but there were some important exceptions as noted below). At no time was the Boer population large enough that subsistence hunting alone would have reduced the numbers of game animals.

The early accounts appear to be in close agreement that when this vast area was first entered by whites it was teeming with large numbers of mammals and birds. Among the most common of the larger mammals was a zebra-like animal called a Quagga (*Equus quagga;* Fig. 2.7). It occupied a large range, more or less coincidental with the major expanses of grassland in southern Africa. The Boers shot the animal for sport, for food for their laborers and for the hides, which were fash-

Figure 2.7 *The extinct quagga.*

ioned into grain sacks. There was also a market for quagga hides in Cape Town. The animal appeared to be so abundant that there was no need to be concerned that it be overhunted, but it disappeared forever during the last quarter of the nineteenth century. Some skins and skeletal portions are now to be found in museums.

At about the same time, the southern subspecies of the African lion (*Panthera leo*) was also eliminated from the region. And in the first quarter of the twentieth century, the list of extinct animals was joined by Burchell's zebra (*Equus burchelli*) and a subspecies of Bubal's Hartebeest (*Alcelaphus buselaphus*). However, the importance of this activity is not the extinction of three or four mammals, but the great reduction in or complete removal of large mammals from the ecosystems of southern Africa. This area of subsaharan Africa leads in the alteration of the mammalian fauna after European entry and settlement. Today, animal reserves are scattered over the region and many of the species and subspecies native to the area have been preserved. But populations of larger mammals outside the reserves are localized and generally only tolerated where there is not some competing interest on the part of Africans or Europeans. Vast reaches of grasslands, which at the beginning of the nineteenth century contained a rich collection of antelopes, gazelles, rhinoceroses, giraffe, and the like, are now devoid of game mammals.

What has happened to these ecosystems as a result of the removal of animals? Essentially, there has been extirpation with replacement. The replacement is by man and his domesticated animals, crops, and introduced pastures. However, the replacement is rather ephemeral and could be largely turned back, if man so wished, to the beasts that previously stocked its expanses.

The second area to feel the European impact on animal life was East Africa, where, beginning in the last half of the nineteenth century, English and German colonists exploited the game resource to a marked degree. There is a particularly rich literature, both English and German, on this period, often written in the form of memoirs by persons who returned home after a number of years in Africa. This literature suggests that upper-class males were possessed of a form of hunting mania. There was a virtual love affair with firearms and interminable discussions about gun cartridge sizes and ballistics. Some of the hunting was for sport, that is, for the satisfaction of being out in the open in a difficult environment and taking physical risks to shoot game animals. However, it is doubtful that this subculture modified the faunal picture to any measurable degree. But close upon them came a group of men whose interest was commercial. They viewed the faunal richness of the region as a means to earn the capital that might provide them with a more permanent livelihood. And for some men, hunting offered an escape from a Britain grown dull and tedious with its preoccupation with textile mills and trade.

The chief focus of this commercial activity was elephant ivory. African elephants (*Loxodonta africana*) were numerous at that time, and there was a rich and all but insatiable market for the tusks. And because both males and females grow tusks, both were shot. The potential economic returns were so great that ivory hunting attracted an inordinate number of men during the early days of Kenya Colony. A colonial governor of Kenya is alleged to have expressed the hope that the ivory hunters would hasten and shoot every last elephant so they might then settle down to the business of developing the colony. However, elephant shooting was not confined to Kenya; it extended into Uganda, the Congo, and Tanzania (Tanganyika)—in fact almost anywhere these great beasts could be found.

To the number of "sportsmen" and ivory hunters were added ever-growing numbers of persons who hunted for the market that developed for hides. The game mammals appeared to be inexhaustible until about 1920, when it became evident that unless given some protection, many species would disappear or become exceedingly rare. Nevertheless, game was still so abundant as to amaze newcomers; they often found it difficult to believe that what they were viewing was only a remnant of what was once present.

Here it should be pointed out that beginning in the latter part of the nineteenth century—about 1890—Africa was swept from north to south by a viral disease called *rinderpest*. The disease took a terrible toll of wild and domesticated ungulates. The virus, which is not native to Africa, was accidentally introduced by man, possibly by infected cattle through the Nile Valley or by way of Somalia. It is noteworthy that even with this onslaught, game remained so abundant or recovered so rapidly that hunters in the early part of the twentieth century were greatly impressed.

In the East African region, as in southern Africa, a program providing for the establishment of game reserves was followed, but there was one important difference. Game laws were instituted to prevent the extirpation of nonreserve animals. This is not to say that the numbers of large mammals have not declined in the past seventy years. Many parts of East Africa, where forty years ago wild animals were seen commonly, no longer have any large herbivorous mammals, and the predators—particularly lions and leopards—have become only a memory over large tracts of land.

With the advent of game reserves and the development of hunting laws (with limits on the number of animals that might be killed), a licensing system, and other measures, Africans turned to illegal hunting, that is, poaching. The poachers operate either singly or in groups—recently some have become mechanized.

They hunt without regard to season or limit and employ any method that yields high returns. Wire snares are common and so are arrows tipped with a plant-derived poison such as that from the *Acokantha friesiorum* tree (Fig. 2.8). The proceeds of this hunting consist principally of meat, which is dried and sold. Rhinoceroses are killed for the horn, which enters a clandestine trade to parts of Asia. Many animals, particularly the larger cats, are killed for their pelts, which are also exported clandestinely, mainly to Europe and the United States. Even ivory finds a market, especially in India where its chief use is for women's and girls' bracelets.

It is difficult to evaluate accurately the effect of poachers on wildlife numbers. However, those charged with wildlife management in Africa generally agree that poaching has been important in accounting for the sharp reduction of native animal life for several decades. Of course, there is a certain amount of emotional bias in the discussions about poachers, because in the colonial period it was often illegal for an African to possess a rifle and engage in hunting. Moreover, guns and licenses were and often still are far beyond the economic means of Africans.

We noted earlier that when the economy changed from plant-food gathering to plant-food production, there was an increase in population, with a resulting increase in the pressure exerted on the wildlife. We have also noted that the colonial period was marked by an increase in population. Thus, we can assume that this population increase also resulted in an increased dependence on meat obtained from the wild resource. It should be noted, however, that although many Africans have long relied on wild game for their principal source of protein, no general quantitative or qualitative study has been made of this activity. It would probably be more accurate to suggest that if outside influences had not thrown the African's relationship with the faunal resource out of

Figure 2.8(a) *Zebra caught in an illegal snare set by poachers.*

Figure 2.8(b) *Game warden displaying traps and snares used by poachers in East Africa.*

balance, there would not now be clandestine meat hunting. Uncontrolled hunting will most assuredly result in a major reduction of animals in Africa.

Thus far, we have focused on the hunting of the larger animals of Africa and its effect on the numbers of the exploited species. There is, however, another aspect to the hunting story. Whenever animal populations are manipulated, either by direct exploitation or by habitat changes, differences result in how the surviving forms exert their influences on the ecosystems they inhabit. Each species of herbivore tends to be a *selective* feeder, and its numbers in relation to the other members of its ecosystem are held in check by a host of complex interacting factors. Thus, in time, an ecosystem comes to achieve an

equilibrium in which all the plant and animal members together with the abiotic elements function as a whole (see Chapter 1). When this structure has parts of its life forms removed or altered in numbers, the entire system will shift to accommodate the change. This is particularly noticeable in Africa when predators are removed. The result is an increase in the population of animals formerly held in check by the defunct predator. One example of this is the leopard, which has become rare in most of Africa because of the popularity of the hides for coats. In apparently direct response to this reduction, baboon populations have sharply increased in some areas and baboons are being reported where they had not been known previously (Fig. 2.9). The significance of this ecological shift can be realized when we note that baboons are frequently in a contest with man for crops and baboons suffer from some diseases common to man. As of now, however, we have only general knowledge of how African ecosystems function and cannot predict the results if there is a manipulation of the number and species of the wild animals. We only know there will be changes.

Discussions of the relationship between man and the animals of Africa usually are confined to man and the larger mammals, that is, the game animals. However, the in-

Figure 2.9 *Baboon troop.*

vertebrate animals have also been affected by man, and some of the results, in turn, affect the mammals. Of the invertebrate animals man has influenced, the most important to the ecology of subsaharan Africa is the tsetse fly (*Glossina*). This insect is present over an area greater than the continental United States, and because it is the principal carrier of nagana (Trypanosomiasis), a disease of major veterinary and human importance, man has made intensive efforts to eradicate it. Now confined to Africa, *Glossina* once had a much greater range—which may have included North America in the geological past.

Glossina transmits a flagellate protozoan of the genus *Trypanosoma* by feeding on a mammal that has the protozoan in its blood. The fly ingests the protozoan and then injects it into the next beast it feeds upon. The wild mammals of Africa are immune to nagana, but most cattle, sheep, goats, and pigs, if fed upon by an infected fly, will become ill. The fact that the disease has a high fatality rate has been a major factor in preventing the spread of animal husbandry in much of subsaharan Africa. The human form of the disease, known to non-African peoples as African sleeping sickness, fortunately is not common. The rate of fatality among persons infected is very high, and even though the disease is not frequent, its appearance in an area usually produces panic.

With few exceptions, the population sizes of *Glossina* are not great. But tsetse is numerous enough to be a major economic problem in Africa. Apparently, the first control measures consisted only of moving people and stock out of an area when the disease appeared and of avoiding those areas where the people had learned there was great danger. It was not until well into the European period that measures to control this insect by direct and indirect means were instituted.

The first and most widely used measure was eliminating the wild animal reservoirs. This was done to deprive the fly of its major source of food and to destroy the main source of trypanosome infection itself. So-called control shooting was initiated in various parts of Africa, especially in the former Rhodesias and in East Africa. The results did not live up to expectations, however, even though a large number of mammals were killed. In Rhodesia (former Southern Rhodesia), for example, it is recorded that from the decade of the 1930s until the end of 1953, under the program of tsetse control, 667,009 large mammals were destroyed. Pressure exerted by conservationists—plus the fact that the fly still continued to exist in the region—brought that program to a halt. Similar control shooting produced similar results elsewhere in Africa. The toll of wildlife was great—perhaps much greater than the available statistics indicate.

Glossina requires shrubs for protection

from the sun. It spends more time in the shade of leaves than in flying. Thus, in parts of Africa, brush was cut in order to reduce the number of flies. This activity was often limited by a lack of labor and money and was sometimes dangerous—animals such as rhinoceroses resented the intrusions. The danger was almost entirely from the black rhinoceros (*Diceros bicornis;* Fig. 2.10), so these animals were removed by control shooting. The most recent method of brush clearance is herbicides sprayed from the ground or the air. However, the defoliating effect of this procedure not only deprives the fly of needed resting places, it destroys the food sources of browsing mammals and may upset the ecology of other vertebrates and invertebrates as well.

Naturally insecticides are used in the drive against the fly, especially several persistent insecticides of the chlorinated hydrocarbon group (DDT being the most common). These are usually sprayed from aircraft, and experimental results indicate that this is an effective means of removing *Glossina* from large tracts of land at a relatively modest immediate economic cost. The more general effects on the rest of the invertebrate fauna in the sprayed areas are not known. If the results in other parts of the world apply here, however, a major disruption probably occurs in the inver-

Figure 2.10 *Black rhinoceros* (Diceros bicornis).

tebrate populations above and within the soil and vertebrate species whose ecologies are directly related to arthropods may also be affected. Thus, we may find repercussions in the bird life in the areas sprayed.

A still more modern method of control is that of sterilizing male flies and releasing them to copulate with the females. This method has proved successful with other fly species in other parts of the world but it is of limited value with the tsetse fly—probably because of the low population densities.

A review of the methods used against the fly indicates that short of a major alteration of Africa's biotic landscapes it will be impossible to eliminate the fly except locally. Thus, the problem is being increasingly attacked by applying chemical prophylaxis to livestock animals. Animals are innoculated to give them a high degree of immunity to the trypanosomes. If this proves successful, it will reduce the inputs of insecticides and herbicides into African ecosystems. However, it may mean that the rate of vegetation alteration will increase, for vaccinated cattle might invade areas that were previously closed to livestock. Thus, man's interactions with the tsetse fly may contain the key to much of Africa's ecological future.

No discussion of man's influence on invertebrates in Africa would be sufficient without mention of the disease called African schistosomiasis or bilharziasis. This disease of man and other primates is caused by a parasitic flatworm that lives in the venous systems of man and other animals. Two species are widely present in Africa—*Schistosoma haematobium* and *S. mansoni*. During the earlier parts of their life cycles, the flukes enter certain species of freshwater snails. Man becomes infected by being in water in which the free-swimming larval form (*cercaria*) occurs. It easily enters the skin (or mouth, if the water is drunk) and moves ultimately to the veins of the urinary bladder. From that organ, eggs of the fluke are periodically passed out with the urine to continue the cycle. The disease is old—evidence of schistosomiasis has been found in Egyptian mummies. The chief injury to man occurs when the eggs previously laid in the walls of the bladder leave with the urine. As the eggs pass through the wall a great irritation is set up that results in the production of phagocytes, and the walls of the bladder become inflamed. With long-standing infections—and this is very common—the walls of the bladder become thickened and the bladder functions are interfered with. Bilharziasis is not a killer but it is debilitating. The disease is found in almost all parts of Africa where there is still or slow-moving surface water. It is estimated that at least 60 percent of Africa's people are infected with these flukes.

The principal influence man has on the snail (the intermediate host) is the enlargement of the areas of fresh water available to them. Thus, irrigation schemes, which are increasingly frequent in Africa, carry with them the threat of augmenting the range of the disease. Control of the host snail has been successful only locally, such as in the mining areas of southern Africa where copper compounds have been introduced into the water for snail control. Any control effort, however, carries the possibility that other animals will also be injured and that the entire aquatic ecosystems into which the molluscicides have been introduced may experience important alterations.

In brief summary, therefore, we have seen that man's influence on the animals of Africa goes far back in time but probably became important only after agriculture was widely established. The chief effect of agriculture appears to have been to increase the number of "game" mammals; it was this richness that was encountered by the first Arabs and Europeans to enter Africa. The Europeans engaged in a number of practices that, in general, led to the decline of the larger vertebrates of Africa. Moreover, through their efforts to control the ecosystems, they altered many of the previously existing ecological equilibria. The Africans also contributed to the altered faunal conditions but largely as a response to demographic and cultural changes resulting from European influences.

It is not the object of this book to propose conservation measures for Africa or any other part of the world, but the fauna—vertebrate and invertebrate—of Africa's subsaharan ecosystems stand at a crucial point in biogeographical history. The fate of the continent's wildlife now rests largely in the hands of the African people themselves. If certain practices and trends established in the colonial period that are still being followed continue very far into the future, the faunal map of Africa in relation to the larger mammals will include only the game parks and reserves. And, even that may be too optimistic a vision.

2.4 Influences on Surface and Subsurface Waters

When man first disturbed the vegetation cover of Africa, he inadvertently altered the water balances as well. As vegetation alteration increased, so did the alteration of stream flow patterns, water retention in the soil, and, presumably, the watertable levels. In areas where shifting cultivators or herdsmen became so numerous as to reduce the vegetation cover over extensive watershed areas, we may be certain that seasonal flooding became more pronounced. In addition, the seasons of floods were undoubtedly followed by a season or seasons during which the water flow fell below the norms established before human manipulation.

African rainfall nonreliability made these effects even worse. Over much of the region south of the Sahara (and, to be sure, including the Sahara), rainfall reliability is very low. That is, wide annual deviations from the arithmetic mean is the usual condition. Thus when heavy rains occur, their effects are likely to be catastrophic to watersheds that have been laid bare by farmer or herder. Nevertheless, we may assume that until the last century or so a series of man-plant-animal-water equilibria became established in much of this region. If such equilibria existed, however, they have been increasingly interfered with during recent decades. And the prospect for the immediate future is even greater manipulation of Africa's surface and subsurface waters.

From a technological-engineering point of view, the hydrographic changes taking place in Africa at the present time constitute the most dramatic aspect of man's influences in that region. The size and scope of the larger projects of water control are among the greatest ever undertaken anywhere. And these activities are having a profound effect on the aquatic and nonaquatic ecosystems.

Map 2.2 on page 79, which portrays the surface waters of Africa, enables us to make two generalizations. The surface waters are impressive; but they are extremely scarce over large parts of the continent, including the area south of the Sahara. The river that dominates the northern part of the continent is the Nile. It is actually a collection of rivers—Blue Nile, Albert Nile, White Nile—and wends its way across an enormous region. For the sake of convenience, in this chapter we shall discuss only the part of the Nile that is not in the United Arab Republic (Egypt). The area of the Nile in Egypt will be discussed in the chapter dealing with Europe and the Mediterranean. Moving south through the eastern part of Africa, there

are no large rivers until the Zambezi. Like all major African rivers, the Zambezi crosses several political territories. It finally enters the Indian Ocean in Mozambique. Next there is the Limpopo River, which is much smaller than the Zambezi, and then the Vaal. Thence around the Cape to the Orange River, which looks much more important on the map than it is in terms of the water it carries. We then pass over a large expanse of Africa in which surface streams are few and unreliable in their flow. Finally, we arrive at the mouth of the magnificent Congo River, which, by means of a large system of tributary rivers, drains the Congo Basin. So large and far flung is the Congo drainage that one part may be in flood stage while another can be low enough to impede river navigation. Continuing north we pass over several minor rivers and arrive at the delta of the Niger. This extraordinary river begins its journey to the Atlantic close to that ocean but runs inland for hundreds of miles. It passes the fabled city of Timbouctu and makes a great bend to find its direction again. Hundreds of miles later it finally debouches into the Atlantic Ocean. Farther north along the coast we encounter the Volta River, which appears modest lying next to the Niger but is a great river nonetheless. Then, many miles northward, we come to the last of Africa's major streams, the Senegal. The large streams of Africa, examined together as we have just done, appear to be numerous, but you can see from the map that Africa also has large regions, aside from the Sahara, that are not drained by a major river.

The potential of the large and many smaller rivers of Africa as sources for water and power were recognized early in this century, and engineering surveys were conducted in the early decades. However, limited capital and limited demand postponed development. The first demand that was created was for power. Africa is, by and large, poor in coal, and the major petroleum fields thus far discovered have

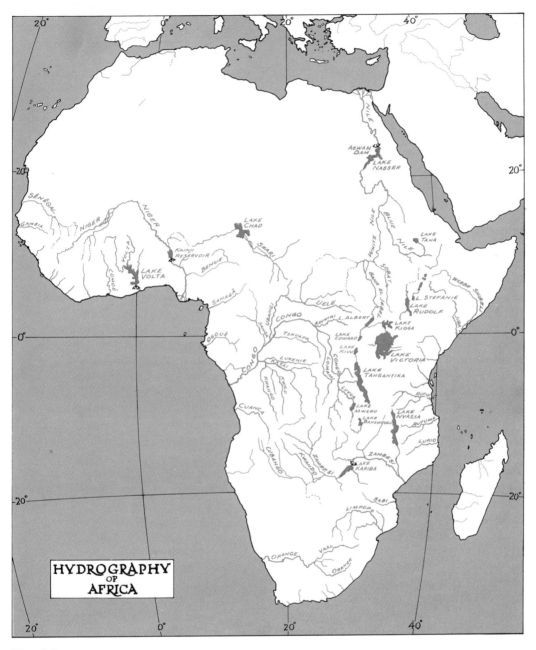

Map 2.2

been exploited for only a brief time. Long before this, a need for power was created with the development of mining and industrial activity, particularly in the southern part of the continent. This demand gave rise to the construction of the Kariba dam on the Zambezi River, in what is now Rhodesia (Map 2.2). This dam impounds a lake that has a surface area of approximately 4300 square kilometers and is approximately 123 meters deep at the deepest place. It is said to be the largest man-made lake in the world, based on waterholding capacity.

A more recent project, also chiefly for the production of power, is the Akosombo Dam in Ghana. The dam impounds a lake—Volta Lake—on the Volta River that has a surface area of approximately 8555 square kilometers. The greatest depth of the lake is approximately 70 meters. This lake is so large it can be shown on a map of relatively small scale. Of even more recent vintage is a dam on the Niger River at Kainji. The lake is about 1280 square kilometers in area and has a maximum depth of 55 meters.

These three dams have all been created within the past decade. Two dams were constructed earlier in what is now Sudan. The oldest of these, the Sennar Reservoir, was completed in 1925 and has a surface area of 600 square kilometers.

It is obvious that such large projects must have produced significant ecological changes. Fortunately, the ecological implications of these huge man-made lakes was recognized quickly, so studies were initiated to determine their impact on the aquatic and nonaquatic ecosystems.

Thus far, chief attention has focused on the Kariba and Volta projects so our discussion will relate chiefly to those lakes. However, what is being learned there may well be applicable to all large artificial lakes in Africa.

The first and most obvious ecological change that took place in response to the Kariba project was the removal of approximately 50,000 persons, mostly Africans, from lands that were to be inundated by water. Some aspects of the ecology of the cultures of these people, who were resettled on land made available for the purpose, have been studied by anthropologists.

Ecological attention first focused on the lake. That is, what would occur when a river was stopped in its flight and turned into a lake? First of all, before the lake was filled, an effort was made to rescue some of the animals that would be drowned by the rising waters. This project, named operation Noah, received much attention in the world press but probably succeeded in saving only a small fraction of the animals in the area.

Damming the river naturally slowed the flow of the water and this offered opportunities to those aquatic organisms—plant and animal—that thrive under such conditions. Furthermore, the great mass of organic (chiefly plant) detritus in the water provided a rich nutrient source for some animals such as fish of the genus *Tilapia*. This was a beneficial effect, because *Tilapia*, as well as some other fish favored by the new conditions, are highly valued for food and supply protein to the often protein-poor diets of the people inhabiting the region. An unfortunate consequence of the new conditions, however, was the marked increase in populations of the snails that act as intermediate hosts for schistosoma species (see previous section). This was worsened by the (now discontinued) practice of clearing all the trees adjacent to the shoreline that would be drowned. These cleared areas provided a habitat for water weeds, which, in turn, provided an excellent habitat for the snails.

The most visible and perhaps most dramatic change was the invasion of aquatic plants. For a time, the surface matt of vegetation was so thick that navigation on the lake was threatened. The plant responsible for this, *Salvinia auriculata*, is a native of South America (Fig. 2.11) and first appeared in the Zambezi drainage in the 1940s. However, it was not a problem then because it did not survive well in the swift water and was found only in the limited areas that had slow-moving currents. Almost as soon as the lake behind the dam began to fill, this weed began its population explosion. In a short time it formed large, floating matts of vegetation, which provided a niche for as many as forty other plant species. At its peak, the *Salvinia* masses covered almost 20 percent of the lake surface. In the last few years a decline

Figure 2.11 *Infestation of* Salvinia *in Kariba lake.*

has been noted, and it may be that the final equilibrium point will result in a less ecologically dramatic and damaging situation. The plant shades the water below it, interferes with oxygen exchange, and creates a habitat very unfavorable to most aquatic animals.

Until recently, ecological study focused on the area above the Kariba dam, but it became evident that possibly far greater ecological changes were taking place below the dam. In many respects, the principles that were discovered in areas downstream from the Kariba dam apply to streams below dams anywhere in the world.

One chief purpose of any dam is to provide a regular flow of water throughout the year. This means that floods, if they occurred prior to the construction of the dam, will no longer occur except under unusual circumstances. In the case of the Zambezi below the dam—the middle Zambezi, as it is known—floods were a regular event before the dam, and this flooding was an integral feature in the aquatic and nonaquatic ecosystems. Between floods, the floodplain was used by various species of herbivorous mammals, which were forced out of the area during the floods because water covered their feeding ground. One effect of the dam is that now the herbivores are able to remain in the

floodplain all year. This means that there is no resting period for the plant biomass and it is no longer sufficient to carry the herbivorous mammals. The species composition of the plant cover is changing. Even without the added grazing-browsing pressure, some of the vegetation that had been adapted to annual inundation would disappear, and this has occurred.

Although floods have been prevented, water is occasionally released from the dam at irregular intervals, and these sudden releases may have damaging effects on the wildlife below. For example, reptiles, which lay their eggs in the soil near the river in what is normally the low-water period, may have their eggs washed away. Nesting birds, adapted to the dry-season conditions, may find their nests, eggs, or nestlings destroyed when there is unseasonal human-induced rises in the river water. However, these rises do have the beneficial effect of flushing unwanted aquatic vegetation down the river. A geomorphic effect that has ecological implications relates to the fact that when the river is kept low, during periods of lake filling, the bottom of the stream tends to be scoured and, conversely, when there is a sudden spate, bank erosion may rapidly increase. In addition, the chemical nature of the water below the dam has probably changed, especially in regard to the reduced silt load. Thus, the nutrient content of the water is lowered and this has consequences for some of the plant and animal life (formerly?) able to live in this stretch of the river.

Ecological changes associated with the Volta project are similar to those of the Zambezi, except that there is relatively little river between the dam and the ocean and no recent history of a complex vertebrate fauna utilizing seasonally dry floodplains for food or other purposes. Thus far, no invasion of Volta Lake by exotic vegetation has been reported, but this may occur in the future because exotic aquatic plants move around Africa with considerable speed. Like the Kariba case, there has been a great increase in the fish life, especially *Tilapia,* in the water behind the dam. Here also, the slowing of the water has encouraged the propagation of snails involved with schistosomiasis.

The ecology below the dam has also undergone changes but they are not so dramatic as those below the Kariba dam. These changes included the drying out of creeks, the reduction of fish life formerly of dietary importance to the human residents, the reduction of agriculture that

formerly used the annually flooded areas after the silt-bearing water had drained away, and the incursion of salt water in the lower part of the river.

Clearly, to build a dam is to manipulate some complex and incompletely understood ecosystems. And there are many signs that other large dams will be constructed in Africa, so the aquatic ecology of the continent will undergo even more extensive changes. It is probable that most of these dams will be constructed in the Congo Basin, which is said to contain a hydroelectric potential exceeding 100 million horsepower—an amount equal to about 25 percent of the world's total hydroelectric power potential. However, we may also look to the upper Nile for projects, particularly in the vast area in Sudan where hundreds of square miles of water are covered by floating vegetation called *Sudd.* Preliminary studies in the Sudd region (Fig. 2.12) suggest the possibility of ridding the area of the vegetation and thus making available large quantities of water for irrigation. If this comes about, the present ecology of this region will surely undergo important changes, which cannot now be predicted.

Turning from the giant man-made lakes

Figure 2.12 Sudd *vegetation on the Nile in Sudan.*

of subsaharan Africa, it is important to examine the opposite end of the scale, namely, small lakes and ponds. These bodies of water are impossible to map, although they are almost everywhere on the African landscape. Being an easy means of providing water storage against dry periods and a means whereby fish, principally *Tilapia,* can be grown to provide protein for human populations, their formation has been encouraged by governmental authorities. Unfortunately, little is known of the ecology of such ponds in relation to the adjacent ecosystems. However, the mere fact of their abundant and increasing existence is worth noting, and they definitely are extending the ecological support base for man in Africa. Of course, where ponds are sufficiently numerous one might expect an increased input into subsurface water supplies, but we have to await the development of more ponds and small lakes before this can be shown.

As noted earlier, surface water is scarce over large reaches of subsaharan Africa, especially during the dry seasons. One of the measures employed to overcome this has been to drill or dig wells—thousands of these have been constructed within the past three or four decades. The chief immediate ecological effect of the wells is to allow man to move himself and his animals into regions that were formerly too dry to permit other than transitory use. By extending the human ecological base, the ecosystems will be subjected to greater influence by man. However, the African population increase has made such measures necessary and thus far little thought has been given to the ecological implications.

This brings us to a discussion of irrigation. One might think that irrigation should have been part of the discussion of the big lakes, but the largest lakes are used principally for the production of electricity and there is only a limited need for irrigation water in the dam areas. On the other hand, the dams in Sudan, for example, have been constructed principally for irrigation (Fig. 2.13). Also, a large and rapidly growing array of dams of lesser scale have been constructed chiefly to provide irrigation water. In terms of land area affected, this aspect of hydrographic change in subsaharan Africa is by far the most important.

There is scarcely a country in Africa that is without irrigation plans. And, as would be expected, the number and areal importance of these plans are in direct proportion to the seasonal scarcity of surface

Figure 2.13 *Part of a major irrigation project in Sudan.*

water for agriculture. As Africa's human population grows, the pressure for the development of irrigated lands will grow. Unfortunately, there is a lack of information regarding what happens to African ecosystems when they have their water budgets sharply altered as a result of irrigation. However, the general features and problems can be examined.

The greater availability of surface water makes it almost certain that schistosomiasis will spread and so too will diseases that have insect vectors whose life cycles are at least partly spent in water. Another problem relates to soil changes that sometimes occur when irrigation water is not managed properly. One result can be great stretches of saline flats that are useless for agriculture and are ecological deserts. Irrigation schemes also encourage greater concentrations of people, which facilitates the transmission of communicable diseases. Of course, there are many economic benefits as well.

Water pollution is not yet a major problem in subsaharan Africa but it certainly exists. In the case of *Schistosoma* you will recall that the eggs are forced through the bladder wall of the human host and voided with the urine. The fact that ponds, lakes, and rivers are favorite places for urination is largely responsible for the widespread distribution and persistence of this ailment. Similarly, it has been shown that Africans are frequently infected with intestinal parasites whose transmission is facilitated when human feces are allowed to enter the water supplies used for domestic purposes. In general, little attempt has been made to avoid contaminating water with human wastes and the people

do not understand that contaminated water should be treated before it is used for domestic purposes. Of more local importance but growing in significance is water pollution due to inputs of chemicals. Two principal sources can be noted—mining and associated operations and insecticides and herbicides.

Mining, which will be discussed in greater detail in the next section, is distributed widely but locally in subsaharan Africa. The principal areas of concentration are located in southern Africa, with the result that streams are frequently contaminated with mine waste. Similar conditions are found in the highly mineralized Katanga mining district in Zaïre. They are also found elsewhere—but on very localized bases.

Insecticides and herbicides have come into ever-growing use in Africa, as we discussed earlier, and these often find their way into aquatic ecosystems. Their effects have received little attention thus far. The few data available indicate that at least for a brief time after application, DDT produces measurable reductions in fish life, although in one case the fish population rebounded to a much higher level than before application and then fell back to normal. Considering the widespread use of insecticides in Africa, there is a clear need to study the short and long run effects of these chemicals in all of the continent's major ecosystems. Herbicides have been used in tsetse control, as indicated in an earlier section, and these chemicals have also been used to control unwanted vegetation around some of the new lakes. The long-lasting ecological effects are not yet known.

2.5 Influences on the Soils and Geomorphology

Human influences on the soils and geomorphology of Africa extend far back into time, but, as we have seen in previous discussions, the greatest degree of impact has come within the last few thousand years. These influences may be divided into two broad categories—influences on the soil cover and influences on the geomorphology associated with mining operations.

Soil. During that long stretch of time when the only human technology in Africa was hunting and gathering, it is unlikely that significant impact was exerted on the soil until fire was obtained. You will recall that we do not know the antiquity of human-controlled fire in Africa but that a figure of approximately 60,000 years has been offered as a reasonable estimate. In any case, we may assume that when human-set fire became part of the dynamics of African ecosystems, the previous rate of soil erosion shifted upward, at least locally. The severity depended on the frequency of the fires, the time of year they were set, and the nature of the slope and soil over which the fires burned.

When agriculture became widely established, there also must have followed an increased rate of soil erosion. This must also have been the case when the herders arrived. With their herds and use of fire, they were in a prime position to accelerate the rate of soil erosion. However, after a time, a certain degree of stability must have been achieved, except for unusual local circumstances. We do not find many allusions to soil loss in the first travelers' accounts (although some of these people would not have recognized soil erosion under any circumstances).

Allowing for a certain degree of ecological myopia, we must accept the fact that soil erosion on a continental scale is a phenomenon of colonial times and the still short period of political independence. During this time, one of the most visible human-induced changes in the landscapes of Africa is the great loss of soil over thousands of square miles.

These changes relate to European influences, perhaps the most important of which was injecting forces that led to an increase in human population. We have seen that this increased population led to the more rapid removal of vegetation. But the increased population also led, as a direct result of that removal, to a dramatic increase in the loss of Africa's soils. One person, viewing an African river at flood stage and noting its red, soil-filled waters, exclaimed, "Look, Africa is bleeding to death!" And the aphorism is well taken, for the soil and water are indeed Africa's life's blood.

No region is free of soil erosion induced by human use of the land. However, some regions have been more seriously affected than others; these include Uganda, Kenya, Tanzania, and vast tracts of southern Africa, particularly Bechuanaland and Basutoland. The chief causes are cropping on too short a fallow (rest) period and too many domestic animals on the available land (Fig. 2.14).

This subject is emotionally charged and has many political nuances, because some of the conditions arose during the colonial period, when Africans were kept on "native reserves." As human population grew in some of these reserves, the traditional ecological ways of life became unsuitable, with the resulting destruction of soil cover.

In addition to the increased pressure on the soil exerted by the growing number of slash and burn cultivators, the introduction of commercial crops, such as cotton in Uganda, resulted in intensive farming with little or no fallow period. In some cases, the soils simply could not withstand this type of use, the result was and continues to be great soil loss.

Figure 2.14 *Extreme accelerated soil erosion in Kenya.*

During this period, the number of livestock in African hands increased dramatically. Moreover, colonial officials prevented the once nomadic peoples from wandering over the territories they had occupied seasonally. The result was that a greatly increased number of animals had less and less land available for grazing. In addition, the remaining grazing lands were all too often ecologically marginal, the better expanses having been preempted by Europeans.

But one must not believe that only an expanding African population caused this soil loss. European practices had similar consequences. This was especially significant in parts of southern Africa, where large groups of cattle were kept in limited spaces (kraaling). After the grass had been consumed and the roots cut and trampled, the herd would be moved to another spot and the process would be repeated. European farming practices in many respects were far less enlightened than those of

most Africans. But the direct effects on the soil were often highly localized because of the insular nature of the European settlements in agricultural areas. The chief exception is southern Africa, where, even though persons of European descent are a minority, they are sufficiently widespread to exert considerable ecological impact over large areas.

In addition to the *physical* loss of soil through accelerated erosion, the chemical and physical changes that take place as a result of land-use practices must also be considered. The most immediate impact on the soil is caused by grass fire. For a few minutes, temperatures in the upper few millimeters of the soil may soar to over 600°C. In one experiment in South Africa, it was found that after a grass fire passed, the surface temperature remained as high as 550°C for six minutes but that at a depth of approximately 5 centimeters, there was barely any temperature increase. Other studies repeat the finding that only the top layer of soil is subjected to an extreme change in temperature with the passage of a grass fire. When wood is piled and burned, as in the Chitemene system, the soil temperatures may go higher and penetrate more deeply. The soil microflora and microfauna are also influenced by the passage of fire, but there are insufficient data to permit a generalization of its long-term import.

More subtle but probably of greater significance than the passage of fire is the change in soil temperature that results when tree or brush cover is removed and the soil is exposed to the sun. In a tropical forest or dense woodland, soil temperatures tend to be nearly isothermal, that is, they experience only slight diurnal variations. However, when such an area is cleared, there usually is a sharply increased diurnal range of temperature extending at least 15 to 25 centimeters into the soil—depending on grass cover, nature of the soil, and exposure conditions in general.

As noted earlier, one benefit of fire to shifting cultivators (in addition to the obvious one of clearing away unwanted vegetation) is that the resultant ash supplies needed nutrients for the crops. This, however, is of transitory significance, and if the fallow period is shortened too much, a general deterioration in the physical and chemical nature of the soil results. This appears to be occurring with increasing frequency in Africa.

In our discussion, we have mentioned the use of insecticides for the control of the tsetse fly and other insects. As yet, research is not sufficiently developed to permit a general statement as to the effects of these insecticides on the soils of Africa. However, we may assume that in those areas where the use of insecticides is sufficient to accomplish the desired changes in insect populations, the biotic content of the soils also must be altered.

Geomorphology. Mining of a sort began when a person first picked up a bit of stone in order to use it for some purpose. However, this activity, which later became the basis of lithic "industries" in

the Paleolithic, could not have left any visible mark on the earth. Moreover, there is no obvious evidence that even the arrival of the Iron Age to subsaharan Africa resulted in any but the most local changes of the geomorphology. The discovery of mineral deposits during the colonial period changed that.

The first major discoveries of valuable minerals were made in southern Africa in the latter part of the nineteenth century. The region has since become known as the most important diamond mining district in the world. Also of continuing importance are the gold mines located in the Johannesburg district (Fig. 2.15). Other mineral deposits have also been discovered and exploited in this region. The result has been the creation of enormous piles of mine refuse ("tailings") which dominate the geomorphology of the local scene. In addition, there are dramatic gouges and holes in the earth—the result of the great volumes of material removed. People's reactions, when viewing the mine dumps, vary. Some feel consternation, but the other extreme was expressed by a person noting the Witwatersrand mining district

Figure 2.15 *Mine dumps in the Johannesburg area, South Africa.*

near Johannesburg where some 60 million tons of mine wastes are dumped each year. He dismissed this by saying that the huge piles of detritus "bring an interesting element of relief . . . to an otherwise rather monotonous and unproductive landscape." Beauty, truly, is in the eye of the beholder.

The Katanga district of the southern part of the Congo Republic is a mining area important for its copper. An open-pit method is used here, which results in the huge scars in the earth typical of this mining method. Mines are scattered over an area roughly 450 kilometers long and 90 kilometers wide, although most of the larger ones are grouped in a relatively small area.

In addition to the great mining regions, there are many smaller mining districts in Africa. Each one produces geomorphic changes as a result of the movement of earth. Although of a local nature, the geomorphic expression has frequently been dramatically altered by man and the changes are likely to remain for centuries to come.

2.6 Influences on the Atmosphere

Weather records for most of the subsaharan region are of such short duration as to make any discussion of trends during the last few centuries all but impossible. Nevertheless, there has been much speculation regarding climatic change in this region during the past hundred years or so, such as a debate as to whether the Sahara is extending southward as a result of increasing dryness. However, we are not yet in a position to determine what, if any, *macroclimatic* changes are taking place in Africa.

There is one area of climatology, however, to which we can address our attention—micro or bioclimatology. You may recall from the discussion of climate in Chapter 1 that the microclimate refers to the part of the atmosphere that is in contact with an organism or object and is, in some degree, influencing the organism or object. Viewed from this scale, we can assert that man has influenced and continues to influence the atmosphere of Africa.

When man alters the vegetation cover, he also alters the microclimates of the area to some degree. These changes may be modest, such as when his livestock graze a Kenyan savanna, or profound, such as when he cuts down heavy forest to make way for crop cultivation of pasture for his animals. In the latter cases, the previously existing conditions of relatively stable air temperature and humidity shift to conditions of broad diurnal changes.

When the fallow period is long, the microclimatic changes are as ephemeral as the period of cultivation; but when custom or circumstances require short fallow periods, the microclimates have greater significance because they persist for greater periods of time. One of the most important of these persistent features is the dehydrating effects of the air near the ground. In the absence of trees, the air moves much more freely, is heated more readily during the day, and acts as a transporting agent for the moisture given up by the soil and the remaining vegetation. In areas where before cutting there was a delicate equilibrium between precipitation and the maintenance of forest, removal of the forest for a long period of time might result in microclimatic changes that would make it all but impossible for forest to become reestablished. It has been suggested that this is what happened and is happening in considerable parts of the savanna vegetation area adjacent to the remaining forests in Africa and within the region of forest-savanna mosaics (Map 2.1 on pages 58 and 59). The same line of reasoning is sometimes used to account for the southward movement of the Sahara each year.

In this latter case, there are not yet available the kinds of data we need to be certain of the causes of this advance. On the other hand, the microclimatic changes associated with forest removal are well known. All that is in doubt is how influential these changes have been in relation to the other major features of the African ecosystems. In the author's judgment, these changes have been important and contribute to the development and persistence of degraded ecosystems in many parts of Africa.

It has also been hypothesized that subsaharan Africa is receiving less annual precipitation because of forest removal. This is based on the view that forests are the source of large quantities of atmospheric moisture, and so when extensive forest areas are removed, precipitation must also be reduced. Whatever merit this hypothesis may possess, it is almost impossible to test it in Africa because of the lack of comparative data. If forests are ever shown to be important in the subsaharan water (hydrologic) cycle as it relates to precipitation, the estimate that only about one-third of the forested area remains (see section on vegetation) might prove important.

The limited amount of industrialization in the region and the even more limited number of automobiles and trucks preclude the possibility of air pollution from such sources except in a few localized areas, mostly in southern Africa. However, during the season of burning associated with slash and burn cultivation, vast areas of the continent are covered with a smoke haze, which, of course, is a form of atmospheric pollution but one that appears not to result in pathology in man, beast, or plant.

This region has not yet contributed to the world atmospheric supply of radionuclides as a result of atomic devices being detonated in the atmosphere. And Africa thus far has not been a major recipient of such materials released by explosions detonated elsewhere in the world.

Chapter Three

Human Influences on the Ecosystems of Asia

The concept of a place called Asia is largely a Western abstraction. Its chief utility is to allow one to lump into one huge entity an area that comprises between 41 and 43 million square kilometers (as compared to approximately 23 million square kilometers for North America). In such a huge area, the ecological and cultural variations are enormous and defy any attempt at discussing them in a few convenient generalizations. And in many ways the culture history and ecology of this vast area remains little known. Though teeming with our species in many parts, it is far from being completely under man's dominion.

3.1 Cultural Background

If one reflects that Mainland Asia includes most of the U.S.S.R., the Chinese People's Republic, the Mongolian People's Republic, Laos, Cambodia, North and South Vietnam, North and South Korea, Thailand, parts of Malaysia, Burma, Bangladesh, Tibet, Nepal, Bhutan, India, Pakistan, Afghanistan, Iran, Iraq, Saudi Arabia, Yemen, Muscat and Oman, and the Trucial States, with a combined human population that exceeds 2 billion persons, it becomes evident that we are confronted with a Herculean task when we attempt to say something meaningful, in a limited space, about the past and present cultures of these people. One naturally seeks first for some unifying element, but there is none.

Man was widely distributed in this region by half a million years ago, and we have evidence of human occupation of caves near Peking at about 400,000 years ago. The Peking sites and others of similar antiquity indicate that *Homo erectus* was able to occupy a diversity of ecological situations, ranging from the hot and humid forests of tropical Asia to areas where winters must have been bitter and long. That *Homo erectus* possessed fire is known from the cave deposits in Peking, but even if we did not have this evidence, we would have to conclude that fire was used because it would be necessary in order to occupy such high latitude sites.

After long study of the cultural prehistory and history of Asia, we can make two general conclusions. Man wandered about over this landscape to an extraordinary extent and he was also extraordinarily sedentary in habit. These conflicting conclusions derive from the fact that sedentary habit and isolation led to much of the cultural diversity encountered in the region today. The isolation was never perfect, however, and the people were not content to stay in place. Moreover, they were repeatedly invaded by outsiders, who usually lent not only their cultures, but also their gene pools to the evolution of the Asian human complex. That these forces are still at work is apparent to even the most casual observer, for many political boundaries, to cite one example, are still in a state of flux.

The prehistory of most of this region is still so little known and understood that we must pass over it entirely. And even the historical period of Asia cannot be completely examined because much basic research still remains to be done. Thus, we shall begin our study at the end of the fifteenth century, when the Western European powers made their first serious attempts to gain a colonial foothold in parts of the area.

China was already thousands of years old and, at this time, was under the Ming Dynasty (1368–1644). Its western border was similar to that of the present, although it was not nearly so well defined. Korea existed as an entity, but much of what is now Laos, Cambodia, and Vietnam was an area that might best be designated as Annam. This was more a hodgepodge of cultures

derived from Indian and Chinese influences than a political entity. Siam had emerged as a visible and viable entity, and Tibet, at least as a region, was known to exist. India, as we know it, did not yet exist, but the cultures of the region had already enjoyed a history only slightly less long than that of China. Persia, which was west of the Indian region, was, in turn, bordered on the west by the Ottoman Turkish Empire. Most of the vast reaches of land that today are the central and eastern parts of the U.S.S.R. were not part of any recognized political unit. Russia, however, existed in what is now the western part of the U.S.S.R.

Eastern Asia had not been a stranger to the visitations of outsiders, and those areas that lay closest to the Mediterranean, such as Persia and the west coast of the Indian subcontinent, had known foreign traders for many centuries before the fifteenth. However, these trade contacts appear to have had no major effect on the Asian cultures.

The change began with the Portuguese voyages of discovery in the last half of the fifteenth century. When it became clear that a feasible sea route to the Indies existed, the Portuguese were followed by the Dutch and the English, all of whom were seeking access to the resources of the region. The fortunes of these European powers were not equal, however, so although Portugal was first on the scene, it never gained the influence that England did (and Holland, in the former Dutch East Indies). British influence became most marked in the Indian subcontinent, especially in the area that was to become India, Pakistan, Bangladesh, and the Malayan part of Malaysia. What began as a commercial enterprise wound up as a colonial undertaking of huge scope.

What was the nature of the cultural influence? We might say that weighed against the region's total history, English influence was negligible. There was a story told shortly after Indian independence

(1947): A newsman asked an Indian peasant what he thought of the partitioning agreement that separated Pakistan from India, but the peasant clearly did not understand enough of the matter to answer. So, to liven up the conversation, the newsman asked, "Well, how are things with you now that the British have left?" To which the peasant replied, "The British, were they ever here?" This story may overstate the situation, but for the masses, the British presence was a distant phenomenon. Although that presence did influence the lives of even the most humble, it was so subtle that it escaped recognition by millions of persons.

What, then, were the principal cultural influences of England in the region? First must be counted language, for if the British did nothing else, they provided the leader elite of this polyglot region with a *lingua franca,* namely, English. To be sure, the polyglot situation still prevails and most of the people cannot comprehend English, but the well educated class does. This is no small matter when one realizes that more than a dozen major languages are spoken in this region. The British also introduced modern surface-transport systems, established forestry departments, began at least a rudimentary program to preserve natural areas, and encouraged the development of dams for irrigation and the production of hydroelectricity. But most important, they influenced the mortality and morbidity of the people by disseminating the materials and methods of Western medicine. This has had and is having the most profound effects on the demography of the region.

Portuguese influence on the mainland was minor, although the Portuguese held longest to a tiny remnant of their once greater colonial empire in India. They were forced out of Goa some years after the British surrendered India and Pakistan (which means Bangladesh as well). The French engaged in colonial adventures, in the latter part of the nineteenth century, in what was once Annam but which later came to be known as Indo-China. They exerted some cultural influence in the region, the most important of which may have been spreading the French language among the educated leadership class. The French colonial government was forced out in this century and ever since, the area has been in great turmoil.

China escaped direct colonial rule, but the influence of foreigners in its affairs was at times very great. The Chinese People's Republic of today encompasses about 9,366,000 square kilometers and lies between approximately 46° north latitude and 20° north latitude—a huge area, which includes many of the major macroclimates recognized by climatologists.

Westerners are overwhelmed by the fact that China possesses such a long, traceable history and that something recognizable as Chinese civilization was present while elsewhere mankind lived in varying degrees of barbarity. This tremendous continuity *in situ* of culture sets China apart from other nations. But China did not exist as a geopolitical entity during all this considerable period of time. In fact, one cannot accurately speak of *a China* until only a few centuries before the Christian era. And even then, it bore little resemblance to the present-day country in terms of area or government.

The emergence of the Chinese state is not unlike the development of states in Europe. Beginning in the northwest part of the region, a center of culture developed that evolved into imperialistic entities characterized by a series of dynasties, each of which sought to expand its control over a larger part of the land. Although the vast distances, poor surface transport, and poor communication facilities made the task of physical control difficult, it seems to have been exercised with some effectiveness and for long periods of time. However, the first effective unification was not achieved until the rule of the Mongols, which lasted for almost a century (1279–1368). Mongol rule finally gave way but China remained a large and viable political unit until events of recent times led to the outsized influence of foreign powers. The influence resulted in the political dissolution of China, but Chinese *culture* was not seriously injured. The outsiders did not leave any languages behind as a common language and the Chinese adopted only those cultural aspects they deemed useful.

In the past two centuries, China has been beset by threats from almost all sides. For a long time, the Russians have been troublesome on the west, and in this century Japan occupied parts of China and the United States has operated naval units on some of its rivers. But in spite of all these forces, it is difficult to identify the outside influences that have modified Chinese culture, except for one thing—the theories of Karl Marx. The Communists successfully took control of China in 1947 and since then have tried to bring all the Chinese citizens under the banner of Marxist belief. Where this will lead China is a question of the twenty-first century, but already some of the effects are evident on the landscape.

Aside from China's political convictions, the size of its human population receives the most attention from foreign observers. The most recent UN estimate (1972) is that the Chinese number close to 800 million, making China the most populous nation in the world. Even allowing only a modest standard of living, the resource needs of so many people stagger one's imagination. Moreover, it is estimated that China's population is increasing at an annual rate of about 1.8 percent.

The other major geographic unit in Asia is the U.S.S.R. Compared to China, Russia came late to the geopolitical scene. But in the vast region occupied by the U.S.S.R. today, peoples of widely varying cultures have existed and moved since early times, and these movements became large scale after the horse had been domesticated and integrated into warfare. The great grass-covered steppes (plains) made it easy for seekers after power and empire to move military forces.

Although never subjected to colonial rule, in the modern period Russia has had more than a little difficulty with foreign political influences. The courts of the Czars were extensions of all that was deemed most civilized in French culture, and the Russian language was considered by the aristocracy to be merely a kitchen language that they used to communicate with servants. Although this was not universally true, as is eloquently shown in the Russian writing of Dostoevsky and others, the general cultural attitude among the nobility was that anything Slavic was crude and unworthy. However, one cultural aspect cut across all levels of Russian society—a profound degree of superstitious mysticism. Although the Russian Orthodox Church has long been the chief religion of Russians, their break with a pagan past was incomplete and this contributed to a pervasive and at times overwhelming gloomy mysticism, which is clearly seen in the writings of many Russian authors.

While other parts of the world were experiencing social revolutions, Russia went on serenely following a way of life that became increasingly anachronistic, especially when compared with the social advances in France. This was shattered with the successful revolutions of this century, which brought the Bolsheviks to power, destroyed the Russian aristocracy, and set Russia on a path that is resulting in, among other things, some of the most profound environmental changes ever brought about by man. In 1971, there were an estimated 245 million people in the U.S.S.R. and the population was increasing at 1 percent per year, or about 2.4 million persons per year.

3.2 Influences on Vegetation

Southern Asia. The area we are calling Southern Asia stretches from the Arabian Peninsula on the west to Vietnam on the east and from the southern tip of India to the middle slopes of the Himalayas. It contains a vast array of ecosystems. In Arabia, on the west, are great tracts where scarcely a living plant can be found; whereas in other parts of the region, dense tropical forests cover hundreds of square kilometers. The area contains some of the dryest parts of the world and also has a rainfall station that may be the wettest in the world.

Although in no large part of this region has the vegetation been unaffected by man, the areas of greater population most easily illustrate the impact of human activities on the vegetative cover. Thus, these places must be where most of the changes took place (Maps 3.1a, below, and 3.1b on page 96).

Fire may have been the oldest means whereby man influenced the vegetation cover in this part of the world, but hard evidence for the great antiquity of fire in most of this region has not been brought to light. However, given the apparent age of human-used fire in China (approximately 400,000 years), one may assume an age of at least many thousands of years for fire use in the entire region. If fire was in general use during the Paleolithic period in this part of the world, it may have altered the natural vegetation cover in certain situations.

However, with the invention(s) and spread of agriculture, we know that fire

Map 3.1a

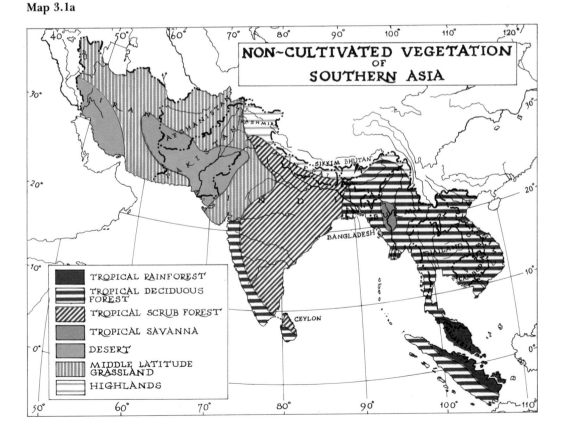

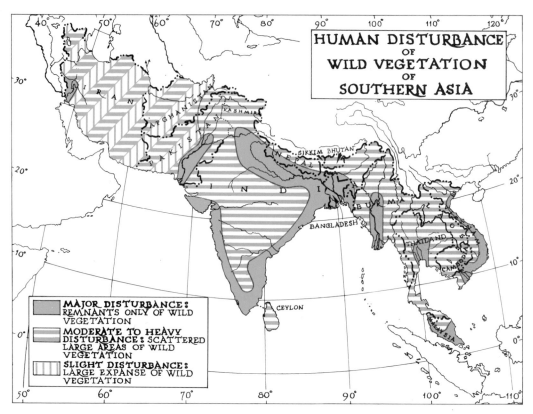

Map 3.1b

played an ecological role almost from the beginning—it was indispensable for clearing away vegetal debris that had been cut by the farmer. If we assume an approximate age of 7000 to 9000 years BP for the first appearance of agriculture in this area, we have an idea of the time span during which fire was of ecological significance. However, in the early period, and perhaps for many years following, agriculture was a highly localized phenomenon. There were probably tiny "islands" of cultivators in the midst of an "ocean" of hunters and gatherers. We do not yet possess reliable chronologies for the succession of agriculture over gathering in this part of the world. Gathering societies persisted even into modern times, but these have now almost all vanished.

We can only conjecture about the ecological importance of fire in times past, but abundant information *is* available on the impact of fire now and in the recent past. And, because some of the observed uses of fire do not appear to have sprung from recent cultural developments, we can suppose that they date from the ancient past.

Fire, to be effective, must be set where there is a large quantity of combustible vegetation. In this part of the world, this condition occurs where a fairly well-developed wet season is followed by a season of drought, or at least a period of markedly reduced rainfall. Because much of the area falls into this category, it is seasonally vulnerable to fire set by the human inhabitants. Of course, using fire to clear plots of cut vegetation in order to plant a crop is a widespread cultural activity. The result is not only major ecological changes on the burned plot but also on neighboring land, because such fires frequently escape if the vegetation is dry enough. However, there are many other reasons fires are ignited in the vegetation of this region.

Fire is employed to clear away vegetation that might offer cover to large predators such as tigers and leopards. This practice is understandable when one realizes that these big cats were once widely distributed in much of this region and sometimes turned to a diet of human flesh. But even where man eating was not a problem, there was always the possibility of attack by these animals. In addition, fire is set as a means of obtaining game meat. An area to be "hunted" is set aflame and, if the fire is sufficiently swift, a number of animals will be trapped and killed by the flames. The "hunters" then only have to go over the charred ground and pick up the animals that were caught by the flames. This hunting method is probably a survival from the earlier gathering stage.

There has long been a market in Asia for the shed horns of deer (as well as the horns of other animals) in the belief that they are possessed of aphrodisiac properties. Members of the deer family usually shed horns each year, but they are difficult to locate in grass and brush. Thus, fires that cover large areas may be ignited just to make the search for shed antlers easier.

The visible results of such indiscriminate burning are several, but the most important for vegetation is that in many areas of the seasonally wet-dry tropics, forest has given way to grasslands. These grasslands, to which the Philippine term *kogón* may be applied, frequently consist of only one genus, *Imperata*, a grass that is well adapted to repeated fires and seasonal drought. It is a coarse grass and has little value to humans because it provides poor forage for livestock. The spread of *kogonales* has emerged as one of the more serious problems in parts of this region. In such places as Indonesia, the spread of *kogonales* is measurable in thousands of hectares per year (Fig. 3.1).

Far more general in distribution, however, are the vast tracts of degraded forest and woodlands that are maintained in their poor condition by the frequent passage of human-set fire. No large areas of vegetation cover in the seasonally moist areas of Southern Asia have entirely escaped from the impact of human-set fire. Even those forests that appear to be climax are more likely in a successional stage before completion of the sere.

The importance of grazing livestock is

Figure 3.1 Imperata *grass covering abandoned farm plot in foreground and a recently-cropped field with dead trees in middle ground. Philippine Islands.*

not equal over this region. In some areas, the ecology and culture favored the development of large numbers of such animals as cattle or goats, whereas in other locations, these and similar animals are almost absent. Unfortunately, where livestock is significant to the economy, overgrazing is usually the rule. This is true not only in parts of India, for example, but also in parts of the western end of Asia. Moreover, overgrazing is undoubtedly of great antiquity. As yet, those who keep herd animals often do not understand that the numbers must not exceed the carrying capacity of the pasture. This lack of understanding is worsened by the fact that in most of the areas where livestock is important, rainfall is highly variable. Good rain years, during which the herds are allowed to increase, are followed by periods of poor rain, during which the stock exert damaging pressures on the limited forage. The result is that such areas often suffer extreme damage to the plant cover. Heavy grazing pressure results in the more palatable plant species being grazed out while less palatable plants come to dominate in the altered niches.

Fire is also an important pasture-management tool. It is common practice for herders to set fire to pasture lands each year in order to encourage new growth of palatable vegetation. Burning for pasture improvements, if carefully controlled, can be useful in maintaining native pastures at optimal carrying capacity. However, as practiced in this area, the result is often the destruction of sparse plant cover over already overgrazed pastures and the spread of the fire into adjacent shrub or forest lands.

In most of Southern Asia, the people frequently do not yet use fossil fuel. Thus fuel for heating and cooking must often come from nonfossil vegetal materials. As yet no good study has been made of the wood fuel needs of this region, but even simple arithmetic shows that the figures must be large. The human population of India alone exceeds 560 million. In many parts of the more densely settled parts of India, particularly along the Ganges Plain, the masses simply cannot burn wood regularly. Here the women and girls make briquettes of cow dung which when burned furnish much of the heat required for cooking (Fig. 3.2). It has often been remarked that use of animal dung for fuel is extraordinarily wasteful of a fertilizer resource. It *is* of great utility as a soil

Figure 3.2 *Making fuel briquettes from cow dung in India.*

amendment since it loosens heavy soils and also helps soil to retain moisture. These expostulations about misuse often overlook the fact that such use is symptomatic of a pressing lack of wood fuel, a lack that has come about through the extremely high wood fuel need that has long prevailed in the Ganges area and in other parts of southern Asia. It requires extraordinary vigilance to protect any woodland or forest vestige adjacent to settlements in India and other Asian countries. In India, it is the village that takes the heaviest toll, for it is estimated that about 600,000 villages exist there.

In spite of efforts to move into the age of steel, wood construction still dominates all but a few urban landscapes. The demand for construction wood has not been equal over the region. In the driest areas, almost all such wood must be imported; but even in those areas that have tree species of commercial worth, not all are useful. The number of tree species for which there is a world market is not great and these tend to be overexploited. The result is that the trees are much reduced in number, the ecosystems from which they are taken are also damaged, and the vegetation cover in

addition to the trees also changes.

Perhaps the most outstanding example of this phenomenon is teak lumbering (Fig. 3.3(a) and 3.3(b)). The teak tree, *Tectona grandis,* is distributed from India eastward into the island world of Indonesia, but it is not continuous over this range nor does it occur in equal densities. The wood is highly valued because of its great durability, resistance to rot, and the fact that it can be worked into fine furniture. But teak trees grow slowly, and it may take 150 years for a tree to gain the diameter required for economic harvesting. It should therefore come as no surprise that almost all teak areas have been overharvested and that there is attendant damage to many aspects of the ecosystems.

The most recent chapter in human-induced vegetation change in this region relates to warfare. Although warfare is not new to the area, its technical attributes have changed over the years and now pose serious threats to several of the region's ecosystems. This problem became evident during the war in southeast Asia.

Three aspects of the techniques of the war affected the vegetation—defoliants, saturation bombing, and bulldozers. Defo-

Figure 3.3(a) (*Left*): *Teak forest in Thailand.* **(b)** (Right): *Working elephants in a teak forest in Thailand.*

liants were sprayed from the air into several major ecosystems in Vietnam to prevent enemy forces from utilizing the forest for protective cover. The mangroves and tropical semideciduous forest were the most extensively effected (Fig. 3.4). Various estimates have been given of the number of square kilometers of forest that were sprayed, but no exact figure is available. However, hundreds of square kilometers were involved and some of this area was sprayed more than once. That there are immediate ecological changes is evident because the object of the spraying is to cause the trees to shed their leaves. The duration of ecological changes induced by

such widespread use of defoliants will not become clear until some years, perhaps decades, have passed. This massive input of chemicals into tropical ecosystems is without precedent and the ecological results may prove disastrous.

Saturation bombing causes massive local ecological disturbance with complete disruption of vegetation at the point of impact. The size of the surrounding area affected depends on the explosive yield of the bombs and the terrain characteristics. Therefore, in those areas where saturation bombing was most effective, from a military point of view, the result was major changes in vegetation. Many bomb craters quickly became little ponds in the wet season and the result will probably be thousands of small hydroseres (Fig. 3.5). The dry areas between craters will also undergo their own successions. Whether or not the input of chemicals at the bomb sites will adversely affect the vegetation is not yet clear.

Figure 3.4 (*Top*): *Mangrove forest in South Vietnam that has not been sprayed with defoliant chemicals.* (*Bottom*): *Mangrove forest in the same region in 1970 after having been sprayed with defoliant chemicals in 1965.*

In order to expose enemy movements, large and often elongated tracts were cleared of virtually all vegetation by bulldozers especially outfitted for the task. Unless such tracts are repeatedly cleared, they will undergo a succession similar to that which occurs on the abandoned plots of shifting cultivators. To be sure, the land in the bulldozer clearings does not have a dressing of ash, but this may prove of limited significance after a short lapse of time.

The ecosystems may be able to rid themselves of many of the chemicals placed in them by military actions simply because much of the area is subject to heavy seasonal rains. The water should in many (most?) instances flush away the contaminants. The question then appears to be, what happens to succession in some of these ecosystems, for example, mangrove, when the climax has been massively altered? Will the same climax return with time or will another kind of climax result? Only time will provide the answer.

Thus far we have concerned ourselves

Figure 3.5 *Bomb craters in South Vietnam.*

chiefly with man's influence on the wild vegetation of Southern Asia. We now shall examine those situations where permanent change is deliberately sought and then maintained, that is, where more or less permanent agriculture is pursued. In those parts of the region where intensive agriculture is practiced, one finds some of the most "humanized" landscapes in the world.

Chief among such landscapes must be the irrigated rice fields, and many of the flood plains of the moister parts of this region are so developed (Fig. 3.6). One is tempted to think of these areas as possessing human amphibious cultures because there is often a blending of dry and wet lands. Even the rice must spend part of its last days before harvest, not in water, but on dry land from which the water has been drained. Here, man is both a land and water creature, spending time wading in the rice paddies he created and living on the rivers or canal waters, but trekking to and from his places of rest and labor on dry land.

Figure 3.6 *Rice paddy landscape in southern Asia.*

But all agriculture is not rice. Wheat, barley, and millets dominate huge areas where there is neither the rain nor soil for rice. In fact, these landscapes are more numerous than rice paddies in most of India and in the area west of India (Fig. 3.7). Extreme parcelization of the wheat and barley lands has often been the case, however, resulting in minute parcels each of which must feed many mouths. One result of this is the suppression of all but the smallest vestiges of noncultivated vegetation.

A distinctive landscape in this region is the plantation. Many different crops are cultivated on plantations, but in terms of area, the most important is rubber. Ecologically, the plantation has a low diversity of plant species. The rubber "forest" that is a rubber plantation retains some of the attributes of the highly diverse forest once present in terms of shade and other physical factors. But the human-induced and maintained ecosystem is simplified biotically. On a map of the entire region, plantations do not appear important, but locally, they are ecologically significant and present distinctive landscapes.

Although in Southern Asia a number of important crop plants, such as rice, bananas, sugar cane, black pepper, wheat, and barley, were domesticated, crop plants from other areas were also introduced. Some of these, which came from the New World within historic times, have become economically important—maize (corn; *Zea mays*), tobacco (*Nicotiana*), chili pepper (*Capsicum*), peanuts (*Arachis*), tomatoes (*Lycopersicon*), and manioc (*Manihot*). There were also introductions from Africa, other parts of Eurasia, and various islands around the world.

Some trees were also introduced, but not yet on a major scale. More than one species of *Eucalyptus* tree have been introduced into several countries, and pines (*Pinus*), both native and introduced, have been set out in small plantings. Exotic trees are frequently seen in areas of human settlement, many of them having been chosen for their bright blooms or for the extensive shade they provide.

Accidental introduction of "weeds" must have been going on for thousands of years, but there has not yet been a careful study of the phenomenon. The water hyacinth, *Eichhornia*, from South America is one fairly recent entrant. It is a pest in many of the wetter parts of the region and little use is made of it, except that Chinese farmers

Figure 3.7 *Wheat fields of small landholders in India. The Persian water wheel lifts water to irrigate the fields.*

around Singapore are reported to feed it to hogs with success.

To summarize: In Southern Asia, a long period of human occupance, a variety of human ecologies, and a large human population in many areas have led to the modification of a major part of the wild vegetation cover. The changes are most easily seen in and adjacent to the centers of high human population, but no large area has escaped some human modification.

China. The enormity of China—in population, area, and ecological diversity—tends to intimidate any ecologist trying to discuss it in a limited space. Moreover, westerners have a considerable ignorance of China, both past and present. However, most scholars agree that man's impact on the wild vegetation cover has been major. Although this deduction is derived from bits and pieces of information from the historic past, there is sufficient scattered evidence to indicate that in centuries past China was not the forest poor nation it is today. Just *how* rich it once was remains a partial mystery (Maps 3.2*a*, below, and 3.2*b* on page 104).

The principal area under question is the now largely treeless north China Plain. It is covered, for the greater part, by deep loess deposits on the upper slopes and deep alluvium in the valleys (Fig. 3.8). Because Chinese civilization is rooted in this area, the first major ecological impact on the plant cover most likely occurred here. There is no agreement as to what the vegetation cover was when man appeared on the scene, at least 400,000 years ago. Fire drives, if used, certainly would have taken

Map 3.2a

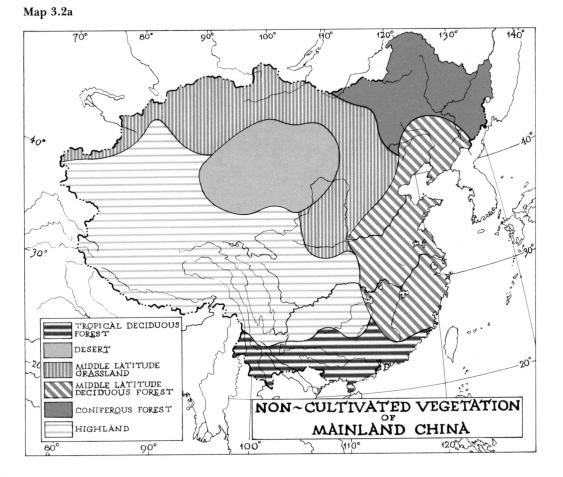

TROPICAL DECIDUOUS FOREST

DESERT

MIDDLE LATITUDE GRASSLAND

MIDDLE LATITUDE DECIDUOUS FOREST

CONIFEROUS FOREST

HIGHLAND

NON-CULTIVATED VEGETATION OF MAINLAND CHINA

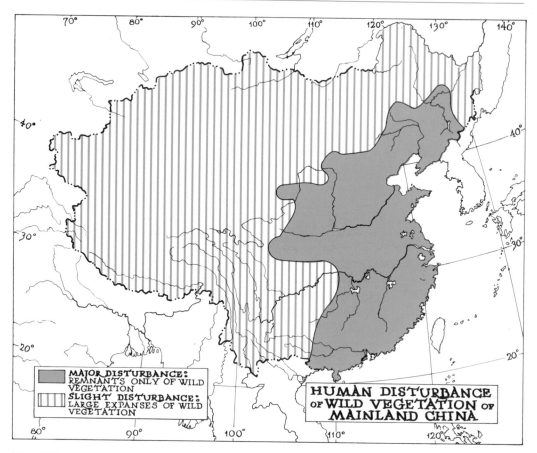

MAJOR DISTURBANCE:
REMNANTS ONLY OF WILD
VEGETATION

SLIGHT DISTURBANCE:
LARGE EXPANSES OF WILD
VEGETATION

HUMAN DISTURBANCE
OF WILD VEGETATION OF
MAINLAND CHINA

Map 3.2b

some ecological toll of the vegetation.

Because of the lack of data, we are forced to skip to more recent times, when agriculture was present. Whether agriculture was autochthonous or introduced is not known, although probably it was introduced. We must assume that the earliest phases were similar to the slash and burn cultivation found in other parts of the world. However, over the centuries, this evolved into a distinctive form of sedentary agriculture to which the term *Chinese hoe culture* has often been given. This term is useful for it indicates the major tool used.

Chinese hoe culture has many notable features: an intensive use of labor, sedentary villages, use of fertilizer (often human feces), field rotation, and an almost complete suppression of wild vegetation. The

farmers suppressed the wild plant cover because it could not be permitted to compete with crop plants; there was an ever-growing need for fuel for heating and cooking; and hiding places for tigers, which were greatly feared, had to be removed.

Thus, the wild vegetation in this part of China was and is being subjected to extreme pressures. Although there has been no careful demographic study detailing the growth of human population, it is known that for many centuries this has been a densely settled area. Today, the density exceeds 100 farm persons per square kilometer. This, plus the fact that the area usually has cool to cold winters (in the northern part) makes the need for plant fuel enormous. In *The Good Earth* Pearl Buck tells most poignantly of the ex-

Figure 3.8 *Intensive cultivation in North China. The dam impounds water for irrigation during dry periods.*

treme shortage of fuel in this region, and travelers almost always remark about the lack of wood and how straw and twigs are esteemed as cooking fuel.

Thus, an interesting question arises. Did the Chinese remove a once extensive forest or woodland and then turn to scraps for fuel? Or did the region lack trees from early times, so that the Chinese who settled there had to use more humble and less obvious fuel of vegetal origin from the beginning? The available evidence suggests that within the period of agriculture, most of the area did not support other than dispersed trees. Possibly there were local exceptions, such as areas adjacent to running waters or where soils were more supportive of trees than loess-derived soils generally are. In the early period of agriculture, the trees may have been sufficient for human needs. However, the expansion of agriculture, the increasing need for fuel, and the desire to suppress plant cover for predators resulted in the essentially treeless condition encountered today. The southern and southwestern parts of China do not show such a marked deterioration of forest cover, although agriculture is often intensive. However, the repeated and excessive pressures of China's people on the plant cover has made China the poorest of the major nations in forest resources. Possessing an estimated 20 percent of the world's population, China has less than 7 percent of the world's forest resource. This statistic is even more significant when one remembers that most Chinese (in China) do not consume fossil fuel but must rely on wood or other nonfossil fuel of plant origin.

One of the more important ecological facts about China is that grazing livestock animals are not important in the ecology of most of the country. Grain is raised in huge quantities but, unlike in the United States, it is fed almost exclusively to humans rather than animals. Thus, one does not find major vegetation removal because of overgrazing, at least not in the areas where most of the people live. The animals that are significant in the ecology of most of the country are small and able

Figure 3.9 *Reforestation project in the People's Republic of China.*

to consume matter that is not suitable for human food. Thus, there are many hogs, because they convert garbage to protein, and fowl, particularly ducks, because they exploit parts of biotopes that are not directly accessible to man.

The agricultural landscapes of China are among the most distinctive in the world and cover the greatest area of any nation in Asia. The suppressed wild vegetation has been replaced by simplified ecosystems, which are the result of centuries of local and regional experimentation to find the crop plant or combination of crop plants best suited to soils, climate, and human needs and tastes.

The northern half of China is devoted to wheats and millets, with soy beans important near the coast and up to 500 kilometers inland between the Huang (Yellow) and Ch'ang (Yangtze) Rivers. Southward, one passes through a broad east-west belt of mixed cultivation, where wheat, rice,

and sweet potatoes are grown. This is followed by the great rice-growing area of China, which is about one-third of the national territory.

Many of China's principal crops are exotics. Rice was most likely domesticated in Southern Asia, although southern China may also have figured in its developmental stages. However, the Chinese developed and selected the varieties suited to their soils and climate. Wheat and millet came from southwestern Asia, as did cotton. China's major crops were established long before contacts were made with the New World. However, a large number of New World crops were adapted by the Chinese as supplements; among these are maize, tomatoes, chili peppers, white potatoes, and peanuts.

The Chinese do not often assemble exotic plants to decorate their landscapes. Indeed, the Chinese have sometimes shown a disregard for nature that is unrivaled even

Map 3.3a

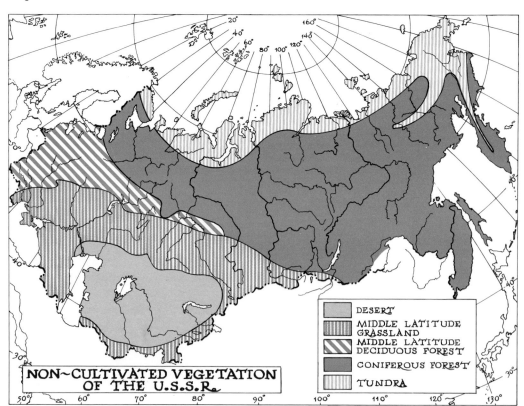

DESERT

MIDDLE LATITUDE GRASSLAND

MIDDLE LATITUDE DECIDUOUS FOREST

CONIFEROUS FOREST

TUNDRA

NON~CULTIVATED VEGETATION OF THE U.S.S.R.

by Western society. In any case, although exotics are present, they are unimportant features of the Chinese cultural landscapes.

The shortage of wood is, of course, well known to Chinese authorities, and over the years there were attempts at reforestation. These usually met with only limited success, because fuel-hungry people removed the growth long before it was large enough to be of ecological or economic importance. The current government has started a large-scale reforestation effort, which may prove successful. But some time must elapse before it can be determined whether a turnaround in the long-term reduction of China's tree cover will occur (Fig. 3.9).

U.S.S.R. There are few more striking cultural contrasts than those between China and the U.S.S.R. China is a unified culture reaching far back in time, whereas the U.S.S.R. is a recent political assemblage of as disparate a group of cultures one can find. Although Russian is the common language, dozens of other languages are spoken. And although fealty must be pledged to the State, a range of ethnic diversity remains that is probably unequalled in any other nation. Obviously we cannot trace a single ecological tradition here; rather, we find a complex of traditions that merge in various ways with the policies of the State.

With China, one must talk about the wild vegetation in restrictive terms because there is so little of it. With the U.S.S.R. this is not the case, although there are areas where man has had a major impact on the plant cover. However, a map of the non-cultivated vegetation of the U.S.S.R. shows that needle-leaf trees cover a wide belt extending from the Baltic on the west to the Pacific on the east (Maps 3.3a, on page 106, and 3.3b below). This somber forest, often located on poorly drained acid soils

Map 3.3b

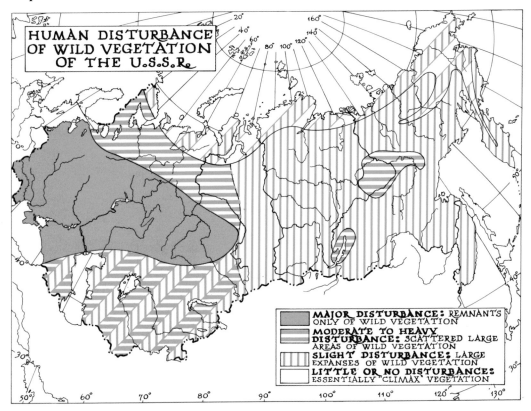

Figure 3.10 *Taiga forest in the U.S.S.R.*

is the taiga (Fig. 3.10). For the greater part, the taiga has not yet been altered by human agency. But in the west and elsewhere where logging is occurring, the forest is being removed. However, so large is this ecosystem that prodigious efforts will be required before alterations on a regional scale become visible.

South of the taiga there is an area of grassland that extends eastward almost uninterruptedly to the Chinese-Russian frontier. Are these expanses of grasses (the "steppes" of Asia) a natural climax or are they the result of human actions? On the west, there is some evidence that the grasslands expanded after the invasion of farmers in the Neolithic period, but no firm answer is yet possible. It may be significant that in North America, areas with the same kind of macroclimate have large forested tracts. There may be no simple ecological reason why grass dominates such a large swath of the U.S.S.R. but the search for causes is a task for future research (Fig. 3.11).

Whatever the genesis of the grasslands, they were and are being altered by humans aiming to extend the agricultural base. And, of course, emphasis is placed on that staple of the Russian diet, wheat. Much of the area of the grasslands, particularly in the west, is suited to wheat production. However, as one travels eastward, precipitation reliability becomes less and wheat farming becomes increasingly chancy. However, this has not deterred the gov-

Figure 3.11 *Grasslands in Siberia.*

Figure 3.12 *Wheat fields in the U.S.S.R. during harvest.*

ernment from converting at least 40 million hectares in the east from wild grasses to wheat—in what the Russians refer to as the "virgin land scheme." Change on such a large scale is only possible because of the availability of machinery and labor—and a state that can push such an effort (Fig. 3.12).

Elsewhere in the U.S.S.R., vegetation changes are chiefly local, and the tundra, which lies to the north of the taiga, has probably received even less human-induced stress and change than has the taiga. In many respects the U.S.S.R. is a pioneer land in which, until recently, man was numerous only locally. Thus, until recent years, maximum vegetation changes were confined to the west, the heartland of Russian culture. Now, with the desire of the state to exploit all its territory, one may expect more changes along the lines of the virgin lands scheme.

3.3 Influences on Animal Life

Southern Asia. For hundreds of thousands of years, man has been inflicting his presence on the animal taxa of Southern Asia; but until recent times, most of the animals were able to withstand the human pressures imposed on them (at least we have no evidence to the contrary). That situation has altered in this century, and in recent decades Southern Asia has emerged as a disaster area for animal taxa. The threat is particularly acute for mammals, but other vertebrate classes do not escape.

Why should this condition have so recently developed? The answer is partly in the explosive growth of human numbers, with the attendant need to exploit the soils and other resources more intensively. This has drastically reduced the size of the area that is habitable by wild animals. In addition, there has been an overkill of many animals for sport or to obtain materials for commerce, or to get rid of animals that are considered undesirable because they offer a physical or ecological threat to the human inhabitants.

Table 3.1 is a list of some of the animal taxa most in danger today. Some of these, such as the Arabian oryx, may be gone from the wild state by the time this book is printed. Most of the others are destined for a similar end, unless measures are taken to reverse the trend. The Indian tiger, for example, will be able to survive only, if at all, in the remote areas where it

Table 3.1
Some Endangered Mammal Species in Asia

Common Name	Zoological Name
Dugong	*Dugong dugon*
Indian lion	*Panthera leo persica*
Tiger (all)	*Panthera tigris*
Indian rhinoceros	*Rhinoceros unicornis*
Sumatran rhinoceros	*Didermocerus sumatrensis*
Asian wild buffalo	*Bubalus bubalis*
Kouprey	*Bos sauveli*
Takin	*Budorcas taxicolor*
Arabian oryx	*Oryx leucoryx*
Orangutan	*Pongo pygmaeus*
Proboscis monkey	*Nasalis larvatus*

is not persecuted by man. There is little likelihood that any of the rhinoceros species will survive outside zoological gardens by the end of the century. These beasts are hunted for their horns and also are shot where they are thought to be damaging crops.

However dreary this tale of imminent extinctions, it is actually surprising that so many animal taxa have managed to survive for so long in a region where human population densities often reached high levels. Considering the teeming millions in India, one might be amazed that any significant part of the autochthonous fauna survives to the present. How can we account for this?

It has been suggested that the reason so much of the wildlife of India survives is that the majority of people are of Hindu faith and this religion inhibits the killing of animals. This must be a factor, for in general, Hindus regard all animals with respect and protect animals to a degree that might astonish even the more avid animal lovers in Western society. But not all persons in this region are Hindus, and so we must look for other causes as well.

In general, we find that hunting has not been an important activity of the masses for a long time. This may have come about simply as a result of sedentary living habits which are focused on the village and its im-

mediate surroundings. Conditions there were not suited to the support of any but a few wild animal species.

The forested tracts harbored tigers and other dangers against which the common folk had little means of protection. Thus, such areas were avoided except in unusual circumstances. This is evidenced by the published accounts of Western hunters (chiefly British) recounting their experiences of hunting man-eating tigers and leopards. Almost without exception, the country folk possessed no means of combating these carnivores, so panic and economic disruption usually occurred until the offending cat was killed by a hunter. Moreover, the big cats have large feeding territories, so one animal could terrorize a large area. The fear of large carnivores is a frequent aspect of folk belief in this area. One writer recounted how a Malay, before entering a forested tract, would call out loudly to His Highness (meaning any tiger that might be within earshot) that he, the man, was only passing through the forest on legitimate business, meant no harm, and therefore respectfully requested that the tiger not eat him (Fig. 3.13).

Southern Asia is important as a source

Figure 3.13 *Bengal tiger* (Panthera tigris).

Figure 3.14 *Plowing with a water buffalo, Laos.*

for domesticated animals. The varied biomes provide a generous array of animals that might be brought into human cultures. In the moist region of the east, the water buffalo was domesticated; today it is one of the most valuable elements in the agricultural complex of rice production (Fig. 3.14). This is also the area where the chicken and the peacock were domesticated. Although the peacock's economic usefulness is limited, its beauty has led to its being dispersed over much of the earth as an ornamental creature. In the arid west, the dromedary camel was domesticated; and although this animal is giving way to trucks, it still remains important in some parts of the desert world. It was probably also somewhere in the western area that the donkey was first domesticated.

Perhaps the most spectacular domestication is that of the Asian elephant, *Elephas maximus*. This animal has been an economic factor in moist tropical Asia for thousands of years, and only recently has its role given way to modern machinery. It was especially valuable in the teak forests and was also pressed into service as a general transport beast. In addition, the elephant often figured in war and in peacetime pageantry.

Although the distinction is often made between wild Asian elephants and domesticated ones, it is difficult to demonstrate a sharp difference between them. Few elephants are born in captivity so it was necessary periodically to round up wild elephants of a proper age to be "broken" into training. This centuries-old practice coupled with the fact that the age and sex composition of the wild herds were often as well known as that of any domestic herd of cattle blurs the time-honored distinctions between wild and domesticated animals. The so-called wild herds had many centuries of genetic selection imposed on them through the capturing process.

China. The long history of dense human settlement in parts of China coupled with the long history of removal of the vegetation cover has had a profound effect on the wildlife. There has been no detailed study of man's impact on the animal life species by species but historical documents provide a general picture. Just prior to the start of major vegetation alteration, China had a rich and abundant fauna, including many horned mammals, a variety of predators among which the Siberian tiger was the largest, many small mammals, and a wealth of bird species.

The Chinese, understandably, were unwilling to allow ecological conditions that could offer cover to the larger and more dangerous beasts. The fear of tigers appears to have been general, and this led to repeated burning or other methods of removal of dense shrub cover that might offer protection to these beasts. In addition, the intensive nature of cultivation in much of China did not allow the preservation of habitats suited to many other animals—at least in the areas of the older, dense human settlements. However, where land was not well suited to agriculture and where human numbers were and are relatively low, many animals managed to survive. The larger mammals now in serious danger are the tiger, *Panthera tigris;* sika (deer), *Cervus nippon* ssp.; and takin (a deer-like animal), *Budorcas taxicolor* ssp.

China is not known to be an area where many significant animal domestications took place. However, the carp, goldfish, silkworm (*Bombyx mori*), and mallard duck were domesticated here and have been widely dispersed by man to many parts of the world.

U.S.S.R. In contrast with China, in much of the U.S.S.R. man has had a comparatively modest effect on the wild animals. This does not mean that all species have escaped, but vast areas of the nation, particularly the taiga region, still have an almost intact fauna. However, those areas where human settlement is densest, such as the west, have shared the European experience of greatly reduced numbers of the larger wild animals. Even in the east, in the endless grassy undulations of Mongolia, the wild horse and wild Bactrian camel may have already become extinct. If any do remain, their numbers are probably too small to assure their future existence.

Table 3.2
Some Mammals Introduced to the U.S.S.R.

Common Name	Zoological Name
Mink	*Mustela vison*
Raccoon	*Procyon lotor*
Muskrat	*Ondatra zibethica*
Raccoon dog	*Nyctereutes procyonoides*
Eland	*Taurotragus oryx*
Chapman zebra	*Equus burchelli antiquorum*
Grant zebra	*E. burchelli boemi*
Nilgai	*Boselaphus tragocamelus*

Animal domestication has been of minor importance in Russia until recently, when the government started to augment the native faunal wealth. However, because much of this is in the form of introduction of animals, it may be argued that this is not domestication but simply human-induced animal introduction. But some of the exotics are being reared in captivity and others are liberated.

In this century no other nation has embarked on such a large-scale introduction of exotic wild animals into areas where they had previously been known but had been exterminated through overexploitation or habitat destruction. In Table 3.2 is a list of some of the animals that have been introduced into the U.S.S.R. within recent years. Note that emphasis is given to furbearers and of these, the most successful introduction was of the Nearctic muskrat, *Ondatra zibethica*. This small, aquatic mammal now accounts for the largest share of the annual fur catch in Russia because it has dispersed widely in the taiga region.

3.4. Influences on Surface and Subsurface Waters

Human modification of the fresh waters of Asia stand second in importance only to human-induced vegetation changes. Although these modifications reach far back in time, it has been in the recent period that the most striking changes have occurred—and they are still occurring. The development of civilization in most of this part of the world is closely tied to human technical achievements in the area of water control.

Southern Asia. The southern part of Asia has some extremely dry regions, so in order to pursue agriculture, it has been necessary since early time to dam and otherwise divert surface waters. In the most arid parts, such as the Arabian Peninsula, water is so scarce that no amount of damming will provide enough water for crop production. But in Pakistan, for example, water is available and this has led to extensive irrigation works. Even in areas where the total annual rainfall appears adequate for agriculture, one finds dams to provide water through the dry periods or to augment rainfall in the years it is not adequate. Irrigation is the first reason man sought to control and modify the hydrography of this region and remains, today, the principal impetus for such modification.

Available evidence indicates that the earliest attempts to modify the surface drainage for irrigation occurred at least 3,000 years B.C. (5,000 years BP). These attempts were made by the Sumerians in the lower reaches of the Tigris and Euphrates Rivers (lower Mesopotamia). They developed an elaborate system of canals, which were extended over the centuries as political rule passed from one power to another.

It seems that irrigation techniques spread from the Mesopotamian area eastward, particularly to the valley of the Indus River, which is the lifeline of Pakistan. Some of the earlier irrigation systems were unsuccessful because they did not provide a method for storing water for the dry periods and the distance the water could be moved from the river onto fields was limited. An important invention to bring water in alluvial fans to the surface was the *qanat*. A qanat consists of a series of vertical shafts each dug to allow excavation of a tunnel which is directed into a water-bearing stratum within the alluvial fan.

The vertical shafts serve to provide air for the men digging the tunnel. The latter is angled so that the water it taps flows out to the surface where it can be used for irrigation and other purposes.

Modern machinery and engineering techniques revolutionized the irrigated landscapes of this part of the world, and there have been decided benefits in increased crop production. However, with these benefits came a problem. Throughout the world, the irrigation of dry lands brings the threat of adding to the mineral content of the soils until they can no longer produce the desired crops. And no large irrigated part of dry or semidry Southern Asia has escaped this effect. It is most marked in parts of Pakistan—thousands of hectares have been rendered useless for further farming and other large areas are far less productive than they used to be (Fig. 3.15). In addition to salinization, irrigation has caused a rise of water tables as a result of percolating irrigation waters. This has reached the point where the land's suitability for the production of such a crop as cotton is reduced or eliminated.

In addition to the fact that the size of the

Figure 3.15 *Salts accumulated at the surface due to bad drainage and waterlogging in an area of irrigated farming.*

area under irrigation has greatly increased in recent decades, the form of irrigation is different. Now, irrigation is mainly perennial, whereas previously, irrigation relied on seasonal floods to provide water. Thus the water-land relationship has been markedly altered in recent times, not only qualitatively but quantitatively as well. And, of course, the increase in irrigated land has caused a sharp increase in the number of persons living in irrigated areas. In one part of Pakistan, for example, when seasonal irrigation was practiced there were about eight persons per square kilometer. Today, with perennial irrigation, there are more than three hundred persons per square kilometer.

India has long been the site of irrigation efforts. While the British were in India, considerable effort was directed toward the development of irrigation and many thousands of hectares were put under irrigated cultivation. It has been in this century, however, and particularly since political independence, that irrigation has been given a high priority in government planning. Today, there are approximately 30 million hectares under irrigation (an area approximately two-thirds the total land area of California). Most of the irrigated land is in the Ganges Valley, but large areas are also in the south and central parts of the country (Fig. 3.16).

In addition to massive dams, extensive canals, and similar engineering works, in thousands of square kilometers in Southern Asia the water is simply lifted a few feet from available rivers and diverted through ditches and canals to the fields. This technique is most common where rice is the prominant crop, and such irrigated landscapes are almost a geographical cliche in parts of the wetter lands of Southern Asia.

No detailed study has been conducted on water pollution in Southern Asia, but the essential facts are well known. Human body wastes is the principal source of water pollution in most of the area, because in-

Figure 3.16 *Irrigation project underway in India.*

dustry has not yet grown enough to be a major contributor. Most people in this area do not understand disease and so virtually all surface water adjacent to human settlement is polluted with pathogenic organisms. Schistosomiasis is widely distributed in the region, and cholera is an almost perpetual threat. The situation is made even worse because practice includes drinking untreated water from certain rivers and streams that are thought to be holy (Fig. 3.17).

Irrigated fields are major foci for waterborne diseases, and the widespread habit of applying human feces directly to the fields as fertilizer assures pathogenic conditions. In no part of Southern Asia has man yet been willing or able to handle his body wastes in a safe manner. Local populations do adjust to the parasitic load they are obliged to carry, but this adjustment is paid for by high infant mortality and a lifetime of lowered energy for those who survive childhood.

Figure 3.17 *Religious pilgrim drinks from the polluted Ganges River.*

China. China has long been the locale of impressive attempts to harness rivers. Irrigation works may be as old as the oldest known irrigation canals in Southern Asia. However, it has been only recently that the people of China have achieved monumental changes in the hydrography of the nation. These changes are still in progress, and most of them are yet to come.

Irrigation projects are generally on a small scale; only recently have mammoth projects come to be significant to agriculture. The most dammed river in the country must be the Huang (Yellow), a stream that wanders over a large swath of northern China. This river has long been a problem because it sometimes produces devastating floods but other times goes dry. The silt load of the river, which is heavy, accounts for its Western name—Yellow River. During the past twenty years or so the Chinese government has engaged in a major project to tame the Yellow River. Many dams have been constructed (with more planned) in an effort to pre-vent floods, provide water in dry years, and generate electricity.

The Chinese intend to tie the Yellow River system of water control to that of the Ch'ang (Yangtze) River to the south by a series of connecting canals. Elsewhere in the country, other large dams are under construction as well as many lesser water control schemes. The result is a major modification of the natural hydrography of the country, which will be even more pronounced if the government continues its efforts. Quantitative data respecting China are often suspect, but it does appear that China's boast of having more area under irrigation than any other nation may be true. In 1960, the UN estimated that approximately 80 million hectares of Chinese land were under irrigation (an area about twice the total area of California). Even allowing for considerable error, there seems to be no question that China is embarked on a giant effort to modify the surface drainage of much of the nation.

Little is known (outside of China) at present as to the ecological changes resulting from such massive changes in the surface waters. It is known, however, that the problems associated with irrigation in other parts of the world are also present in China—increase in soil minerals (salinity) and waterlogging resulting from raised water tables or inadequate drainage. But other problems associated with introducing large quantities of water into ecosystems not adjusted to them must also occur. One might expect, for example, that disease has increased in some areas, particularly in the almost tropical south. On the other hand, the water may offer favorable habitat to waterfowl and other desirable aquatic organisms. Many interesting questions regarding the water schemes in China cannot be answered at present.

Although data are almost nonexistent, it is reasonable to suppose that water pollution is general in the country and waterborne disease is common. Most of the pollution is probably the result of untreated sewage; but in the more industrialized parts of the country, effluents from factories are undoubtedly also present. The Chinese, as we have seen, have not shown marked sensitivity to their environments and, in fact, have been quite destructive of their ecosystems over the centuries. Although the current regime is engaged in some conservation practices, levels of pollution are tolerated that might cause Westerners considerable alarm. Another major problem, as in southern Asia, is the use of untreated human feces for fertilizer.

U.S.S.R. The Russian view of nature seems to be best exemplified by the manner in which they seek to modify the drainage systems of the country. Whereas China has the greatest area under irrigation, Russia has the most major alterations of the natural drainage of large regions.

As is often the case, nature and the requirements of man are not in synchrony in Russia—at least insofar as the placement of major rivers is concerned. Most Russians live west of the Ural Mountains, but the rivers there, such as the Dnieper, Volga, and Don, are relatively modest when compared with the rivers of the east. The Volga has the largest annual flow of the western rivers, but it has a little less than half the volume of the great Yenisey-Angara system. However, all but one of the great rivers of the east (Ob-Irtysh, Yenisey-Angara, Lena) drain northward to the Arctic Ocean and pass through some of the wildest and least settled lands on earth. Thus, most of the modification of the natural hydrography has occurred in the western region, mainly to develop water transport systems (canals, rivers, and lakes) and to produce electricity.

The Volga River has been repeatedly dammed and channelized on a scale that is apparent even on small maps of the U.S.S.R.—and the large-scale removal of water has led to one of the most profound changes in Russia's hydrography. This river is the chief source of water for the Caspian Sea—a huge lake with no surface outlet—in the south of Russia. The level of the Caspian has been falling on such a scale (about 3 meters in the past forty years) that some ports have been closed. The problem was partly caused by a long period of reduced precipitation in the Caspian region but it was made acute by man's interference with the Volga. The result is the alteration of one of the world's largest hydrographic features in area, depth, and probably mineral content. Similarly, the rivers draining into the Aral Sea have been diverted for irrigation, causing the level of that body of water to fall also.

The lowered water level in these lakes has had an important impact on the aquatic life. In the Caspian, for example, it has become difficult for some fish species to migrate into the Volga for the annual spawning process. Perhaps more important, many of the swamps and marshes that formerly were places for fish spawning and nesting sites for huge numbers of waterfowl have all but disappeared. Here certainly is a major example of recent negative ecological change induced by man. The government is well aware of the problems but thus far has come up with plans that will only worsen the situation.

In the future, the eastern part of the country will see most of the modifications. Already, large lakes have been created in this region, and plans have been laid to make even larger ones. There have been suggestions for turning rivers around to make them run south through huge canals. If such schemes are ever accomplished, the ecologic effects might be global.

Water pollution is a widespread problem in Russia but is most marked in the west, especially where there are concentrations of people and manufacturing. Not only manufacturing effluents contribute to pollution; more than half the sewage dumped into Russian waterways is not treated at all and much of the rest requires more treatment to be medically safe. The Caspian Sea has been much injured through the dumping of petroleum and the pumping

Figure 3.18 *Lake Baykal, U.S.S.R.*

of ships' bilges. But the most dramatic pollution threat is occurring in Lake Baykal.

Lake Baykal is one of the world's most unusual and ecologically priceless treasures. Almost everything said about the lake must be stated in superlatives. It is the world's deepest lake, the lake with the largest volume of fresh water, the oldest lake, and so on. It contains a biological treasure without equal, and about three-fourths of the organisms are endemic to the lake. The faunal and botanical variety is great and ranges from algae and protozoans to a seal that is found nowhere else in the world. The water quality itself is extraordinary because it is judged to be extremely clean. Eutrophication has scarcely begun— an astonishing fact considering that it is the oldest lake in existence (Fig. 3.18).

Lake Baykal is now threatened by paper factories that have been built on its shore.

Few industries can more quickly pollute clean water than a paper mill. According to recent information, the government has taken some measures to prevent damage to the lake's water, but it is not known if they are effective. Moreover, areas that are watersheds to the lake are being logged off; this may cause an increase in the sediment input into the lake, thus hastening the process of eutrophication.

The current and recent impact of man on the fresh waters of the U.S.S.R. demonstrate that economic dogma does not matter at the industrial level, insofar as the impact on ecosystems is concerned. It is not important who owns the tools of production. The preoccupation with production appears similar to those of countries run under the private-enterprise system, and the impact on the exploited ecosystems is essentially identical.

3.5 Influences on the Soils and Geomorphology

Given the long history (and prehistory) of agriculture in Asia, it is not surprising that this is the source of the major changes produced by man on the earth's crust in that area. The changes are chiefly associated with accelerated soil erosion, and no large part of the region used for crops or livestock has escaped them. Often dramatic, although far more local, are the changes brought about by minor or major engineering works or by warfare.

Southern Asia. Wind and water erosion are natural processes here as almost everywhere in the world. However, human modifications of the vegetation cover for agriculture or grazing have often altered the natural tempos and created situations that lack any visible ecological balance. The dryer parts of the region are subject to winds that lift off exposed soil and transport the dust over great distances. The moister areas are subject to floods (as also are the dry areas on occasion) that, in a brief time, can also transport much of the exposed soils. No accurate survey of accelerated soil erosion has been published for this part of the world, but investigators agree that it is often serious. The poverty of most of the farming population plus a long history of land parcelization through inheritance have fostered conditions conducive to the deterioration of the soil cover over large areas.

However, in certain river valleys in both the wet and dry areas, man has created, through diking, terracing, and, for centuries, adding organic materials to the fields, generally desirable conditions. Indeed, foreigners find some of the human-created geomorphic expressions the most esthetically pleasing of the landscapes in southern Asia. Although the ecology in such areas has been altered to a major degree, the new ecosystems are apparently well balanced and carefully tended by man.

In the section in which water was discussed mention was made of the ingenious wells called qanats (Fig. 3.19). In addition to significantly altering the local hydrography, these wells exhibit a considerable geomorphic expression—through deposits at the shaft surface of rings of debris hauled from below. Moreover, the areas in which the qanats occur are so arid that the rings persist for many decades (perhaps for centuries). They have even been mapped as though they were natural geomorphic features.

Warfare is not new to this region, but the ecological scars were usually faint. Recently, however, the technology of warfare caused many ecological alterations not at all difficult to discern. Mention has been made of the bombing in Vietnam and the hundreds of thousands of water holes thus created. The holes are usually surrounded by a ring of earth cast up by the explosion. It is not known how long these craters will last, for they are mostly located in areas subject to torrential monsoon rains. At present, however, they are the most conspicuous minor geomorphic feature over hundreds of square kilometers in the war zone.

China. A long history of forest clearance and vegetation removal has resulted in such accelerated soil erosion that it constitutes one of the most dramatic aspects of human-induced modifications in China. Although never the object of a detailed investigation, accelerated soil erosion is so widespread and often of such serious proportions that few writers have failed to note it. A significant aspect of the phenomenon is that natural conditions lend themselves to major human disturbances. For example, the extensive loess surfaces of parts of northern China are, at best, fragile and extraordinarily sensitive to erosive action which is triggered by the removal of vegetation or by unsuitable farming practices. Thus, in the loess region one sees many dramatic examples of accelerated soil erosion. But this area is not unique insofar as major soil changes are concerned. The Chinese agricultural population has long exerted a crushing ecological burden on the land and it has been only through ingenuity and willingness to settle for a harsh existence that many people have managed to survive.

The story of change is not all negative. The Chinese have long used human wastes and other organic materials to maintain soil fertility and even to improve it. Further, much of the more settled nonurban portions have been terraced or otherwise altered to slow the loss of soil. In some of the oldest farmed regions, the present soils are probably very different physically and chemically from the soils people first

Figure 3.19 *Air view of a qanat in Iran. The circles of excavated earth can be seen around the edge of each vertical shaft.*

began to farm thousands of years ago. And some of these soils may well be a decided improvement in terms of agricultural utility over the soils encountered by the first farmers.

U.S.S.R. The recent drive on the part of the Soviet government to increase the area under cultivation has resulted in extreme cases of accelerated soil erosion in some areas. The erosion is often caused by wind transport after the previously protecting layer of grasses has been turned under for grain crops. It was particularly acute during the Khrushchev years when the government, through the virgin lands scheme, sought to open up new areas to wheat production. The scheme was par-

tially successful in terms of increased grain production, but in some areas it backfired and created dustbowl conditions that adversely affected millions of hectares of land. It was reported that in some extreme instances, crop lands were covered by dust up to one meter in depth, all but destroying them for any productive use in the near future.

Quite apart from the soil movements associated with efforts to increase the agricultural lands, accelerated soil erosion has long existed in different parts of the Soviet territory. In some instances, the damage has been major; deep gullies developed and valuable farm lands were lost, more or less permanently.

3.6 Influences on the Atmosphere

As will become more apparent in later chapters, there is a close correlation between the status of industrial development and the level of atmospheric modifications in a particular region. Thus, one must expect a wide range of conditions in Asia as far as human modification of the atmosphere is concerned.

Southern Asia. Most of Southern Asia is too poor economically to modify, on a macro scale, the atmosphere of the region. The minor use of fossil fuel and the generally modest industrial development assures that this region is not yet a significant source for atmospheric contamination. However, where there are concentrations of heavy industry, such as the Tata steel works in India, the air can be polluted. And in the larger urban centers, air temperature and dust content are increased, as is common in cities the world over.

Although there are few studies to document the deduction, local microclimates have surely been altered throughout this region. These alterations are chiefly associated with vegetation changes, but they might be of such a scale as to have altered some aspects of the macroclimates as well. The biological and general ecological implications of such widespread changes in microclimates are not understood at present, for they have been little appreciated and hence little studied.

China. Until recently, China must have been only a minor contributor to air pollution, but with the rapid increase in industrial development and the detonation of nuclear devices in the atmosphere, it is achieving the dubious ecological status of a "major power." The Chinese government publishes little that would provide a basis for evaluating the impact of their society on the atmosphere, but insofar as general air pollution is concerned, this impact has been increasing rapidly in the last few decades. The most dramatic demonstration of this was the detonation of a nuclear device in a remote desert region. Radioactive materials became a part of the general atmospheric circulation in the Northern Hemisphere.

The total consumption of fossil fuels is not known but it may be reaching a point where the CO_2 input is large enough to be measured in the world's atmosphere. If this is not yet the case, it will surely be so in the near future as China moves toward ever greater industrial production.

The major aspect of atmospheric modification to date, however, must relate to microclimates. We have already seen that the Chinese people have enormously altered the vegetation cover over the cen-

turies; it follows that this has resulted in microclimatic changes over vast areas. These have not been studied in any detail, so their total ecological significance cannot be evaluated.

U.S.S.R. A major concomitant of the recent decades of industrial growth in the U.S.S.R. has been air pollution due to the gases and particulates emanating from factory chimneys. In the most industrialized areas, such as Magnetogorsk, the air is so contaminated at times as to exceed almost anything that exists in North America. Until recently this condition has been accepted as the necessary price of industry, but air pollution has reached such high levels in certain areas that the government has become concerned.

Until the Russian people become much more affluent than they are at present, automobiles will remain a minor source of air contamination. However, increasing quantities of petroleum are being consumed for the production of energy, and this will result in increased sulfur and nitrogen oxides in the atmosphere, at least locally.

The massive agricultural schemes are, of course, causing changes in the microclimates and there are suggestions, though no adequate proof as yet, that macroclimatic changes are also occurring. Some public concern has been voiced about the possible climatic consequences should the Soviet government carry out proposed schemes to divert southward one or more of the great north-flowing rivers. It has been suggested that were such plans executed, global weather changes might result, but there is little agreement as to the direction of such changes or their magnitude. However, the scope of Russian engineering imagination and the level of technology make for possible future climatic changes of a scale that would not only be impressive but possibly very damaging to the ecology of large regions inside and outside the U.S.S.R.

Chapter Four

Human Influences on the Ecosystems of Europe and the Mediterranean

4.1 Cultural Background

The complexities of European history and prehistory make a simple summation difficult. Thus, what is presented here is a broad view of some of the events that produced the European and Mediterranean ecological scene we see today.

Almost everything written about the earliest period that man was present in Europe will soon require modification, because knowledge of the prehistoric period is being increased almost constantly. However, it does appear certain that *Homo erectus* was in parts of Europe by at least 300,000 years ago and in the Mediterranean region even earlier. *Homo erectus* was replaced by *Homo sapiens* not suddenly but by a slow process of evolving characteristics, which was accomplished by about 60,000 years ago. It was this species that was to spread over much of the earth's surface.

During these hundreds of thousands of years, the cultural scene appears not to have changed to any great degree. Hunt-ing and gathering technologies probably increased in sophistication, but essentially man, throughout this long time period, was a nonproducer of food.

In the eastern end of the Mediterranean region, farming was well established, at least locally, by about 9000 years BP. It appears that barley and wheat were the principal crops. The routes and timing of the westward dispersal of domesticated plants, animals, and agriculture are still not completely known, but evidence that farmers were in the Danubian Basin by around 7000 years BP has been found. Until recently, scholars believed that the Danubian region was the staging area for the diffusion of Neolithic culture into Western Europe. However, recent archaeological discoveries and new interpretations indicate that although the Danubian Basin was important to the development of Neolithic culture, Western Europe experienced developments in agriculture independently.

Agriculture did not immediately sweep aside the old gathering ways. After the first fairly rapid spread of agriculture, the areas known today as France, the British Isles, most of the Iberian peninsula, Scandinavia, northern Italy, and Switzerland were not much affected. However, the enormous advantages of food production over food gathering led agriculture to prevail, so by 3000 years BP, gathering as a way of life had all but vanished from the regional scene.

The evolution of agriculture and agricultural systems in this region is a complex subject. Here we can draw attention to only a few elements. Slash and burn cultivation, often including domesticated animals kept for food, became the first basic agricultural system. This system, with a number of local variations, persisted until the next major change—the invention and geographical diffusion of the animal-drawn plow.

It appears most likely that the plow was invented in Egypt or a little to the northeast of Egypt. Hundreds of types of plows have been used since the first invention, but we shall use the term *plow* to stand for all of them. The plow-animal complex diffused from its place of origin westward (and eastward but at a later date) and appears in the archaeological record in Britain by about 4000 years BP. It seems to have penetrated into the Scandinavian region slightly later. But for a number of reasons, the appearance of the plow did not immediately mark the death knell of shifting cultivation. Most important was the fact that much of the region was still in forest or fairly dense woodland. This condition favored slash and burn cultivation because plow cultivation requires more or less permanent clearings. Thus, there was a time lapse of many centuries before the plow emerged triumphant on the Europe-Mediterranean agricultural scene. Even today, one may see local surviving examples of slash and burn cultivation.

As agriculture was undergoing changes in distribution and technology, so were other aspects of the material cultures of the people of this region. Chief among these changes was the move away from stone tools to metal tools—first bronze and later iron. So significant were those developments that it is common for historians to speak of the Bronze Age and the Iron Age. There was also an ever-greater tendency for the people to congregate in villages—a trend that was underway in some parts of the region even during the early phases of the spread of agriculture.

As local areas became more densely settled and agriculture changed from shifting to sedentary systems, land tenure (ownership) altered. Under shifting cultivation, private landownership was rare or nonexistent, but under sedentary agriculture, private ownership became the rule. This is a complex subject and has been the object of study of many scholars. For our concerns, it is important to note that there are usually direct and indirect relationships between land tenure and ecological changes.

But during this time people were not content merely to stay in one place. On the contrary, there were many movements of people over the regions, and these movements resulted in the spread of various technologies and ways people viewed and used the land. The result was the creation of complex, culturally modified landscapes throughout Europe and the Mediterranean. Even today, it is this complexity that attracts visitors to the region from all over the world.

Toward the end of the eighteenth century, a new economic (hence, ecological) era began to emerge in Europe and the rest of the world. Generally known as the *industrial revolution,* it is still spreading around the world. The ecological implications are enormous.

In the latter part of the nineteenth century, Europe was again the staging area for another "revolution"—the *medical revolution,* that is, a revolution in death control.

Although this is still being diffused to much of the world, it has already become so influential as to be a root cause of the rapid increase in human numbers.

The demographic history of Europe and the Mediterranean region is known only in part. We know essentially nothing about the numbers of *Homo erectus* that occupied the region, and for *Homo sapiens sapiens,* we have accurate head counts for the entire region only for recent years. However, within historic times, the general trends of the population have been well studied; it is clear that over these centuries, there has been much shifting of the principal population centers. For example, during the Roman Empire, the population of the Italian peninsula was concentrated in the southern part; today, the center is in the north and the south is a poor region with a relatively sparse population. It appears that Europe and the Mediterranean area have been particularly well suited to the needs of man.

During the past 8000 years or so, the human population in Europe usually increased, but the increase occurred at differing speeds and, for brief times, was even reversed. The major reversal of which we have knowledge occurred during the Middle Ages, when plague repeatedly swept the region. Of course, in local areas, disease took a regular toll, as did famine and war. In spite of local and regional upsets, however, man survived and increased. The major increases in population appear to be associated with (1) the invention and spread of agriculture (2) the spread of the plow (3) the industrial revolution and (4) the medical revolution. Populations more than once have exceeded the carrying capacities of the region. The most recent example is the potato famine in Ireland, which occurred in the last half of the nineteenth century because of a disease that struck the plants. Hundreds of thousands of people died from starvation and hundreds of thousands more fled to England and the United States. Other parts of the European region have been unable to provide for all the population at times and the result has been mass migrations—mainly to the Western Hemisphere during the past few centuries. Even today, some parts of the region, such as Spain and Italy, are unable to provide employment for all. Again, the result has been the migration of thousands of people, mostly young men, into those parts of Europe where there is a market for their labor.

The history of human impacts in Europe and the Mediterranean is so complex that it defies any attempt to generalize in a small space. Nevertheless, the main impression one carries away is that of a region of largely "humanized" landscapes, because it has long been under the influence of man. In this fact there is much hope for the future of our species, for in spite of the fact that man used and reused the ecosystems in a determined fashion for so long, the "earth abides." It not only abides, but together with mankind, it often creates some of the most lovely and esthetically pleasing landscapes to be found anywhere in the world.

4.2 Influences on Vegetation

There is no other region of comparable size where one may so easily and repeatedly view the effect of man's impact on the vegetation cover. To understand the vegetation cover of this region is to understand the history of man's geographic dispersals there and his technical and other cultural achievements over thousands of years.

The story is not only one of destruction. On these great expanses of land, which were reshaped to meet the changing needs and esthetic desires of our species through time, some have emerged as artistic achievements. There must be few persons who do not feel uplifted when viewing a long line of slender cypresses against a blue Italian sky, a sense of exaltation when passing over the polders of Holland, or bitter-sweet emotions evoked by the plains of La Mancha in the searing summer sun. Transformation of the earth by man does not necessarily mean destruction nor must it imply that the results cannot be esthetically pleasing.

The Mediterranean. For the purposes of this discussion, the Mediterranean area includes Spain, Portugal, southern France, Italy, southern Yugoslavia, Greece, Albania, southern Turkey, Lebanon, Syria, Israel, extreme northern Africa from Egypt to and including Morocco, and the associated islands. So uniform is the plant cover over most of this area that it is customary to refer to it as "Mediterranean-type vegetation." The suggestion is often made that this type is more or less the *direct* result of prevailing climatic conditions. However, although climate is a significant factor here (as everywhere else), it is not the key to understanding the vegetation patterns we see today.

Mediterranean shrub vegetation is not the only type present, as you can see on Map 4.1*a* on page 125. There are also extensive areas of desert vegetation, needle-leaved trees and mixed forests of broad- and needle-leaf trees—but the so-called Mediterranean type does predominate. The overwhelming feature is the preponderance of shrubs. These are interspersed from time to time with grassy areas and palms or with stands of trees (ignoring, for the moment, the great extent of farmed lands). Some of the commoner elements in the Mediterranean shrub lands (also known as maqui, matorral, garigue, jara) are rock roses (*Cistus*); wild olive (*Olea*); asphodel (*Asphodelus*); lavender (*Lavan-*

dula); oleander (*Nerium*), in the canyons and seasonally dry water courses; Mediterranean fan palm (*Chaemerops humilis*); certain oaks (*Quercus*); aleppo pine (*Pinus halepensis*); and a host of aromatic shrubs, of which the above-mentioned lavender is one (Fig. 4.1a, 4.1b, 4.1c).

There is no question that the plants of the Mediterranean region are adapted to the climate and soil conditions. But to what degree is the present *distribution* of the Mediterranean vegetation the result of human actions? The evidence is overwhelming that man has played a major role (if not *the* major role) in the evolution of the post-Pleistocene plant geography of the Mediterranean area.

It is increasingly assumed among scholars that man controlled fire in pre-Neolithic times. Thus, it appears reasonable to assume that by means of fire, *Homo erectus* and, in the Paleolithic, *Homo sapiens* produced extensive modification on the plant cover of the area.

With the spread of agriculture, our task of relating human acts to vegetation change becomes somewhat easier because modification of the vegetation was an inescapable corollary. Although we lack a detailed chronology, we do know that the Middle East was the staging area for agricultural dispersal westward. However, we know almost nothing about the ecology of Neolithic farming in this part of the Mediterranean region.

Judging from the available carbon 14 dates, the dispersal of farming westward in the Mediterranean was not rapid. There is evidence that agriculture was being practiced in the Iraq area by about 11,000 years BP, but the *earliest* suggested date for Spain is 7000 years BP and for a site in southern France, 7500 years BP. One must regard this information with extreme caution, however, because there are enormous gaps in our knowledge respecting the archaeology of this region. Nevertheless, the available information suggests that farming moved westward at a relatively leisurely

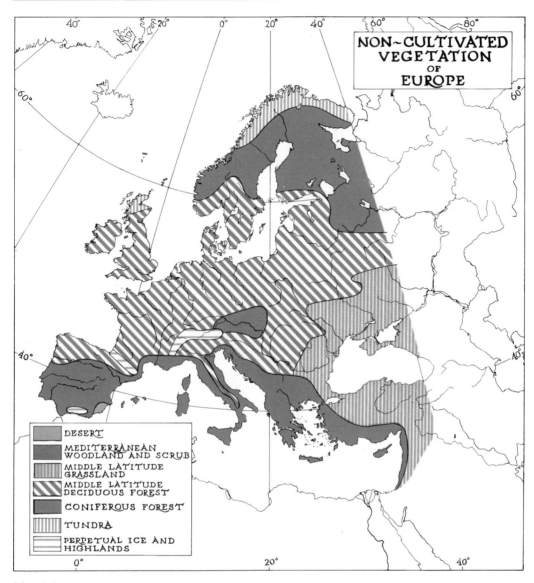

Map 4.1a

pace, perhaps along both the northern and southern rims of the Mediterranean Sea.

If the positions of the known (or supposed) Neolithic sites in Spain offer an accurate clue, there was limited continental penetration by Neolithic farmers in the Mediterranean area. Thus, we might assume that the impact on the vegetation was confined largely to coastal sites or short distances inland.

With the passage of time, agriculture became a more complex activity in this area and especially after the plow was adopted. Even in places where the plow was slow to be accepted various forms of farming developed that required intensive use of human labor on the same lands year after year. These changes, accompanied by an ever-growing human population, led to profound alteration of the vegetation of the most intensively farmed areas. Such areas were no longer confined largely to the coasts; they extended generally over the land, especially in river valleys where

Figure 4.1(a) *Oak-grass vegetation in California.*

Fig. 4.1(b) *Mediterranean fan palm.*

the alluvial soils were (and still are) especially attractive to plow farmers. In some particularly favored locations, such as along parts of Spain's east coast, farming using chiefly human labor developed and persists to the present day. This latter comes down from at least Moorish times and perhaps, in part, from the earlier Roman period. On these lands, the wild vegetation has been all but obliterated and this remains true for some of the areas to this day (Fig. 4.2). In other areas such as in the valley of the Río Guadalquivir in Spain, lands that had provided agricultural surpluses in Roman and Moorish times fell into disuse and were partially reclaimed by brush and fan palms. This process is now being reversed as agriculture once again spreads up and down this valley (Fig. 4.3).

Of all the Mediterranean landscapes that have been humanized, none exceed and few equal in degree of humanization those of Italy. From the harsh and bony hills of Sicily in the south to the narrow confines of the Adige River valley in the north are varied vegetation expressions caused by farmers and herders over thousands of years. But if Italy often offers great beauty in its changing vegetation expression, the same cannot be said for much of Greece, Albania, southern Turkey, and the eastern end of the Mediterranean, and the North African rimland. Although not all the starkness is attributable to the farmer and herder, much of it is. The story continues to unfold today, although usually it is one of attempted restoration rather than further degradation.

Interwoven with farming almost from the beginning is the keeping of grazing and browsing animals, most notably sheep, goats, cattle, horses, asses, and camels.

Fig. 4.1(c)
Mediterranean shrub vegetation on the east coast of Spain.

Goats and sheep—if the archeological record or its interpretation is accurate—were the first of this group to be domesticated, and no other domesticates have produced greater changes in the vegetation. Both these animals appear to have been domesticated first in the Middle East and then to have been taken with man to many parts of the Old World (and later to the New World also). It is possible that goats had to be reckoned with earlier than sheep in the alteration of the vegetation, for goats are admirably adapted to finding food in the poorest of shrub. They do so largely by browsing, although they will graze on grass as well. These small, nimble-footed beasts early provided their keepers with milk, meat, and hides. Being of modest size and strength, they fit in well with almost all the agriculture systems that appeared (Fig. 4.4).

Because the goat was such a useful and easily *mis*managed beast, it was allowed to destroy a significant part of the ecological patrimony of the region. Many of the slopes laid bare to the sun of summer and erosion from the winter rains are barren because, for thousand of years, careless herders burned the vegetation. This was done to encourage an ever more impoverished shrub to put forth new growth to feed flocks of goats. Of course, the goat is not directly at fault. Man was unable or unwilling to control goat numbers and to refrain from the annual burning. Thus it is to fire and goats that we may correctly attribute much of the current geographical distribution of the Mediterranean shrub vegetation. Although at the policy making level recognition of this fact has come only recently, one now finds governments controlling goat grazing, outlawing burning of

Figure 4.2 *Irrigated huertas near Valencia, Spain.*

hillside pastures (although in Spain, the herders persist in spite of the law), and offering financial inducements to engage in alternate uses of the land. One notable example of this in recent years is the island of Cyprus. The government pays a bounty to goat owners who give up their animals to engage in some form of desired crop growing. The result has been the replacement of shrubs by trees on some parts of the island.

In earlier centuries, sheep were valued chiefly for milk and meat. Relatively late in the story sheep started to be kept for their wool. One of the most famous wool sheep breeds, the Merino, attained importance in Spain, although the breed may have been introduced there by the Moors. Sheep raising achieved an almost unbelievable importance in Spain from the thirteenth to the eighteenth centuries. Indeed, sheep raising is still significant in Spain, although it no longer enjoys its once preeminent position. In 1273, the sheepmen of Castile organized themselves into a powerful organization called the Mesta, which was given the personal protection of the crown

of Castile, itself deeply involved in sheep raising. The Mesta came to be a power almost unequalled in Spanish history because it was answerable almost to no man.

The ascendancy of wool production in Spain marked a renewed attack on the woodlands that remained, especially those that lay in the areas most favored by sheep. The country was laced with a network of broad sheep roads (cañadas), and the drovers secured rights to extensive grazing areas, often to the direct detriment of agriculture. Even allowing for a degree of exaggeration of conditions because of the friction between farmers and sheep raisers, major destruction was wrought on the vegetation cover by overgrazing and the removal of trees and brush by fire to enlarge the pasture areas.

Although most dramatic in Spain, sheepherding was also important in many other parts of the Mediterranean area and often resulted in great destruction of the vegetation. Moreover, there were also horses, cattle, asses, and, in North Africa, in addition to these, camels. The North African rim has long been an ecologically

Figure 4.3 *Young orange grove in the Rio Guadalquivir Valley, Spain.*

fragile area where the delicate balance between plants, soil cover, and sparse precipitation is easily disrupted by human actions. Overgrazing has resulted in a heavy ecological toll here and the forests and woodlands that remained up to Roman and Moorish times are no more. Only the careful detective work of a few scholars tells us that the barren rock piles so frequently encountered are often the result of human actions over the past 2000 years.

As though it were not sufficient that farming and herding laid toll to the Mediterranean vegetation, trees were taken for fuel and lumber. We tend to forget that until recently, wood was practically the only fuel used. So impressed are we with our achievements with metal (Bronze Age, Iron Age) that we overlook the "age" of overwhelming importance to our species— the "wood age."

We may never be able to obtain a detailed quantitative statement of how much wood was burned for fuel during the hundreds of centuries before coal and petroleum became available. But the use of the woodlands for fuels must have vied with herding as a cause of vegetation changes.

The Romans, for example, were fond of hot-water baths and, as they extended their empire around the Mediterranean rim, they laid heavy toll on adjacent wood-

Figure 4.4 *Herd of goats, Greece.*

lands and forests to supply the great quantities of fuel needed for these baths. It has been suggested that forest destruction along the Mediterranean coast of Turkey reached its high point during the Roman period and was a direct result of woodcutting.

Later, the Moors in Spain tended to concentrate their populations into relatively large settlements, which wrought havoc on the forest (wood) resources of the immediate hinterlands. It has been suggested that Córdoba attained a population of almost 1 million persons (although that figure appears high) during the height of the Moorish occupation, which lasted from the beginning of the eighth century to the end of the fifteenth century. During that time, the Sierra Morena, whose slopes run down to the edge of Córdoba, were exploited for the enormous quantities of wood the city required. Another great center was Granada, and it was from here that the last of the Moors was "officially" expelled from Spain in 1492. But during their long stay, the Moors almost completely stripped the north-facing slopes of the mighty Sierra Nevada (Mulhacén) of tree cover. These slopes were converted into bare rocks capped on the upper levels by snows (Fig. 4.5).

Figure 4.5 *Deforested slopes of the Sierra Nevada, Spain.*

The need for wood increased beyond that for cooking and heating because wood was also the source for the manufacture of charcoal. Today, when charcoal suggests a backyard barbecue, it is easy to forget that charcoal making is one of man's great technical achievements. Without it many of the metalurgical discoveries and advances of the centuries before coal was used probably could not have occurred. Charcoal makers were active for many centuries and we can only guess at the toll exerted on the trees. The needs of Roman smelters, of Moorish tile factories, and of other kinds of factories requiring the hottest fuel available caused many hills to be stripped bare.

Wood was also a prime construction material for buildings and ships. In those days, ships were largely of wood and naval strategies were often directed toward acquiring and maintaining access to the forests that produced the most desirable kinds of lumber for seagoing craft. Other scholars have shown how great was the overutilization of the more favored species of trees, not only for shipbuilding, but for all types of construction.

The opposite side of the coin of vegetation removal is the introduction of new species and the attempts to replace the forests that were destroyed. Introductions of plants began with the spread of agriculture (perhaps earlier, but we have no evidence). In about the ninth century, through Arab influences, sugar cane and certain citrus fruits arrived in Spain. The tree known in English as Saint John's Bread (*Ceratonia siliqua*), native to the eastern end of the Mediterranean, was dispersed widely by human means, as was the fig, grape, and olive, all of which are native to different parts of the region (Fig. 4.6(a) and 4.6(b)).

These and other crops gave rise to distinctive new ecosystems dominated seasonally or perenially by one or more plants. Perhaps the most durable and distinctive of these man-made ecosystems is the olive grove. Olive trees cover hundreds of thousands of hectares in this region. Olive trees are capable of yielding fruit for many centuries, and they are perfectly adapted to the harsh summers, thin stony soils, and steep terrain on which many of them are planted. Although lacking the taxonomic diversity of the wild thickets in which wild olives occur, an olive grove does retain attractive features for some of the animals that would have found a place in the wild shrub—if it were still present.

The most recent major influence on the crop-plant complex of this area developed as a result of the discovery and introduction, after 1492, of many plants from the New World. These include potatoes, maize, tobacco, peanuts, agave, green peppers, beans, and a host of lesser items.

The tree flora of the region has been added to from early times. For esthetic reasons, man has moved Mediterranean species around the region, well away from their points of origin. Among this group are cedars of Lebanon, Atlantic cedar, Lombardy poplar, many of the pines, and some species of fir. These trees often appear in parks, along roadways, and on private lands. Several pines, acacias, and *Eucalyptus* species (from Australia) have been introduced for wood or other uses.

In the gardens of the private homes and in the numerous parks, a broad array of plants from all over the world is found. The Mediterranean region is hospitable to many exotic plant species.

The latest and most positive aspect of human influence on the vegetation is reforestation. Most of the government in the

Figure 4.6(a) *Winter view of vineyard, Italy.*

Figure 4.6(b) *Olive grove, southern Spain.*

Figure 4.7 *Pine reforestation, Spain.*

Mediterranean area have become alert to the need to replant some of the land to permanent tree cover. The most frequently planted trees are pines, which are adapted to many of the soil-climate conditions of the area. Perhaps the most favored species are the maritime pine, *Pinus pinaster,* and the aleppo pine, *Pinus halepensis.* The maritime pine not only yields wood and lumber but is often a valued source of naval stores (Fig. 4.7). In recent years, *Eucalyptus* trees, native to Australia, have been planted extensively because they grow not only under difficult ecological conditions but at a fairly rapid rate as well. These trees are grown principally for cellulose. In the south of Spain, where there are large areas set out in *Eucalyptus* trees, the landscapes present a decidedly alien appearance. One is often struck by the paucity of other living things in these man-created forests.

The reforestation efforts in this region are exemplary and, in general produce highly desirable effects. There are, however, some negative impacts on the wildlife which will be discussed later in the chapter.

Western and Northwestern Europe. For the purposes of this discussion, Western and Northwestern Europe refer to the following countries or parts of countries: Great Britain, Ireland, Norway, Sweden, Finland, Denmark, Germany (West and East), Belgium, Netherlands, Luxem-

bourg, and France (excepting the Mediterranean area). A glance at some maps depicting the vegetation cover of this region would lead the uninformed to conclude that it is essentially one great unbroken stretch of forest. That is definitely not the case. Although there are considerable areas now in forest, even greater areas are not and are given over to many other uses. On Map 4.1b, on page 133, you can see that Great Britain, Ireland, and large parts of France have almost no forests and that the principal areas of forests are in Norway, Sweden, and Finland. The chief reason for this condition is human actions over the past several thousand years, particularly in the last 1000 years.

Traveling over this region on the autobahns and other roads, one can easily be misled by the many tree-covered tracts through which the roads pass. But the roads are usually lined with trees only to a shallow depth; even where this is not the case, the forest has often been planted by man. The result does not lack esthetic qualities, but we should be aware of the *cultural* input and the fact that the landscapes are, to a major degree, artifacts.

Another misleading aspect is that much of the reforested lands are in conifers, whereas the natural tree cover for many of these areas is broadleaf trees of several genera and species. With the possible exception of areas in the north, it would be all but impossible to discover any area ex-

Map 4.1b

ceeding a few hectares that has a vegetation cover free from significant human influence.

Because we know almost nothing about the ecological relationships of early man in this region, we must move to the time (rather late in the period of human occupance) when farming appears in the record. The penetration by Neolithic farmers into this region appears to have been chiefly by way of the Danube River corridor. Some penetration may also have come by way of southern France, but there were not many paths available in early times.

Over the centuries, these farmers occupied much of the region, and they brought about changes in the vegetation where soil and other conditions were best suited to their style of farming. So basic was the use of fire to reduce cut vegetation to ashes that the term *brandwirtschaft* (fire economy) has been applied to it. The term aptly describes the importance fire had for these people. It also indicates that we confront an ecological force of great significance. The picture that has emerged is that there was no *general* clearing of forests or woodlands over large areas and that at any given time extensive areas of relatively undisturbed forests remained. Although there must have been variations, it appears that when a plot was farmed for, say, two years, it would be abandoned and left undisturbed for twenty or more years. In some areas, the fallow period may have been shorter and this may have slowed or even prevented succession back to a forest or woodland climax.

The plow was present in this region before the invasions of Julius Caesar. However, its major ecological impact was delayed until it was modified and the general technique of plowing the often heavy European soils was developed to a degree that permitted its wide adoption. By at least 400 AD the plow was making its presence felt throughout most of the region; and during the next several hundred years, a massive clearance of the forests and woodlands took place. The changes were dramatic. Where once almost unbroken forests of oak and other broadleaf trees stretched from the Schwartzwald to Dover Channel, a thousand years later only remnants remained. And even then the process of removal did not stop. But forest removal caused by the spread of plow culture is only part of the story.

From almost the beginning of farming, man had a number of domesticated animals, such as cattle and pigs. These animals played a role in modifying the vegetation, and our knowledge of their habits tells us that in areas where man and beast were numerous it would have been all but impossible for many forest species to regenerate.

In addition there was the need for fuel. The need was even greater here than in the Mediterranean area, for short, cool summers are followed by long and often bitterly cold winters. And the need for fuel was not only for heating and cooking. In the early centuries of technological development of mining and manufacturing ever greater quantities of wood and wood converted to charcoal were required. So great were these needs that their impact on forests may have equaled that of the farmers.

Just as there was a great demand for lumber in the Mediterranean area, so too there was such a demand in Europe. As the emerging nations developed their merchant and naval fleets, they needed the species of trees that yielded lumber suited to the marine environment. Especially hard hit were the oaks, for they could be made into long straight timbers for the keels of vessels, and the firs, which being tall and straight, could form the masts of sailing ships. By the seventeenth century, England had become permanently short of the timber required for its fleet of merchant and naval vessels. It has been said that this man-caused shortage influenced British foreign policy for a long time because Britain had to seek timber abroad. It was acquired from Europe, the New World, and India.

It is difficult for us to imagine what it was like to live in a world where wood and lumber were *the* natural resources. Yet, it was in such an environment that Europe struggled up from barbarism; and the energy released from the burning of wood and the devices contrived from wood made possible much of the success of that struggle. One aspect of that struggle was the gross modification of the wild vegetation.

Given the centuries-long history of tree removal over this vast area, it is not surprising that efforts were finally directed toward reforestation. Some attempts may be traced back to the eighteenth century, but no significant progress occurred until more recent times. Even now, large areas remain almost denuded of trees, except for those few allowed to remain—and these are almost never mature stands. The favorites for reforestation are the conifers because they have a rapid growth rate, particularly when compared to the oaks. Thus, in Germany and France extensive tracts are covered with conifers rather than with the hardwoods of an earlier time. However, a coniferous forest is ecologically different from a mixed broadleaf forest, so reforestation in this or any other region does not mean that the previous ecology has been restored.

The vegetation changes in Central and Eastern Europe are nearly the same as those in the areas we have already discussed.

4.3 Influences on Animal Life

Europe and the Mediterranean area are part of the Palearctic zoogeographic region (see Map 1.4 on page 28). Most of the larger, economically important animal domestications occurred in this region, and it is there that the numbers and geographic distributions of many native animals were most modified by man.

Animal exterminations. Some scholars have suggested that pre-Neolithic societies were responsible for exterminating some of the larger mammals that were present in parts of this region toward the end of the Pleistocene. However, the evidence is almost entirely circumstantial—the animals seem to have disappeared in a relatively brief span of time, which includes the period when man-the-hunter was wandering over these landscapes. Whether these events are linked cannot be proved now, but future research may allow a better interpretation of the data.

When one considers that Europe has been densely settled since at least classical times, it is surprising that few animals were exterminated through human acts. Only six mammals are recorded to have completely disappeared and not even one land bird is known to have been exterminated. This extraordinary fact cannot be easily explained except by stating that land birds of the region were able to adjust to the ecological conditions created by man.

However, although there were few *regional* exterminations of mammals and none of land birds, the distribution and number of many of the larger faunal elements were grossly modified. Many animals are now so reduced in number and geographic range that their continued survival is questionable. The European brown bear (*Ursus arctos* ssp.), for example, once ranged widely over much of Western Europe but is now confined to a small area in northwestern Spain, two small areas in Italy, and somewhat larger areas in Yugoslavia and Greece. In prehistoric times, the brown bear was associated with man in Britain, but it seems to have disappeared before written records begin. Similarly, the wolf (*Canis lupus*) once enjoyed a much wider range than it does at present. Actually, it is surprising that this large animal has managed to survive in the wilder parts of Spain, because strenuous efforts have often been made to rid the area of this predator. Even in France and Germany, occasional reports appear that a wolf has been sighted—impressive evidence that this partial progenitor of the domesticated dog has unusual powers of survival. (Bears and wolves continue to survive in comparatively large numbers in many parts of the U.S.S.R., however, and their demise there is not imminent.)

Private game parks are one factor that allowed the survival of the mammals most sought after by hunters. These parks are looked after by gamekeepers long noted for their determination to protect the birds and mammals from poachers (illegal hunters), and they have been responsible for preserving some of the game birds and mammals of this region down to the present day.

The North African rim has experienced some animal exterminations within the recent past, including the Barbary lion (*Panthera leo leo*) and the Atlas bear (*Ursus arctos* ssp.). There is no record of exterminations of land birds native to this area. Although the detailed faunal history of the area is not known, we can assume that the extensive alterations in vegetation had a profound effect on the number as well as distribution of many of the native vertebrate animal species.

Animal Domestications. In Chapter 1, we discussed animal domestication and pointed out that it is often difficult to determine if the bones of an animal found in a site of past human occupance belong to a domesticated animal. The difficulty arises, in part, because often we do not agree what a domesticated animal is and cannot be certain what osteological (bone) changes are associated with a long history of breeding in captivity. Thus, any attempt to establish the time and place such animals as goats, dogs, sheep, cattle, horses, and other economically important domesticated animals were first taken into human culture will seldom if ever be successful. However, the rich archeological sites of the Mediterranean-European region at least allow us to see that from early in the Neolithic some, and later all, the animals just mentioned (plus a few others) were intimately associated with human societies.

Which of these animals were first established in human societies in the region we are examining? We can only answer tentatively at present. It seems that goats, sheep, asses, and possibly cattle were originally domesticated in or near this region.

Figure 4.8(a) *Muskrat* (Ondatra zibethica).

Figure 4.8(b) *Grey squirrel* (Sciurus carolinensis).

Figure 4.8(c) *Coypu* (Myocastor coypus).

The European wild pig, *Sus scrofa,* may also have been domesticated here. The question of the original domestication of the dog has filled many pages, but the general consensus is that the dog was developed chiefly from the wolf and simultaneous domestications may have taken place over the Eurasian continent.

In addition to the fundamental importance of human-controlled breeding of these animal species was their human-controlled dispersal over almost all of this region. In some instances this began in pre-Neolithic times. Man has substituted his domesticated animals for many of the preexisting native animals in the ecosystems of this area.

Animal Introductions: deliberate and accidental. We have mentioned the introduction and spread of domesticated animals in this region. But another aspect of deliberate introductions begins in relatively recent times. The most dramatic example is the introduction of the North American muskrat (*Ondatra zibethica*) into Central Europe in 1905 (Fig. 4.8a). A private Czech citizen imported only five animals, but within a few years, the species had spread into Germany and Austria. Moreover, its advance was aided by repeated introductions to other parts of Europe, and soon its durable pelt was much sought after. This prolific marsh dweller found conditions to its liking and is now a conspicuous part of the European mammalian fauna.

Across the Dover Channel, in England, two introductions appear to have satisfied few persons. In 1890 the native grey squirrel of Eastern North America (*Sciurus carolinensis*) (Fig. 4.8b) was introduced. It quickly spread over England and Scotland and people thought it drove out the native red squirrel (*Sciurus vulgaris*). However, it has been noted that just prior to the introduction of the grey squirrel, the red squirrels suffered an epizoötic (an epidemic disease among animals) that reduced their numbers almost to the point of being rare, so the exotic squirrel merely occupied a niche that had been recently partially vacated. Whatever is the true story, the result is that the American squirrel became a conspicuous part of the English landscape.

Within this century, the coypu (*Myocastor coypus*), an aquatic rodent native to southern South America was introduced into England and several places on the Continent (Fig. 4.8c). So far it gives no sign of explosively increasing its numbers, but the authorities who know well the problems associated with it in North America, are watching the animal closely.

Sportsmen who desire to hunt or fish for animals not otherwise available to them have been a stimulus for introductions. As a result, a number of native animals have been reintroduced to parts of Europe and the Mediterranean area where they had been hunted out or where, for a time, the changed ecological conditions had made it impossible for them to live. In this way, red deer (*Cervus elaphus*), wild boar (*Sus scrofa*), ibex (*Capra hircus*), and game birds such as the red legged partridge (*Alectoris rufa*) have been returned to areas they once occupied. In addition, some nonnative animals, particularly certain species of fish, have been introduced. Among these are rainbow trout (*Salmo gairdneri*), brook trout (*Salmo fontinalis*), and black bass (*Micropteris* sp.), all from North America.

4.4 Influences on Surface and Subsurface Waters

The varied culture histories and the diversity of water quality and quantity in this large region assures great diversity in the kind and degree of hydrographic manipulation and alteration by man. We have no information for the thousands of years before recorded history, but we may be sure that at least by the time Neolithic farmers were altering the vegetation cover, there were concomitant changes in the runoff of streams—and perhaps other hydrographic changes as well. But because of a lack of supporting evidence we can state only that man's influence on the non-marine waters of this region probably began early.

With the historical period usually referred to as classical, we are able to begin documenting this aspect of human activity. The preeminent culture of that time was that of the Romans. We have seen that Rome spread its influence around and far beyond the Mediterranean area and had much impact on the vegetation because of an energetic development of agriculture. However, the rainfall in the Mediterranean area is usually limited and concentrated during a brief period of the year. Therefore, they also found it necessary and desirable to alter the natural water conditions. Fortunately, from their point of view, the Romans possessed great engineering skills for the construction of aqueducts, underground canals (*cuniculi*) and various kinds of drains.

Roman aqueducts are such visible and frequently preserved features that almost everyone knows of them. They were constructed mainly to lead water from often distant sources in hills or mountains to where it could be used for domestic or agricultural purposes. In addition, some aqueducts, especially in densely settled areas, appear to have been used chiefly for drainage. The most famous of these was the "great cloaca" of Rome; it may have been the first significant attempt to provide a large city with an adequate sewer drain.

Although irrigated agriculture predates the Roman period by many centuries, the Romans developed, for that time, elaborate irrigation systems dependent on the water provided by aqueducts. Unfortunately, few persons in those times grasped the relationship between plant cover and a dependable water supply. The result of this ignorance can be shown, for example, by the aqueducts that reach back into the now barren slopes of the Atlas Mountains. These once tree-clad slopes provided a runoff of sufficient reliability and magnitude to justify a large input of labor and capital in the construction of the now almost useless aqueducts.

The Romans were also capable of a certain amount of canal building. In many respects, the engineering problems of simple canal construction are less complex than those of either aqueducts or cuniculi. The major requirement (in the absence of machinery) is for a reliable and large supply of human labor.

During the period when Moorish influences were important in the Iberian Peninsula, approximately 700 to 1492, the management of water for irrigation purposes became particularly refined. This was most notable in the area around Valencia, where even in Roman times an intensive irrigated agriculture developed. The special skills of the Moors made possible the use of all available water sources, even quantities that we today might consider to be no source at all. The carefully constructed works for the diversion of surface water onto fields are still common features in many southern and eastern Iberian landscapes. Moreover, many of the old techniques are still in daily use (see Fig. 4.2).

However, the removal of tree cover by Romans, Moors, and other cultures led to a more rapid runoff of water immediately after a rainfall. Some of the seasonal torrents in the Mediterranean area occur, at least in part, because of the removal of the soil and vegetation "blotter" which, if in place, would allow for a much slower, more prolonged runoff.

By the Middle Ages, there had developed increased uses for water and often greater emphasis on traditional uses as well, which, in turn, influenced the hydrography of various parts of this region to a marked degree. Perhaps the most significant development was the increased number of canals constructed for moving boats and cargo across the grain of the land or at least across the grain of the prevailing drainage. Water transport was discovered to be relatively inexpensive and, in the general absence of highways, water transport made economic sense. The Iberian Peninsula and the Mediterranean rimlands in general took little interest in canals because of the scarcity of fresh water and the accessibility of marine transport and river systems. Canal construction developed first in France and the Low Countries and later spread to much of Western Europe and Great Britain. In these areas, an essentially anthropogenic hydrographic net developed that gave a distinctive new expression to many of the local regions.

As the European demand for luxury agricultural products rose, greater interest was shown in draining swampy lands that appeared as though they would be profitable to farm (Fig. 4.9). Thus, drainage became a significant activity from the late Middle Ages onward. Later, when the relationship between mosquitoes and malaria was established, there was further drainage to remove the habitat of these disease-carrying insects.

We are able to discern efforts to construct dams from ancient times; this technical achievement was only a little more recent (in parts of the Old World) than the invention of agriculture itself. However, it was not until comparatively recent times

Figure 4.9 *Farm land in the Camargue (Rhone Delta), France.*

that engineering skills and the availability of machinery made possible the massive hydrographic changes we associate with modern dam building. And, as might be expected, it is in the water-scarce world of the Mediterranean that some of the most dramatic landscape transformations have taken place because of this activity.

Even allowing for the long history of dam construction here, this century is the period during which the major water impoundment has taken place. Virtually every country around the Mediterranean has constructed dams in recent years, but Spain has succeeded more than all the others in altering its surface hydrography. Most of Spain is troubled by unreliable precipitation and torrential runoffs when rain falls or snow melts. Now, many dams hold back these waters and make irrigated agriculture possible throughout all or most of the year.

The lakes that have been created provide habitats for fish, including exotic species such as black bass from North America. Besides fishing and other recreational activities, the impounded waters are often used to generate electricity before they are led into canals and on the fields of crops. So widespread are dams and the associated construction features that many of the freshwater ecosystems of Spain have been changed. Although little is known of the ecology of the rivers before damming, it is safe to state that the biological productivity has been increased, accelerated soil erosion has been reduced, and in general, a healthier ecology has resulted. (One must remember that the water impoundments usually do not represent an alteration of pristine conditions, but are attempts to remedy the long history of ecological abuses.)

The most recent major change in the surface and subsurface waters of this region is related to pollution. Water pollution is by no means something new in this region, as we can recognize merely by looking back to the fetid cities of the Middle Ages, in which water was drawn from the open sewers that were the depositories of all manner of wastes. However, in terms of magnitude, we should focus mainly on the last century, particularly the last few decades.

An examination of a map of Europe can suggest some of the major reasons that so much of the fresh waters of the continent are polluted. For example, large urban and industrial centers are located far from the ocean, so the people must dispose of wastes in areas far removed from the relatively efficient mixing ability of marine waters. Because of this, some of the rivers of this region are little more than sewers but serve also as major sources of water for domestic and industrial uses. One would find it difficult to locate any river in this region that is not polluted to some extent, but we shall focus our attention on three—the Rhine, the Seine, and the Thames.

The Rhine River begins its journey to the Atlantic at Lake Konstanz in Switzerland. It flows in a generally northward direction, passing the following major cities in West Germany: Mannheim, Frankfurt on the Main (actually Frankfurt is located on the Main River, a tributary of the

Rhine), Wiesbaden, Cologne, and Dusseldorf. Eventually, the Rhine passes into Holland. An examination of a large-scale map of the Rhine Valley will give you an appreciation of the number of people that populate it and thus use the waters of the Rhine. In recent years, there have been spectacular losses of fish in the Rhine due to pollution and concern for the river's future has been growing. If the Rhine were destroyed, human settlement along this corridor would be impossible, so a campaign was recently begun to clean it up.

The fabled Seine begins in the low hills of Langres in southeast France. However, after meandering through Paris, its condition is quite different from when it arrives. In Paris, it receives the waste products of millions of persons and their pets and the froths, foams, and untold chemicals of all the factories along its banks. Men still fish along the banks of the Seine in Paris and one cannot but wonder at the optimism of the fisherman and the rugged constitutions of the fish—mostly carp—able to live in this substance that often bears only a slight resemblance to water. However, the government has become aware of the effect of Paris on the river and there is reason to believe that the Seine may regain its health soon.

Few rivers of Europe can evoke the mood of the end of the nineteenth century as well at the mere mention of its name as the River Thames. Nor are there many rivers whose waters have been so intimately bound up in the affairs of man. It was on the broad waters of the Thames estuary that the Romans sailed in A.D. 43, when they established Londinium. And in later centuries it was on these waters that England sent forth its traders and adventurers to all parts of the world. But it was also this river that man so polluted that it was often a source of such odors as to make, as one observer noted, the taking of tea on the terrace of the House of Commons in the summer almost impossible. The Thames, particularly where it flowed past London, achieved the distinction of being one of the dirtiest rivers in the world. The mess finally became so apparent that the government initiated efforts to clean the river and these have been very effective.

4.5 Influences on the Soils and Geomorphology

Man's influence on the soils and geomorphology has been closely related to his population size and level of technology. In general, the degree of impact has been greatest in the areas experiencing long, dry summers. However, one can scarcely move any appreciable distance in any direction in this region and not encounter examples of human-induced modifications of the soils and landforms.

Until farming made its appearance, it is fairly certain that man had little effect on the soils and geomorphology of the region. But with the establishment and spread of agriculture and domesticated animals, changes began to occur in the soils and even in certain of the landforms. We cannot determine whether or to what degree the early farmers perceived that their practices induced accelerated soil erosion, but some scholars believe that a few terraces may have been constructed for the purpose of retarding soil loss. However, another possibility is that the human-made terraces may well antedate the invention of agriculture and may have been used to direct runoff water onto fields in an effort to increase the yield of wild plants. In any case, the terraces are old in the Mediterranean part of this region; whatever their purpose during the early phases of agriculture, they had the effect of altering the natural geomorphic expression and reducing the loss of soil on farmed or grazed slopes.

In the classical period, however, we find extensive evidence of human modification of soils and landforms in this area, but we must remember that the evidence had a long history before this time. The Romans, through their desire to extend agriculture, modified the land surface in many parts of their empire. Agricultural terracing was greatly expanded, and the massive removal of vegetation from slopes resulted not only in the already mentioned flash runoffs but also in accelerated soil erosion. Pliny drew attention to the relationship between stripping slopes of their tree cover and the loss of soil, but his concern appears to have had little or no effect on Roman behavior.

The Romans, like the Phoenicians before them, were interested in mining. It was they who opened many of the mines in the Mediterranean region, some of which are still worked. Ore smelting, which was frequently a part of this activity, demanded large quantities of wood fuel. The resulting deforestation, alluded to earlier, led to catastrophic losses of the soil cover over virtually all the inhabited parts and, to lesser degrees, in the more moist areas of the Mediterranean region. Today, when one drives over winding roads through some of the mountain regions of the Mediterranean and sees expanses of gray limestone or glaring crystalline rocks, it is difficult to recall that these slopes were often forested not too long ago. Now they are the habitats of scrambling goats and a few other creatures (including man) who are able to find a means of surviving under these reduced circumstances.

Although the most important geomorphic effect of mining has usually been the result of fuel demands and the attendant wash of the soil, the mine sites themselves must have been, even in classical times, distinctive features on the land. But these

Figure 4.10 *Polder land in Holland.*

were local phenomena and lent only a local flavor of change to the geomorphic expression.

From the beginning of the Middle Ages (approximately A.D. 400) to the present, the more important human modification of landforms and soils in this region has generally shifted northwestward from the Mediterranean. Although modification of these environmental elements has not ceased in the Mediterranean, the major modification in the more recent period was in Western Europe, specifically, Holland (Netherlands).

There is an old saying that "God made the earth but the Dutch made Holland." This is an allusion to the monumental efforts of these people to reclaim, by diking and draining, land that lay beneath ocean waters. At present, about 40 percent of Holland's dry land lies *below* sea level. From the time when farming was first attempted near the Atlantic shore, the inhabitants of this area had to combat the fury of nature, because high tides and fierce winds drove the salt water over the farm lands. The first reclamation efforts were directed toward holding back the sea, during storms, from land that ordinarily was above the high-tide line. But later, attempts were made to push back the sea

and create new lands, protected by dikes and drained by windmill-operated pumps.

The lands thus reclaimed are called *polders* and are used for dairy farming and some crops are grown. As Dutch engineering skills and machinery improved over the years, the amount of polder land was increased. Ambitious projects are presently underway to reclaim additional large parcels. Few landscapes in the world are so much the result of human effort and skill as are the Dutch polders. One is awed by the work that has been expended here over the centuries and by the dogged persistence of a people to wrest its national territory from an even more persistent ocean (Fig. 4.10).

Not only have new lands been created but new soils as well. In fact, the task of making the land suitable for agricultural production has required almost as much ingenuity as the act of drainage itself. Careful experimentation and patience were required to leach salt from these lands so that they could be more than just poor pasture.

The soils of most of Western Europe where agriculture has been longest practiced have been transformed since man first sowed seeds there in the Neolithic. Most of the soils originally were developed under forest cover and climatic conditions that produced acid soils of limited agricultural productivity. Even the Neolithic farmers appear to have recognized this fact, for the early farming in the Danube area was located chiefly on loess soils, which were easily worked and had higher potential productivity.

The wholesale removal of the tree cover over most of Europe ended the part of the soil forming processes dependent on forests. In place of this, man began not only to work the soil, but over the years and with increased intensity in the last century or two, to introduce fertilizers, organic wastes, and other materials that altered the chemical characteristics and the tilth of the soils. One major problem was the widespread distribution of heavy soils that were difficult to work (poor tilth). It is said that some breeds of draft horses developed in Europe were in response to the need for powerful animals to pull plows through these heavy soils. The repeated plowing, fertilizing, and additions of organic materials created a soil cover that is easier to work (improved tilth), less acid, and far more productive.

Similar comments can be made for limited parts of the Mediterranean area, where intensive cultivation over many centuries has also led to the creation of a soil distinctive from that which man first farmed. This must be particularly true of

Figure 4.11 *An autostrada in the hills of Italy between Bologna and Florence.*

some of the intensively farmed and irrigated tracts in Spain and Italy.

Alteration of the geomorphology has proceeded at an increasing pace in the past few decades, because of the desire for major automobile routes and the availability of earth-moving equipment. Vast quantities of earth and rock have been and continue to be moved to provide the cuts and fills of highways across the landscapes of all but the economically poorest parts of this area. The scale and geographic spread of these efforts more than rival the engineering efforts of the Romans (Fig. 4.11).

4.6 Influences on the Atmosphere

The first influences on the atmosphere were local and probably began when fire was used to alter the vegetation cover. Such alteration modifies the *albedo* (reflectivity) of a land surface, which in turn, alters the local energy flux and the local microclimate. It also raises the possibility that the regional climates may have been altered.

We are on sounder ground when we examine air pollution. This is not a new phenomenon in this region; it began when fire came into use and smoke pollution became a part of human existence. However, the chemically complex mixtures we usually associate with atmospheric pollution do not predate the widespread use of coal in this region. That is, "modern" air pollution

Figure 4.12 *Factories in the Ruhr industrial area.*

is largely a creature of industry and its products, such as the internal combustion engine.

The places of most serious air pollution are the major urban centers and the industrial centers. All the large cities of this region—Rome, Madrid, Paris, London—are periodically subjected to extremely polluted air. Until recently, most of this pollution came from burning soft coal, but this practice has been increasingly restricted by government edict, especially in London. Taking the place of coal or adding to coal-smoke pollution are the emissions from engines burning hydrocarbon fuels. It is now usual to encounter almost unbreathable air in the central cities during "rush" hours. And the situation is frequently worsened because poor quality fuel is burned in some vehicles with the result that every car and truck sends forth a cloud of odorous smoke. However, even where the best grades of fuel are burned, smog from petroleum fuel is widespread.

Most of the heavy industry of this region depends on coal for fuel. At best, burning soft coal has often produced extreme conditions of atmospheric pollution (both particulate and chemical), but the problem is worsened in many parts of this region by the burning of low quality coal that emits a large amount of smoke. The area that has long been the most guilty of this practice is the industrial region known as the Ruhr. Here, in an area of some 5000 square kilometers, is located one of the world's major industrial complexes—and one of the world's principal sources of atmospheric contaminants (Fig. 4.12).

Impressive international and national efforts are being undertaken to reduce some of the worst aspects of atmospheric contamination. However, thus far there has been a greater toleration of air pollution than in parts of the United States. In some Mediterranean countries, it is believed that air pollution is simply part of the price that must be paid for industrial progress. Because this was also the prevailing view in the United States until recently, we may expect that ultimately air pollution control measures will be instituted in those countries not bothering with them now.

Chapter Five

Human Influences on the Ecosystems of Australia

5.1 Cultural Background

The condition that has had the greatest influence on Australia is its physical isolation. Until recently, Australia was a kind of cultural and biogeographical attic in which items existed that had long since vanished from most of the rest of the world.

Archeology in Australia is presently an active field of research, so statements made now about the earliest arrival of man in that area will probably be proved inaccurate to some degree. Less than a decade ago the oldest known dates were about 8000 years BP, but since then carbon materials exceeding 20,000 years BP have been dated. The now emerging belief is that *Homo* may well have been on the continent of Australia by at least 30,000 years ago, and evidence of even earlier dates may be found.

It is generally assumed that man arrived first in the north of Australia and then dispersed southward to Tasmania, which, during the Pleistocene was a peninsula of Australia because of lower sea stands. New Guinea, now a large island, was a northern peninsula. There has been much speculation as to how early man crossed from Java to New Guinea. It is clear that he did—only his method remains in doubt. Also in doubt are the places the pioneers first touched in Australia and how many pioneer units were involved. One hypothesis suggests that there were three principal human gene pools that blended to form the Australian aborigine and that the ancestors of the now extinct Tasmanians arrived in Tasmania when it was a peninsula of southern Australia. These people were later cut off from the continent by the rise of the ocean level and did not participate in any subsequent human genetic additions to Australia.

There is no reason to believe that the first humans in Australia did not possess the technique of making fire, and it may be assumed that fire was used for a variety of purposes, including hunting and the preparation of food. There is not the slightest evidence that agriculture was known in Australia before the arrival of Europeans—the economy was strictly hunting and gathering. The material culture of the Australian aborigines, although extremely limited, was highly effective in terms of the needs of the people. Throwing sticks and spears were common, but bows and arrows were unknown. However, the aborigines possessed spear throwers (which greatly increased the velocity of a spear), and nets and traps. They also knew how to prepare poison, which they used in hunting—although this form of hunting was somewhat limited in Australia.

The individual population units were small, probably not often exceeding fifteen persons. They were parts of larger societal entities held together by ties of kinship, but the larger units met only occasionally to celebrate important functions. The food supply appears to have been chiefly responsible for the size of the ecological units. Each unit wandered over a well-defined territory, never staying long in a given place. In fact, extreme nomadism within defined territorial limits appears to have been typical.

When the Europeans first established themselves on the continent, there may have been approximately 300,000 aborigines. This figure is the one most frequently mentioned in print and gives us a picture of a human population dispersed over nearly 7.7 million square kilometers, which is the area of Australia. It has been argued that the size of the population would have been insufficient to exert any ecological impact on the prehistoric ecosystem of Australia. However, these were highly nomadic people so they could spread their influence over much larger areas than could a sedentary culture.

The first recorded European discovery of Australia was in 1605, but the period of significant contact and interest began in 1770, when parts of the east coast were visited by Captain James Cook. Until Cook's voyage, the foreboding coast did not entice colonization by the European powers. Even Cook's generally favorable report did not at once stimulate the British government to establish colonies there. However, the loss of the American Colonies later in the same decade stimulated some interest in this southern continent, particularly as a place where the English government might confine criminals. Thus, in 1788, the first group of convicts arrived; they were established, after a fashion, at the present site of Sydney (New South Wales). Further shipments of convicts as well as free persons followed in succeeding decades. After 1853, the transport of convicts ceased and by then they were greatly outnumbered by free persons. In this respect (as well as others), Australia's early settlement by Europeans resembles that of North America before the American Revolution. The greatest growth spurt in the European population in the nineteenth century came with the discovery of gold in Victoria, Western Australia, and elsewhere, which brought in a rush of gold seekers from many parts of the world.

The establishment of British settlements marked the beginning of massive changes in the use of land in Australia, because agriculture and domesticated animals were introduced almost from the beginning. In succeeding years, industry and mining, particularly mining, were also added and had important effects on the landscape.

The comparative smallness of the human population in Australia has long been a factor in the politics of the country, because in terms of numbers Australia has a very low density. However, only a relatively small part of the continent has proved satisfactory for European settlement; that is, most of the area is ecologically unsuited to the land-use practices of Europeans. From an ecological point of view, Australia is *not* sparsely settled. However, this is difficult for some persons to realize because a population of slightly less than 13 million for 7.7 million square kilometers seems like a vast empty land of equally vast settlement potential.

To help fill her spaces, Australia has been encouraging the migration of Europeans, and the result is a population of widely varying cultural backgrounds. In recent years many persons from the United States also have migrated to Australia. It is too soon to assess the impact these varied peoples with their varying attitudes toward the land will have on the ecosystems of Australia or even to know what cultural ideas and practices they have transplanted.

5.2 Influences on Vegetation

The idea that the aborigines could have exerted any significant influence on the vegetation of Australia has long been rejected. The basis for this rejection is the modest number of people thought to have been present before European entry and the view that only "civilized" man is capable of altering the earth's ecosystems. But, as we have seen, this is an invalid argument. In recent years, an increasing number of Australian scholars have been investigating the possible aboriginal influence on the ecology of Australia in prehistoric times, particularly their influence on the vegetation cover.

This leads directly to a consideration of the aborigine's use of fire. It is clearly established that fire was frequently used by Australian hunters to drive animals. Fires were also set to kill animals trapped by the flames and to open up areas to greater visibility. In fact, so general was the use of fire in modifying the vegetation cover that one Australian author characterized the aborigines as pyromaniacs. Most of Australia experiences prolonged periods of drought each year and during these times, the effectiveness of burns was much greater than they would have been in a more humid environment (see Maps 5.1a on page 148 and 5.1b on page 149). Thus, human-set fire was an important force in the ecosystems of Australia before the arrival of Europeans. We may even hypothesize that some of the large ecosystems of Australia owe their present "natural" vegetation cover in part to a long history of human-set fires. As research continues and more ecologists become aware of the significance of this aspect of Australia's ecology, we will be able to be less tentative when assessing the aborigine's impact on the vegetation.

Although the settlement of Europeans in Australia dates from fairly recent time, the impact on the vegetation has been great.

Agriculture. The initial settlement, at Sydney, was located on relatively poor soils, and most of the early colonists had little or no knowledge of farming. Thus for some years, agriculture, such as it was, was limited to small patches of land. The first crop tried appears to have been wheat. After great difficulty and many decades, wheat was so well established that it became one of Australia's principal exports. Eventually, all the British crop plants and trees were taken to Australia and many were grown with success. Because human population in the historic period clustered chiefly in the southeast part of the country, it is here that the greatest ecological changes of wild and cultivated vegetation took place in the first decades of settlement. The size of the area planted to wheat has fluctuated over the years in response to local and world market conditions. There have been times when the world market was such as to encourage farmers to push wheat farming beyond the safe ecological limits, which

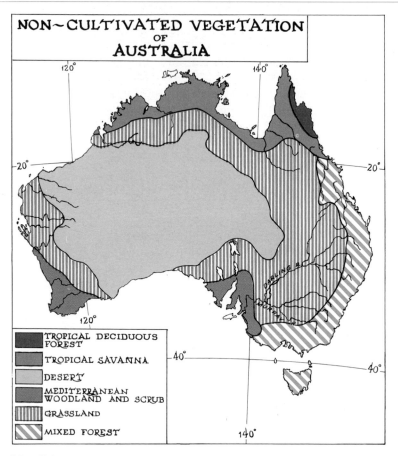

**NON~CULTIVATED VEGETATION
OF
AUSTRALIA**

TROPICAL DECIDUOUS FOREST

TROPICAL SAVANNA

DESERT

MEDITERRANEAN WOODLAND AND SCRUB

GRASSLAND

MIXED FOREST

Map 5.1a

sometimes resulted in great damage to the ecosystems.

Today, the principal farm landscapes are in the Darling River area (see Map 5.2 on page 150), parts of Western Australia, and local parts of Queensland. As you can see, vast areas of Australia are not planted to crops. For its limited area, Tasmania has a relatively large percentage of land in some kind of crop.

Livestock. Although livestock animals were introduced in the latter part of the eighteenth century, it was not until the early nineteenth century that livestock became a significant element in the changed ecological scene. The animal overwhelmingly associated with this change was the merino sheep.

Sheep, as we have repeatedly seen,

require careful management if they are not to become highly destructive to vegetation. Unfortunately, almost no attention was given to the possibly damaging aspects of sheep with the result that poor grazing practices degraded large parts of the vegetation cover. We may never know with exactness all the changes, because little attention was given to the study of the floristic composition of the wild pastures before the sheep were put on them.

Although sheep were far and away the most important livestock animals brought to Australia, cattle, horses, camels, and the usual assemblage of smaller mammals and fowl were also introduced. Cattle, because of their habit of selective grazing and the stockmen's desire to increase the area of available pasture, exert significant pressure on the vegetation cover. For example,

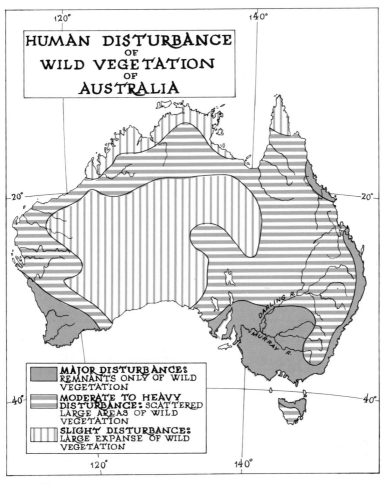

HUMAN DISTURBANCE
OF
WILD VEGETATION
OF
AUSTRALIA

MAJOR DISTURBANCE: REMNANTS ONLY OF WILD VEGETATION

MODERATE TO HEAVY DISTURBANCE: SCATTERED LARGE AREAS OF WILD VEGETATION

SLIGHT DISTURBANCE: LARGE EXPANSE OF WILD VEGETATION

Map 5.1b

in northern Queensland, it is the practice to set fire to large areas to clear them of woody vegetation in order to encourage the growth of grasses for the pasturing of cattle. This is extremely wasteful as far as the vegetation cover is concerned but may be only a continuation of a practice (burning) long established in the area before Europeans arrived.

Exotic plants. We shall begin our discussion of exotic plants with a mention of some that escaped from areas of cultivation and spread over parts of the wild lands of Australia.

The first plant introductions occurred with the first European settlement. Most exotics were introduced in the states of New South Wales, Victoria, and the southeastern part of South Australia—that is, the area of longest and densest non-aborigine settlement, which has contained the largest share of farms and the greatest number of sheep and cattle. Here, man's influence on the vegetation has been extreme and his additions to the list of plants number in the hundreds. In 1930, it was reported that exotic plant species comprised 18 percent of the flora of Victoria, that is, there were 500 exotic plant species, of which 100 were grasses. These plants were often introduced casually, probably chiefly by the accidental inclusion of seeds with shipments of grain or other desired

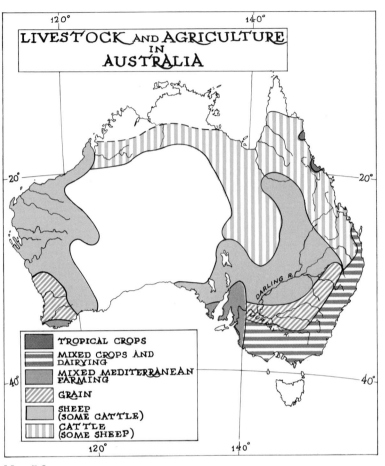

Map 5.2

plants. Sheep played an important role in spreading the seeds of some of these exotics, and disturbed soil (from farming or grazing) often provided a habitat favoring their establishment.

Many of the introduced plants caused great ecological changes and economic loss, but none had these effects as much as a cactus from the Western Hemisphere. Known as the prickly pear cactus, two species, *Opuntia inermis* and *Opuntia stricta,* were introduced in Australia from the Americas in the last century. The results were so devastating that they may well be the most outstanding example in the world of what can occur when an exotic plant is placed in an optimal situation (Fig. 5.1).

Opuntia inermis may have been introduced by 1839. It was later spread into

new grazing areas of south Queensland and New South Wales, where it was planted to form hedges. It is thought that *O. stricta* was introduced into central Queensland about 1860. Almost unnoticed at first, the plants escaped the sites where they had been planted and began to colonize adjacent fields. The rapidity of the spread was startling; by 1900 there may have been about 4 million hectares partially covered by these plants. The peak of the phenomenon was probably reached in 1925, when more than 24 million hectares are said to have been contaminated. Of this vast area (equal to approximately 93,000 square miles), almost half was so heavily covered as to make the land useless for grazing. It is said that in some areas one could view continuous stretches of

Figure 5.1 Opuntia *infested area in Australia before the introduction of the cactus moth.*

hundreds of square kilometers where the infestation was complete. The worst situation was in Queensland but in New South Wales, conditions were also serious.

Naturally, government and private individuals attempted to eradicate this scourge. After many false starts they succeeded in dramatically reducing the area of infestation by introducing a moth (*Lepidoptera: Cactoblastis cactorum*), native to South America, whose larvae feed on these plants. This is one of the most dramatic instances of successful biological control ever attempted.

Trees. The history of men and trees in Australia since 1788 has been one of de-struction in the early years and reforestation and the introduction of exotic tree species in more recent times. Australia is known for its impressive collection of trees of the genus *Eucalyptus* (Fig. 5.2). Many of the members of this genus were assigned vernacular names in Australia that are somewhat misleading, such as mountain ash, snappy gum, yellow box, and coolabah. In addition to the *Eucalyptus*, other valuable tree genera include *Acacia, Cedrela*, and *Araucaria*, to mention a few. In the first century or so of European settlement, the forest areas were ruthlessly mismanaged either by wasteful exploitation for timber or to make space for pasture or farms.

Figure 5.2 Eucalyptus *forest in a sandstone area of Australia.*

Of the several valuable timber trees, the red cedar, *Cedrela toona,* received the brunt of overexploitation. The greatest concentration of these trees occurred in coastal parts of New South Wales, particularly in the Sydney region and northward into Queensland. As is true of its tropical American relatives, the wood of this tree lends itself to cabinetry and its aromatic properties are highly prized. In spite of the great effort required to log an area in the early days, by 1820 permits were required to cut cedar trees around Sydney because the trees had already become scarce. The cedar cutters in Australia played a role similar to that of the fur hunters in the formative stages of North America—both groups were often the first Europeans in a new district and usually were followed by farmers or livestock raisers with their sheep and cattle. The hunt for cedar had some of the aspects of a gold rush, and there was a rapid alteration of the forests affected. Not only was the cedar mined out but the crude methods used prepared the way for forest fires. Eventually, the cedar trees were so reduced in number as to cease being a significant factor in the economy of Australia.

There is reason to believe that *Cedrela* concentrations may also have been influenced, at least in part, by aboriginal land-use practices. Given the long prehistoric use of fire in Australia, the question arises as to a possible relationship between the once relatively high concentrations of cedar and the fires set by the aborigines.

Far more destructive than cutting trees for timber was the practice of ridding areas of their tree cover so the land might be used for farms or pastures. A common method was to ring-bark the trees, that is, to make an incision all around the tree through the cambium layer so the sap would be cut off, causing the tree to die (Fig. 5.3). Great areas of fine timber-quality *Eucalpytus* and trees of other genera were thus destroyed. The attitudes of the earlier colonists were similar to those

of the colonists in the United States. In both countries, the people thought that the forests were unlimited and that trees must be gotten rid of as soon as possible so the land could be "developed" for agricultural or pastoral pursuits. Unfortunately, the timber lands in Australia were quite limited in area; at their peak they comprised only a small fraction of the total area. Tasmania fared better than the rest of the country at the hands of the early settlers, but unfortunately, its forests were ravaged by timber cutters and farmers in the nineteenth century. Now only a vestige of the tree-covered areas of the late eighteenth century remains.

Although once possessing a fairly impressive, albeit limited hardwood resource, the continent has always been lacking in softwoods. Most importantly, Australia does not have any native trees of the genus *Pinus*—the trees that are known as native pines belong to other genera. Perhaps the most useful of the native softwoods is the so-called hoop pine, *Araucaria cunninghammi.* However, the tree is not particularly abundant and has been much abused by the practices described above. Thus, interest developed early in introducing exotic trees, particularly true pines, to provide a supply of soft woods. However, this economic interest came after pines and other exotic tree species had been introduced for aesthetic reasons and wind-breaks.

A complete list of tree introductions into Australia would be lengthy because parks, streets, and private lands all contain exotics. Among the early favorites, however, were pines (*Pinus*) from the Mediterranean region of Europe. For decorative reasons, the stone pine (*Pinus pinea*) was introduced during the latter part of the nineteenth century into parts of south-eastern Australia. Then came cluster (or maritime) pine (*P. pinaster*) and aleppo pine (*P. halepensis*). These three are found in many situations—gardens, parks, pastures, windbreaks—from Perth to the

Figure 5.3 Eucalyptus *trees killed by ringbarking in Australia.*

northern part of New South Wales. The canary island pine (*P. canariensis*) which is so popular in parts of California as an ornamental, is not so well favored in Australia. However, it too is found there.

In the latter part of the nineteenth century, the Monterey pine, *P. radiata*, was introduced from California and achieved considerable commercial success in parts of Southern Australia. It was with this pine that Australians first attempted serious commercial plantings, because it is a fine timber tree and grows rapidly under Australian conditions. The greatest efforts to establish plantations occurred during the first few decades of the twentieth century. Several thousand hectares of Monterey pines were set out each year. There were also some efforts to establish plantations of Mediterranean pines, but they were of limited and local significance. The total area in all pines now exceeds 300,000 hectares, located chiefly in the southeast of Australia.

The human influence on the vegetation of Australia, therefore, has been profound in the period since the first European settlement in 1788. In the areas of densest settlement today, the vegetation cover reflects the hand of man to an important degree. In fact, the southeastern part of Australia has one of the most modified vegetative covers in the world.

5.3 Influences on Animal Life

Given at least a 30,000 year history of hunting before the arrival of Europeans, there must have been disturbances and alterations in Australia's wildlife long before 1788. Nevertheless, this idea has not yet been generally accepted. However, the evidence is increasing that the aborigines exerted a significant influence on the animal life of the continent.

Before reviewing this evidence, it would be useful to examine some of the prominent features of the zoogeography of Australia. Australia, New Guinea, and, traditionally, New Zealand, form the Australian zoogeographical region. However, this chapter relates only to the country of Australia, including Tasmania. New Zealand is discussed in Chapter 8.

The most singular features of the Australian fauna are large numbers of marsupial mammals; the only egg-laying mammals known; and a relatively sparse collection of native placental mammals. In addition, the area has a remarkably rich bird fauna, including two large ratites (flightless birds that lack a keeled sternum)—the emu (*Dromiceius* spp.), and the cassowary (*Casuarius* spp.). Also noteworthy are many members of the parrot family, some of which, such as the cockatoos, are extremely beautiful, and a group called bower birds. Male bower birds often construct elaborate structures of vegetation as an adjunct to their equally elaborate courtship dances.

The reptile fauna is rich and varied and enjoys the dubious distinction of possessing more poisonous than nonpoisonous snake species. Crocodiles occur in the northern part of the country and lizards of the genus *Varanus*, known colloquially as goanas, are distributed widely in the less arid areas. *Varanus* lizards are also widely distributed from Africa to Southern Asia. As might be expected, over much of Australia the amphibian fauna is not rich, but in the moister habitats there is an extremely interesting and often beautiful collection of frogs (family Hylidae and others).

Because of a general lack of surface streams, the freshwater fish fauna is not rich. The Queensland lungfish, a kind of "living fossil," is one of the most interesting elements of the fish fauna.

The fauna of Australia gives evidence of the continent's long physical isolation from Eurasia and even from the island complex on the Sunda Shelf, which includes Sumatra, Java, and Borneo. Man arrived in Australia rather late, although he appears to have been able to reach Java by at least 300,000 years ago.

The most conspicuous placental mammal, aside from man, before Europeans arrived was the dingo, *Canis* (Fig. 5.4). It has been much debated whether this dog was native to Australia or introduced by man. The evidence is overwhelming that it

Figure 5.4 *The dingo.*

was brought in by man—no dingo remains from before 20,000 years ago have been found. Thus, there is no basis for continuing the debate—man brought the dog with him and it was from this one or more introductions that the feral populations derived.

In the enthusiasm to emphasize the size and variety of the marsupial population of Australia, the placental mammals sometimes appear to be rare. This is not true, however. There are many placentals, although far fewer distinct taxonomic forms of them than of marsupials. For example, there is an extremely interesting bat fauna that includes the so-called flying foxes (family Pteropidae). (No evidence exists that marsupials ever successfully colonized the air.) There is also a rodent fauna that may be derived from a single successful colonization occurring well before the arrival of man.

What influence, if any, did the aborigines have on the native animal life? It has long been customary to believe that the aborigines lived in a close rapport with nature, but the accounts of early explorers and of anthropologists suggest that this view is false. What does appear to be true is that the aborigines expended considerable effort to obtain all the animal food possible and developed an impressive array of techniques and devices to accomplish this end. Foremost among the hunting techniques was fire, which we have already mentioned.

To try to determine the extent of the aborigines' influence on animal life, it is necessary to have some relatively clear idea as to how many aborigines were present in Australia when the Europeans first arrived. By extrapolation, we would thus have an idea of how many must have been present for a relatively long period before European colonization. As stated earlier, the most frequently used figure is 300,000. However, at present there is no sound base to support that figure. Concerning vegetation, it was indicated that a few nomadic persons using fire over a large area could, with time, greatly modify the vegetation cover. In the case of animal numbers, however, there may be a much closer relationship between the numbers of humans and their impact on animal life.

In recent years it has been suggested that aboriginal man in Australia may have been responsible for the disappearance of many larger marsupials toward the close of the last major glacial period. The evidence brought forward is largely circumstantial and we may never know whether this is true. If man was responsible for the disappearance of these animals it may have been from oblique rather than direct causes. The very generalized pattern of aborigine hunting suggests that no one animal species was long the object of intensive effort but that hunting was dispersed among many animal species—mammals, birds, and reptiles principally, but often including fish and a host of invertebrates. It may be that changes in postglacial climates coupled with the degrading influence of man-set fires interfered with the habitat requirements of some of the animals.

Less dramatic but perhaps of equal significance is the introduction of the dingo into Australia. This highly adaptable carnivore may have exerted a strong influence on the smaller mammal component of Australian ecosystems. It has long been in a feral state, so it must obtain its food by hunting. Little is known about the ecology of feral dingoes, especially their effect on populations of native mammals. An interesting aspect of the dingo population is that it appeared to be relatively modest when the Europeans first settled in the area. Later, however, sheep were introduced, and when the dingo discovered them, there was an increase in dingo numbers. Dingoes eventually became a serious threat to sheep raisers. Whatever influence the aborigines may have had on the native animal life, however, it was probably modest in comparison to the European impact since 1788.

Deliberate Introductions of Animals. The "first fleet" of Europeans carried not only human passengers but also a small collection of livestock to help the colonists develop a food base. The composition of this initial introduction was not recorded but from an ecological point of view it appears not to have been significant. It was the repeated shipments of animals that led to an ever-increasing array of exotic animals—sheep, cattle, pigs, chickens, cats, dogs, and so on. The earlier sheep breeds were of poor quality; not until the late eighteenth century were the fine, fleece-bearing merino sheep brought in. It was quickly determined that merino sheep raised in Australia produced an excellent quality of wool, and at that time the textile industry in Britain was developing rapidly. The number of sheep quickly increased, and the animals and their drovers soon spread into much of southeastern Australia. By the middle of the nineteenth century, wool production reigned supreme in the economy and sheep numbered in the millions. The herds have continued to grow—by 1969 there were 176,200,000 sheep in Australia (see Map 5.2 on page 150). They are still concentrated in the same areas they were over a century ago and it is easy to imagine the effects on the ecosystems of over 150 years of grazing. The pastures are now very different from what they were before the introduction of domesticated *Ovis aries*.

Although cattle (*Bos taurus*) also were introduced early they never assumed the economic importance of sheep (Fig. 5.5). Their ecological significance, which was also great, was not so much due to the effects of overgrazing or selective grazing, but more to the fact that cattlemen destroyed large tracts of trees to provide pasture. In fact, in parts of Northern Australia, fire-clearing is the principal means at present of opening up pasture lands for cattle. Occasionally, cattle have escaped and established themselves in a feral state. In northern Queensland there are many feral cattle, but the ecology of these beasts has not yet been studied.

Perhaps the most singular introduction was that of the camel (*Camelus dromedarius*). After several unsuccessful minor attempts, success was achieved in 1866 with a shipment to a port in South Australia near Adelaide. This group became the foundation stock for the herds that were later developed. Given the great arid and semiarid expanses of Australia, it appears a reasonable introduction, but one usually does not associate camels with British farm colonists. Thus, camel herders—chiefly Afghans—were also brought in because

Figure 5.5 *Driving cattle in northern Australia.*

the Australians had no knowledge of the animal. Camels played an important role in the development of Australia because they provided transportation in regions where no other animal could have served as well.

The history of camels in Australia was ably summarized by McKnight (see list of readings); he estimated that at their peak, which was in the early 1920s, domesticated camels numbered around 20,000. After that time, other means of transport made the animal less economically attractive, so the number of domesticated (that is, controlled) camels declined; today only a few hundred may be included in this category. On the other hand, the number of feral camels increased; McKnight estimated their number to have been between 15,000 and 20,000 in 1966. Of these, 40 percent were in Western Australia, 30 percent in Northern Territory, and the remainder in South Australia and parts of Queensland. The ecology of the feral camel in Australia has not yet been studied systematically, but it appears that the animal exerts only a minor impact on the desert ecosystems for it feeds mainly on plants that are not significant elements in the diet of other mammals. The chief problem appears to be that the animals frequently destroy parts of fences that were erected to contain rabbits or dingoes, resulting in considerable economic loss. Another problem, especially where livestock is raised and water is scarce, is that camels foul water holes. For these and other reasons, the feral camel is frequently classified as vermin, which encourages persons to destroy them. Because they are the largest animal inhabiting the Australian "outback," it would not be difficult to destroy all but a small number, should this be necessary. Although the evidence is not complete, it appears that this addition to the Australian fauna was less damaging than most of the others and had, for a time, important economic benefits. It should be given more study by ecologists.

In dramatic and painful contrast with the camel introduction was the introduction of the European rabbit (*Oryctolagus cuniculus*) (Fig. 5.6). This introduction and its ecological aftermaths have come to serve as *the* horrible example of what can occur when introductions of animals are made without prior tests.

There was once some confusion as to when the first rabbits were taken into Australia and when the first escape occurred, but it is now fairly well agreed that the first significant liberation of rabbits occurred in 1859 in Victoria. A farmer set free a small number of rabbits and, within three years, they had so multiplied that it was obvious they should be controlled. However, they were not controlled and their numbers exploded. This army of nibblers spread at a rate that appears incredible—about 110 km per year, according to one account—and exerted a devastating impact on the ecosystems they invaded. The economic impact on wool production was also great, because rabbits and sheep often contested for the same plant food. In spite of the seriousness of the situation, however, little systematic study was conducted on the ecology of the wild rabbit in Australia, so the measures taken to control the animal were almost always unsuccessful. And these measures were expensive. They included the construction of fences and the digging of ditches hundreds of kilometers long, and attempts to poison the animals out of existence. Finally, it was discovered that a virus native to Argentina that caused myxomatosis was particularly lethal to European rabbits. After a period of experimentation, this virus was introduced into Australia. During the early period of this introduction (1950s), the effect of the virus on the rabbit population in Australia was phenomenal, with almost 100 percent mortality among those infected. These results threw into sharp relief the economic and ecological toll the rabbits had been taking because the wool clip of 1952–1953 increased by approximately 32 million kilograms.

Unfortunately, however, not all the rabbits infected with the virus died. There is now a growing population of myxomatosis-resistant rabbits which promises new ecological and economic threats ahead. It may be that as time passes, the rabbit will fit into the ecology of Australia and be only a minor problem. But that time has not yet arrived, and the ecological cost of this ill-advised introduction will long provide a warning to approach animal introductions with great care.

One other mammal introduction deserves mention. This is the introduction of

Figure 5.6 *A typical rabbit concentration at a water hole in Australia before the introduction of myxomatosis.*

the European red fox (*Vulpes vulpes*) into the southern part of Australia some time during the nineteenth century. In the absence of native foxes, this small carnivore, which is the object of "the hunt," was deemed an excellent animal to introduce. It has since become well-established in parts of southern Australia and exerts significant pressure on the local ecosystems. It is said that the fox plays a role in holding down rabbit populations, but it also plays a role in holding down the populations of a number of native marsupial species, which probably cannot withstand such predation.

There have also been introductions of exotic birds, in addition to those destined for farms or aviaries. Most of these introductions took place in the southern part of the country, notably in the state of Victoria, and relatively few have become ecological or economic problems. However, the house sparrow (*Passer domesticus*), European starling (*Sturnus vulgaris*), and Indian myna (*Acridotheres tristis*) have become conspicuous elements in the local avifauna in parts of Australia, and occasionally have been agricultural pests. Apparently, their populations have not yet exploded to the extent that they are major ecological or economic hazards.

Although Australia possesses some of the finest saltwater game fishing in the world, many of its freshwater streams are not so well endowed. (the Murray River, however, contains some excellent game fish.) Attempts were begun in the nineteenth century to introduce game fish from other parts of the world with the greatest attention being given to salmonids. Thus, there were attempts to introduce species of trout and at least one salmon species. The salmon did not succeed but the brown trout (*Salmo truta*) from the British Isles (Palearctic) and the rainbow trout (*Salmo gairdneri irideus*) from California (Nearctic) were successful. These introductions were chiefly in New South Wales, Victoria, Tasmania, and the extreme southwest of Western Australia.

Any impact they might have had on the ecology of the streams into which they were introduced is not known. In addition, European perch (*Perca fluviatalis*), several of the carp, and the tench (*Tinca tinca*), all from the Palearctic region, were introduced. The carp were later discovered to be undesirable because their feeding habits degrade aquatic ecosystems below the needs of many other fish species.

Other important introductions included invertebrates. The uncontrolled patterns of crop plant and animal introductions of the nineteenth century assured that many invertebrates would also be inadvertently introduced. Thus there is now in Australia a collection of crop pests that annually causes a large economic loss. Most important, in terms of economic loss, was the accidental introduction, near the end of the nineteenth century, of the blow fly (*Lucilia cuprina*). This fly lays its eggs on sheep, and the sheep, if not treated, are fed on by the fly larvae. Millions of dollars are lost each year because of blow flies. Other exotic flies also cause damage. They have been the object of intense ecological study in recent years in an attempt to rid the country of them or at least bring them under control.

A list of all the known invertebrates introduced into Australia would be quite long, and many of these invertebrates have exerted unwanted influences on the ecology of the region. The entomological geography of Australia has been significantly modified by man since 1788.

Partial and Complete Extermination.
The marsupials of Australia are a threatened fauna. At least six marsupial species are considered to have become extinct in the past 100 years, and some of the forms on the list of endangered species are so limited in number that they are almost nonexistent.

The extinction process in Australia occurs in various ways: (1) direct overexploitation, (2) direct competition between native and introduced forms, (3) habitat destruction by forest removal, and (4) planned destruction of "vermin."

The most dramatic example of overexploitation in Australia is probably the koala bear (*Phascolarctos cinereus*), a medium-sized, thick-furred arboreal marsupial that once had an extensive range in Australia (except Tasmania) in the states of Victoria, New South Wales, and Queensland (Fig. 5.7). All accounts indicate that this harmless and attractive little animal was extremely abundant as recently as the early years of the twentieth century and it was even more abundant before a disease greatly reduced its numbers in New South Wales in the latter part of the nineteenth century. However, koala skins became popular during the nineteenth century, and large numbers were taken every year. In 1889, 300,000 skins were sold in London; in 1924, over 2 million and, in 1927, 600,000 skins were exported from Queensland. Obviously, this could not long continue without serious consequences to the species and finally the animals were given complete protection. It is unlikely that koalas will ever occur in their early twentieth century population size,

Figure 5.7 *Koala female and young* (Phascolarctos).

but extinction appears to have been averted.

Currently, there is commercial overexploitation of some kangaroo species, which are being shot and processed for pet food, and of marine and freshwater crocodiles, whose hides are in demand for leather.

We mentioned earlier the ecologically destructive aspects of sheep raising in Australia and the impact the sheep have had on the vegetation cover during the past 150 years. It follows that there has also been a major impact on the native animals, especially those that are forced to compete with the sheep for available plant food. Although there is little documented evidence to indicate the time or place that a given native animal began to be scarce as a result of such competition, ecologists agree that this has taken place on a large scale and continues to be an important aspect of Australia's ecology. The direct effect of food competition has sometimes been masked by other pressures that were exerted simultaneously on the wild animals by man.

Perhaps even more important than sheep in the competition for available food was the introduced rabbit. We have noted the dramatic increase in the wool clip shortly after the first major reduction in rabbit numbers, which was brought about by the spread of the virus causing myxomatosis. It follows that other animal species must have been affected by the presence of the rabbits and the later release from their pressure but, aside from noting the increase in wool production, little attention has been given to this phenomenon.

The most poignant aspect of biological competition was the pressure exerted on aborigines by invading caucasoids. In the early days, the aborigine was easily swept aside. He was generally considered a minor irritation, unworthy of serious concern. In comparatively recent times, this attitude has been reversed, but it is too late to preserve the aborigine's culture, al-

Figure 5.8 *Dingo fence gate at the Queensland–New South Wales border. The signs on the gate emphasize the importance of keeping the gate closed and post the fine for noncompliance.*

though, as a biological entity, he may survive for many years. Far more final was the effect on the Tasmanians, a hunting and gathering people who were the object of vicious and criminal treatment in the nineteenth century. It is recorded that they were sometimes hunted as "game" and were run down by hunting dogs. In the latter part of the last century, after belatedly being giving official government protection, the last of these people died.

The destruction of natural habitats in Australia has become a major ecological phenomenon in those places where human settlement or raising livestock is best developed. The destruction of the tree cover has greatly reduced the numbers of forest-dwelling species and has put some forms on the list of endangered species. We can make only general conclusions about this aspect of Australian ecology, because for the most part, the phenomenon has been described only in general terms. However, there is no question that many of the smaller marsupials and many birds have had their survival seriously threatened by habitat alteration.

As European man spread out on the landscapes of Australia, he came to contest more frequently with the native animals for the available space, available plant food, and general use of the ecosystems. The dingo was the first animal to receive attention for it had shown itself partial to sheep meat. Various efforts were made to control it—efforts that resulted in large sums of money being paid for rewards for killed dogs, for poison, and for the construction and maintenance of dog fences (Fig. 5.8). Some of the dog fences, built to exclude these canids from specific areas, extend for hundreds of kilometers; however, they are only partially effective. Poison is used liberally but the method by which it is used results in the loss of other animals too.

There is widespread belief in Australia that some animals are vermin and should be controlled or eliminated completely (the dingo is one of these). Unfortunately, also classified as vermin are some native Australian animals—the emu, many of the parrot family, some of the larger marsupials, such as the red and gray kangaroos, and smaller marsupials including some wallabies and wombats. These native forms are included in the lists of vermin, because they compete with man for the crops he produces or the pasture on which he wishes to feed his stock.

Although few animals have become extinct in Australia since 1788, the future of much of the continent's wildlife has reached a crisis condition. The future hinges largely on an approach that is not devoted to the furtherance of economic interests at the expense of the ecological patrimony of the continent.

5.4 Influences on Surface and Subsurface Waters

Most of Australia is arid or semiarid. Surface water is scarce in almost all parts of the country, and in those limited areas where water is relatively abundant, there are cyclical variations that include prolonged periods of drought. Rainfall variability is a problem everywhere, except in parts of Tasmania and a limited area in the extreme northeast (Cape York Peninsula). This is likely the most vexing ecological problem that has confronted man since he appeared in this region at least 30,000 years ago.

It is Australia's fate to be so placed in the world as to miss almost all chance of abundant and reliable precipitation and to lack sufficient high mountain ranges to compensate by orographic precipitation. The highest point in Australia, which has an area of 7,770,000 square kilometers, is only a little over 2000 meters and elevations above 1500 meters are rare. Thus, water control and manipulation offer some of the more dramatic examples of human manipulation of the Australian environments.

In pre-European time, the aborigines made little conscious effort to modify the hydrology. It has been hypothesized that they occasionally enlarged some water holes, and it has also been said that they poisoned other water holes in order to kill emus that came to drink. The vegetable poison used was not supposed to have been toxic to man or other mammals, but this has not been carefully studied.

Probably the greatest effect the aborigine had on the hydrology was from side effects of the burning of vegetation. Thus, one might expect that floods occurred and stream flows were changed where the vegetation cover was most often and effectively modified by fire.

Soon after the settlement, in 1788, of the first European colonists, it became evident that the water supply might be a problem. Although the first settlement was in a relatively humid region of Australia, the long, hot, dry summer gave the colonists considerable concern. Later, when the dry interior of the continent was invaded by sheep drovers and farmers, the full impact of the ecological limitation was felt.

Thus, there soon were schemes to manipulate the surface water for farming and livestock. The southeastern part of continental Australia contains the only major river system, the Murray-Darling, and even the Murray River becomes so diminished in the summer that it is dry from bank to bank at more than one place.

In the latter half of the last century, it was discovered that Australia possessed vast artesian basins. Initially, these gave promise of compensating to some degree for the lack of surface water. Artesian, that is, free-flowing, wells (called *artesian bores* in Australia) were drilled, and they produced large quantities of water, but the salt content was too high to be used for crop irrigation (Fig. 5.9). However, it was learned that cattle and sheep could often tolerate the salty water. The result was that the livestock industry spread into large

Figure 5.9 *Flowing artesian well in Australia.*

areas previously lacking surface water. However, because almost no control was exercised over the number of wells drilled, and there was almost no understanding of where the water came from, the flow of water soon diminished. Some wells ceased to flow altogether, and pumps had to be used to bring water to the surface. Moreover, only about 10 percent of this water was actually used when it reached the surface because the rest was lost to evaporation and seepage. Even though the underground supplies have diminished greatly in the areas where drilling was most pronounced, there is still little agreement as to how long it might take to replenish the supply.

Not all wells in Australia are artesian. In some wells, which are classed as subartesian, the water comes part way to the surface but must be pumped the rest of the way. And other wells simply tap into the water table and the water must be pumped the rest of the way to the surface. However, in almost every area where wells, of any type, have been used, there is a history of lowered water tables and well failure.

As important as are the subsurface waters in Australia, the supply of surface water is even more important. It is from this source that much of the water for irrigation and almost all the water for home and industry is obtained. In the nineteenth century, after a succession of crop failures due to drought, farmers in the Murray valley area decided that their only road to economic security lay in developing irrigation works. Their efforts were hampered, however, by a general lack of experience with arid lands. For example, their legal system contained, insofar as water law was concerned, laws suited to the moist climate of the British Isles and thus were not related to the special water problems imposed by aridity. The legal problems were further complicated by the fact that until 1901, Australia was a collection of colonies, each with its own regulations, rights, and privileges relating to natural resources.

With the emergence of the Australian Commonwealth on January 1, 1901, some of the difficulties were overcome and a more unified and rational approach to water use became possible—even though for several decades water manipulation remained the prerogative of the individual political territories of the commonwealth.

The central government has become increasingly influential in the development of large-scale plans for the modification of the hydrography of Australia. Understandably, most of this activity is concentrated in the southeastern area where there are the most people and the most surface water. Irrigation projects have been developed in the Murray valley area on a relatively large scale, and, equally important, dams have been constructed to provide water for the urban centers of southeast Australia—Adelaide, Melbourne, and Sydney. Almost completed is the Snowy River scheme, the most ambitious and costly of all the efforts thus far put forth to modify the surface hydrography.

Essentially, the Snowy River scheme captures the waters of the Snowy River and by a series of tunnels redirects the water through a mountain range. Thus, instead of flowing directly to the ocean over a relatively short course, the water is diverted into the Murrumbidgee and Murray Rivers. Before the water is released into the rivers, it will be made to generate a large quantity of electricity. The cost of the project is almost 1 billion dollars, which is a very large investment for a nation of approximately 13 million inhabitants. When the Snowy River scheme is completed, approximately 810,000 hectares will be in irrigated pastures and crops. Although the economic advantages are said to be great, there are some ecological aspects that merit attention.

All the changes in the affected ecosystems will not be known for many years but some have already become evident. The Snowy River, below the dams, will be quite different from what it was before. The removal of so much water is an important ecological alteration but it appears to have elicited little attention. Conversely, the addition of enormous quantities of water to the two rivers on the other side of the mountain will also be significant. One must assume there will be changes in temperature and the chemical and biological composition of the waters flowing in the two rivers.

Perhaps the most important change (aside from the changed hydrographic pattern) is the placing of some 8000 square kilometers under irrigation. Aside from the economic benefits, there are some negative aspects.

For one thing, irrigation has created an ecological condition that is suited to the common liver fluke of cattle and sheep (*Fasciola hepatica*). The newly irrigated lands are favorable to certain aquatic snails, which are the fluke's intermediate host, and to the fluke larvae. Thus, this fluke has spread widely and has caused serious economic loss to those who seek to increase their incomes through the new irrigated lands.

Certain mosquito species have also pro-fited by the changed water conditions; some spread a virus disease, known as Murray valley encephalitis, which is a zoonosis, that is, it affects man and other vertebrate animals. Thus, the great area of surface water that has been made available to these insects has increased what has long been a serious health problem in the region.

Freshwater pollution is of limited areal significance in Australia, chiefly because the human population is concentrated in a relatively small area and only a limited quantity of water can be polluted. It does appear, however, that the Murray River is subject to pollution, chiefly because chemicals and insecticides are used in the agriculture of the area, but there are few published data on the subject. During the past five or six decades, agronomists have learned more about the soil deficiencies of the area and the input of chemical fertilizers in the ecosystems of the Murray valley region has increased significantly. It is certain that some of these materials are washed into the surface drainage. Industrialization is relatively minor in this river valley, so the opportunity to contaminate the water supply is limited.

5.5 Influences on Soils and Geomorphology

In the pre-European period, the Aborigines undoubtedly influenced the soil mantle of Australia to varying degrees—depending on the frequency of human-set fires and the nature of the vegetation, and the nature of the underlying soils. But it is not possible to determine the degree or exact nature of this influence. However, the Europeans did not enter on pristine edaphic (soil) conditions.

Almost from the first moment of settlement at the Sydney site in 1788, the Europeans made their presence felt on the soil. The result was the catastrophic alterations of soils in many of the ecosystems of Australia, particularly those in the southeastern quadrant of the continent and in parts of Tasmania. Just as the settlers were confused about the nature of water scarcity because of their cultural and geographical background, they were often in a quandary as how to best use the soils. They tried to apply lessons learned in a different environment. But once past the early, difficult years on the coast, the settlers gradually expanded their plantings of wheat, and this crop became second only to wool in economic importance. It also became sec-

Figure 5.10 *Clearing land with aid of machinery, Australia.*

ond only to wool production in terms of the ecological damage associated with its production.

Wheat has been a "gamble" crop for many years because the size of the area planted to wheat is chiefly determined by world market conditions. And wheat farms, unless checked by law, almost always try to expand their plantings into ecologically marginal areas during the years when grain prices promises to be high—or so low that they hope to recoup by increasing production. At best, wheat farming is usually practiced at the margins of areas where precipitation is sufficient for nonirrigated crop production.

Wheat, to be profitable, usually requires a large area in which nothing else is grown. This leads to the uprooting, overturning, or burning of the previously existing vegetation (Fig. 5.10). The result has been a great ecological cost in Australia and elsewhere, mostly in terms of destroyed soil.

For many years after the first signs of accelerated soil erosion were noted in Australia, there was a great reluctance to admit the evidence in plain view. Erosion is most pronounced in the areas of oldest and densest European settlement, that is, in the southeastern part of the continent. Almost all this area has been subjected to some accelerated erosion; in parts of this region, the process is so well advanced that the land is essentially ruined for any economic purpose.

The principal cause of this soil loss is vegetation removal for crop production or pastures (called *paddocks* in Australia). The hydrographic effects of that removal were already noted. The next ecological element affected is the soil. And as the soil washes away, the ecological picture is further and more seriously altered.

The most important cause of erosion is moving water, but wind is an important agent in areas with prolonged dry periods when the ground is laid bare. Thus, in recent years, great dust storms have blown

Figure 5.11 *Dust storm, Australia.*

Figure 5.12 *Accelerated soil erosion, New South Wales, Australia.*

up. Though not covering as large an area as those in the United States in the 1930s, these storms are nevertheless of dismally large proportions (Fig. 5.11). Water and wind have long had their way with the soils, sands, and dusts of Australia but never to such a degree as recently, since man laid bare the surfaces of areas where drought is the rule and rainfall reliability only a futile hope.

Thousands of hectares of the most valuable topsoil of Australia have been washed to the sea, and these cannot be replaced until the injured ecosystems heal themselves in the centuries ahead. Perhaps more than any other effect of the technological inputs made after 1788, this loss of soil illustrates how extraordinarily fragile the ecosystems of Australia are (Fig. 5.12).

Since the middle of the nineteenth century, mining has been a significant part of the Australian economic scene. At first, mining was focused on gold in the state (then colony) of Victoria. Later, other gold strikes were made, the most notable being at Kalgoorlie in Western Australia. As the world's needs for metals increased, mineral exploration and exploitation increased and the end is by no means in sight. Thus, many local Australian landscapes are dominated by the geomorphic changes associated with mining. Sides of hills are scarred with mine shafts, and large and small dumps of waste rock mark the presence of man's burrowing. Moreover, because so much of the continent is arid or semiarid, these effects will surely last for hundreds of years.

Changes in the earth's crust associated with road building are concentrated in the southeastern part of the continent, the southwest corner, and a few limited areas elsewhere. Many of the longest roads are simply dirt tracks across dry stretches of land. Modest as these may be, they will remain evident for hundreds of years, even if not maintained because the land through which they pass is so dry and slow to change.

5.6 Influences on the Atmosphere

There have been few indications that man has altered the macroclimates of Australia to any measurable degree, but this lack may reflect the lack of attention thus far directed to this aspect of human environmental modification. Nevertheless, Australia appears to fit into the general Southern Hemisphere picture of limited human modification of the macroclimates. This is chiefly because the human population is limited and thus the ability of that population to introduce foreign matter into the general circulation of the atmosphere is also limited. This does not mean that no particulate matter gets into the atmosphere from Australia industry or through the burning of fossil fuels. However, up until now, these inputs have been extremely modest, particularly so when compared with the inputs from northwestern Europe and North America.

Local climates and microclimates, on the other hand, have been subject to human modification, but we have only a general idea of what these modifications are. In the larger cities of Australia, most of which are in the southeastern part of the country, smog is an increasing common phenomenon. It is attributable chiefly to automobile emissions, waste products of factories, and general commerce. In rural areas, which dominate the Australian landscape, we may assume that microclimates have been altered since early times, that is, almost since man appeared on the continent. This alteration would be the result of fire and the alteration of microclimates has been important, especially where forests have been removed to make way for stock raising or agriculture.

Chapter Six

Human Influences on the Ecosystems of North America

6.1 Cultural Background

The exact date of the arrival of man in North America has long been the object of study and debate. Over the years, as more archeological finds have been made and interpretation has become more sophisticated, aided by such dating methods as the carbon 14 techniques, the date has been "pushed back" from approximately 3000 years BP to perhaps 50,000 years BP.

The earliest human invaders are believed to have been hunters and gatherers, and their style of life prevailed until comparatively recent times—when agriculture was adopted in some parts of the region. It appears that in the Western Hemisphere, agriculture originated in Middle and South America (see Chapter 7) and spread from there to parts of North America. At the time of the first recorded European arrivals in North America in the sixteenth century, agriculture was distributed in parts of eastern and southwestern North America, but was nearly absent elsewhere.

The agriculture varied from one part of the region to another. In the East, it was based on corn (maize, *Zea mays*) and squash. There is still some disagreement about the techniques used, especially about how much time a given field was kept under cultivation. It appears that in some parts of the East, almost sedentary agriculture was practiced, because the individual fields were kept under cultivation for ten years or more. The use of fire for clearing trees and brush that had been cut down with stone tools was general.

In the Southwest, various forms of irrigated agriculture were developed, as well as dry farming techniques that relied largely on water storage in the soils. The crops grown were similar to those in the East, with corn predominating. A complete list of all the crops known to these cultures would be impressive, but relatively few of them played a significant role in the diets of the people.

However, even where agriculture was best developed, the people still relied heavily on wild animals for food and many un-cultivated plants were gathered as diet supplements or for other purposes.

In the nonagricultural areas, hunting and gathering were, of course, important. The degree of technological sophistication varied greatly from one region to another, even within local areas. For example, when Europeans first came to California in the sixteenth century, there was no agriculture except along the lower Colorado River and in northern California, where one or two groups casually sowed wild tobacco seed. Some culture groups, such as the Pomo in central California, had an elaborate gath-ering and processing technology devel-oped around acorns. The ecosystems oc-cupied by these and several other ecologically similar people contained nu-merous oak trees (*Quercus*), whose acorns contain many calories. In order to make the material edible, however, it is necessary to grind the acorns into flour and leach the flour of certain bitter components. The Pomo and other groups had developed techniques that made the leaching process effective. The ecological carrying capacity of some of the oak-woodland areas com-pared well with areas of shifting cultivation elsewhere in the world. In addition to acorns, other plant foods were harvested, and a varied array of animals, from deer to small rodents, were also used for food.

Not far distant from the "acorn cul-tures" were other groups that had only a limited technology and appeared to live a marginal ecological existence. Although they harvested many varieties of plant food and included many small animals, such as hares (jack rabbits, *Lepus*) in their diets, they had only a limited ability to ex-tract a living from their ecosystems. As a result, the populations of these groups were modest in size.

Hunting was best developed in parts of the middle and northern regions of North America. It is not necessary to discuss here the many types of hunting groups in this region around 1600 nor the varied hunt-ing methods. It is important to note, how-ever, that in most cases, gathering of plant foods was also an important aspect of the human ecology. The chief exception to this was the Polar Esquimo, who inhabited an extremely difficult environment along the west coast of Greenland. This group of people subsisted on an almost pure meat diet because vegetable food was all but nonexistent.

Thus, the human societies in North America around 1600 were varied and complex, and each type of economy ex-erted an impact on the environment.

The first significant entries of Euro-peans into North America were made in Florida and parts of the Southwest. Later, Europeans established themselves along the Atlantic Coast and penetrated into Canada. Primarily, it was the British and French who settled in the eastern area and they often competed for territory and trade. Although revenue was obtained through exploitation of some of the wild products, especially furs, the settlements were based to an important degree on ag-riculture. Moreover, in the northeastern part of the British Colonies, industry, which was to grow later to enormous pro-portions, was established in response to the needs of the settlers.

Through a chain of events, the southern parts of the British Colonies developed commercial agriculture to the most impor-tant extent, which led to the "southern plantation" landscape. Over the years, various crops figured in the landscape, but with the invention of the cotton gin in 1793, cotton easily assumed the dominant position. The result was that a large quan-tity of labor was required, and the slave trade flourished. The southern plantation system reached its height of economic sig-nificance in the first half of the nineteenth century but was in economic difficulties when the Civil War began. And, by that time, the Northeastern United States had emerged as an important industrialized area. Following the war, the new nation rapidly expanded westward, at the expense of American Indians and other cultural or national groups. American Indians fared somewhat better in Canada, primarily because the French did not have the degree of racial prejudices as did many of those of English extraction.

Partly in response to the rapidly growing industry, which had an almost insatiable need for human muscle power, a major human migration from Europe to the United States was set in motion shortly after the Civil War. The "pull" was jobs and the potential of acquiring land and other goods and the "push" was an economically and socially stagnant Europe, which was experiencing such ecological cataclysms as the Irish potato famine. The "American dream" meant wresting a fortune—or at least financial security—from what was believed to be an endless cornucopia of natural resources. And this was the attitude of the European immigrants as they approached the forests, soil, wildlife, and other resources of the New World. As you can see from our discussion thus far, the brunt of the ecological impact of man in North America during the past century or so fell on the United States. In the following discussion, therefore, the regional designations apply to the conterminous United States (the 48 conterminous states) unless noted otherwise.

6.2 Influences on Vegetation

There is no question that *Homo sapiens* was in North America at least early enough to witness the last major glacial period (the Wisconsin) and its retreat. We still know rather little about the changes that took place except that there were warming trends and trends toward greater drought in some parts of the region. Major lakes in the western part of the region dried up, and warm, moist air masses from the Gulf of Mexico penetrated deep into the heartland during the summer.

The climatic changes accompanying the retreat of the ice affected the survival of plant species. The sequence of vegetation changes during the past 12,000 years or so is currently the object of intense study and some disagreement. However, we do know that around 1600, most of the eastern half of North America was covered by trees. Coniferous forest, called taiga, dominated the northern part of this region, and the central part contained a great mixture of deciduous trees, such as oak (*Quercus*), maple (*Acer*), beech (*Fagus*), birch (*Betula*), and hickory (*Hicoria*), and conifers such as white pine (*Pinus strobus*). The southern area contained great expanses of pines and many hardwood species. In the United States, this forest is said to have extended almost unbroken from the Atlantic to the Mississippi River and, in some cases, beyond.

At the western end of the forest was an enormous grassland—the prairie—which extended from the Gulf of Mexico into Canada. To be sure, some islands of tree-covered lands occurred and, where rivers crossed, there were ribbons of poplars and willows, but these were dwarfed by the great sea of grass.

Farther westward, the Rocky Mountains with their coniferous forests, gave way to the Great Basin, in which isolation from moist air masses created a vast region of limited precipitation. Here great areas of grass again occurred, as well as large areas covered with Great Basin sage (*Artemisia tridentata*), other shrubs, and some small trees. Part of this region, especially in the south, is a desert. West of the Great Basin and the deserts, the pattern is too complex

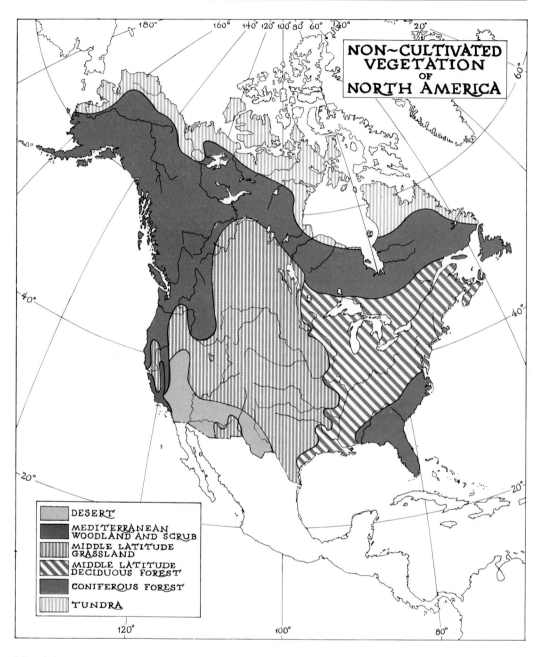

Map 6.1a

for generalization—there were forests of conifers, deserts, vast stretches of oak and brush covered hills, and extensive areas covered by grass (see Maps 6.1a, above, and 6.1b on page 171).

We still do not know if this great vegetative complexity was the result of nonhuman forces alone or if the Amerinds (American Indians) played a role in shaping it. However, there is a growing appreciation of the ability of so-called primitive man in North America to have modified the vegetation cover.

The Great Plains have long been viewed as a mature ecosystem that is largely maintained by a grassland climate (Fig. 6.1).

Map 6.1b

This view is still widely held, but data exist that indicate that some of the assumptions are incorrect. It has been shown, for example, that there is no *climatic* reason for trees not to occur over much of the Great Plains region. The evidence is that trees do occur locally in rocky, *fire-sheltered* locations in much of the area. Moreover, thousands of farmers have planted trees and many of these grow without artificial irrigation.

However, we are still only slightly closer than in earlier years to answering the question about the effect of early man on the vegetation cover of the Great Plains. Some scholars have made the jump from evidence that fire was ecologically significant in the Great Plains to assuming that all or

Figure 6.1 *Midwestern prairie in the Spring.*

most of the fires were of human origin. But fires were also set by electrical storms. Unfortunately, we cannot determine from the charcoal record which agency was responsible for a given burn in times long past. Therefore, we can state only that there were at least two significant sources of fire in this region and that both of these exerted sufficient influence to modify the vegetation toward grass cover. We should recognize that man was a potential ecological force in this region from the time he first arrived.

Although hunting and gathering persisted over large areas until the arrival of Europeans, agriculture spread into limited parts of North America before their arrival. It appears that agriculture had become established in parts of eastern North America by approximately A.D. 300. It was chiefly based on maize (*Zea mays*) but included other plants as well. Although the form of agriculture was slash and burn cultivation, individual fields were sometimes farmed for ten or more years before being abandoned. The effects of such land use on the wild vegetation in this area has been little studied, but probably did not differ significantly from the effects in other parts of the world in similar latitudes. Perhaps one reason why only limited attention has been given to this question is the myth that all the ecosystems of North America were

pristine and virgin until the white man arrived. However, slash and burn cultivation had an important effect on the vegetation of the eastern area, so at least some of the "virgin" forests reported in the seventeenth century by European colonists were, in fact, secondary forests.

In parts of the Southwest, maize and squash were the mainstay crops, but many other plants were also grown. With few exceptions, these crop plants were not native, their origin being south of the North American area. Indian agriculture in this area probably exerted only slight impact on the wild vegetation except on a local scale. Little clearing was needed, and because irrigation was practiced, there was a tendency to farm the same plots for many years. The ecological disturbances caused by these practices seem to have been minimal. The fuel needs of the desert dwellers may have imposed a greater impact on the wild vegetation, especially where there were dense Indian populations. However, even this did not cause major ecological changes.

Although there was a great diversity of human technologies in North America at the time of the European entry, the Amerinds were essentially in ecological equilibrium with the regional ecosystems. There is little evidence however that these equilibria were achieved through con-

Figure 6.2 *California pasture dominated by the exotic grasses–mostly oats,* (Avena).

scious acts; that is, the people probably did not consciously choose to act in ecologically sensible ways. Moreover, the Amerind was but one of a number of biological and physical forces exerting influence on the plant cover.

In the sixteenth century, the Spaniards obtained a foothold in Florida, but they had little ecological impact there. However, in the Southwest, they penetrated in greater force and established major permanent settlements. Much of the area from Texas westward (south of Canada) was under the control of Spain and later Mexico until 1848 when the United States obtained the region by treaty (1819 for Florida).

It was through plant introductions that the Spaniards had the greatest impact on the vegetation cover. Almost as soon as they entered a new area, they introduced a host of crop plants, particularly grains, to provide a food base for settlers. Wheat and barley were the most important introductions but other grasses appeared as accidental voyagers. Among these was oats (*Avena*), which in California eventually invaded great expanses of the land and virtually replaced the native grasses in much of the state (Fig. 6.2). Many other accidental introductions occurred during this period, and some of them, like the oats, became established over vast areas of non-cultivated land. Other deliberately introduced crops included grapes (*Vitis vinifera*), many Old World fruit trees, and vegetables. Crops of Latin American origin were also introduced and established. However, the introduced crop plants tended to remain highly localized because much of the region is arid or semiarid and there was a general lack of abundant surface water for irrigation.

Another force acting on the vegetation cover was introduced livestock. All hooved domesticated livestock were unknown in this area until introduced by the Spaniards. Cattle were deemed of particular value because they required little care (under then existing ranching techniques) and produced hides and tallow, which found local and exterior markets. (Raising cattle principally for meat came much later in this region.) Cattle herds were allowed to multiply without restraint, the only checks being those imposed by predation, disease, and the limitations of grass and water. Like most grazing animals, cattle are selective in their choice of food and do not graze all species of grass equally. This selectivity produces changes in the frequency of many plant species on grazed lands. If a pasture in the West or Southwest is continuously overgrazed, it can lead to the dense establishment of cactus, usually of the genus *Opuntia* (Fig. 6.3).

Figure 6.3 *Various species of* Opuntia *on western rangeland.*

The European impact on vegetation in the East was different from that of the Spaniards in the West. First, the ecosystems were different, because most of the eastern region receives sufficient precipitation to support a heavy and more or less continuous cover of trees and shrubs. Although the nature of this vegetation cover had been influenced by Amerinds, it was richer in tall, woody species than that of the West. The vegetation was not unlike that in the ecosystems from which many of the first settlers had come. The cold winters, moist summers, and tree-clad landscapes were reminiscent of the less disturbed parts of the British Isles and Continental Europe.

Of greatest ecological significance is the fact that most of the early settlers were farmers and the rest had some limited experience in retail trade. Thus, the great forest often appeared less a valuable resource than something that had to be removed to make way for plow agriculture. Only second in importance ecologically is that the settlers brought the plow and thus a system of land use markedly different from anything the region had known before. As we noted earlier, plow agriculture often results in the permanent clearing of land. Thus, the same drama of forest and woodland clearing that occurred in Europe at a somewhat earlier date was repeated in the region between the Atlan-

tic Ocean and the Mississippi River (see Chapter 4). The colonial period laid down the style of how the land and its vegetation cover should be modified, but it should be noted that much of the major forest removal in this region came *after* the colonial period—when the United States was expanding west of the Appalachians and southward (Fig. 6.4).

The story of the colonists' dependence on the Amerinds for food during the first years has been told often. It was during this period that certain of the Indian crop plants, chiefly maize (corn), became accepted into the European diet. Of course, other crops were brought from Europe, and later, crops of diverse origins added. Of these, cotton and tobacco were to emerge in the Southeast as major ecological, economic, and social factors.

Early colonial agriculture was largely of a subsistence nature, with a barely sufficient surplus to feed the townspeople. Efforts were made early to establish crops that could be exported at a profit. Experimentation led to the attempt to grow various plants under plantation systems. Tobacco was the first crop to emerge as a commercial success because the taste for this Latin American plant had become

Figure 6.4 *Clearing trees on the American frontier in the late eighteenth century.*

widespread in Europe. Although highly localized, tobacco-growing makes a severe ecological demand on the land, and the wild vegetation cover gave way completely to tobacco fields where the plant was grown. Soils sometimes became exhausted because they were planted too long to tobacco and were abandoned, and this allowed the areas to be invaded by weeds and brush species of which some were exotic (Fig. 6.5).

Gradually, from the time of the first tiny settlements to the War of Independence and to the beginning of the American Civil War, a variety of agricultural landscapes were developed in this region. Large tracts once thought to be promising farmlands were abandoned to the processes of secondary ecological succession and other, even greater, areas were put permanently to the plow or planted to permanent pasture. One can view this either as a great human drama capped with success or as the destruction of much of the region's ecological patrimony. In any case, the vegetational landscapes became largely humanized, except for the higher, difficult terrains which retain even today much more than a vestige of their former wildness.

Thus far, we have not discussed the middle part of the continent. Curiously, this area, which today contains a large share of the agricultural wealth of Canada and the

United States, was looked upon by the Europeans as a nuisance that stood in their migration path to the Pacific Coast. This attitude was most marked in the gold seekers and farmers before the Civil War. The great ocean of grass appeared at first to offer limited possibilities and was still occupied by various Plains Indians. (The only areas settled by Europeans were in the extreme south—notably New Orleans, settled by the French, and Texas, which had passed from Mexican hands to those of European descent.) The greatest handicap to farming was not the lack of fine soil but the lack of surface water away from the rivers and streams. This not only made farming difficult but even made using the land for livestock grazing difficult.

After the Civil War, however, settlers on their way to the West Coast invaded this region; with them came the forces of the United States government to protect the travelers and to obtain rights and privileges from the Plains Indians. It was also after the Civil War that the open-range cattle industry became significant. This occurred first in Texas, where herds had multiplied during the war years. A market for these cattle existed in the East, so they were rounded up and driven north to the nearest railroad shipping point. However, as the railroad was rapidly extended to the West Coast, these cattle drives became un-

Figure 6.5 *Accelerated soil erosion in a tobacco-growing area in the southeastern United States.*

necessary. Gradually, a more stable form of cattle industry evolved, with herds and herders moving northward to occupy grasslands from which the Indians had only recently been driven. Our interest here lies in what influence, if any, the spread of cattle had on the grass cover. Only a general answer is possible. The grazing habits of cattle (and sheep) differ from that of the bison (American buffalo) that had occupied the Plains before the cattle. Thus it may be assumed that the new grazing activity altered the floristic aspects of the grass cover, but we will probably never be certain of what changes occurred (Fig. 6.6(a), 6.6(b)).

In any case, it was not the invasion of livestock into the Great Plains that had the principal effect on the vegetation cover. It was the invasion of plow agriculture which became spectacularly successful after the invention and dissemination of barbed wire and the windmill. Barbed wire made possible the inexpensive fencing of the cattle range, which kept livestock out of cultivated areas (and also greatly changed the livestock industry itself), and windmills made possible the inexpensive exploitation of subsurface water, through the use of wind to pump water to the surface (Fig. 6.7). Thus, because water was available for domestic use as well as animals, human set-

Figure 6.6(a) *Herd of sheep, Colorado.*

Fig. 6.6(b) *Cattle roundup in Idaho.*

Figure 6.7 *Windmill and waterhole, Texas.*

tlement in isolated areas became possible. Summer rain and winter snow permitted nonirrigated agriculture in a large part of this region, so over huge areas the grassy cover was turned under by the plow. Wheat fields took the place of gamma and other native grasses in the dryer parts (Fig. 6.8(a)) and the more humid areas came increasingly to be planted to maize (Fig. 6.8(b)). The complex of crops and the livestock raising that evolved culminated in the highly mechanized agriculture of that region today. Only the areas where precipitation was too unreliable did the wild grasses remain, but even these areas were modified by heavy grazing and by the accidental and deliberate introduction of exotic grasses.

Turning from the humanized landscapes of the Great Plains to the forested areas of Canada, we encounter a part of North America where the hand of man, past and present, is not always evident. Although the taiga in the east and the great pine and spruce forests in the west are not all pristine, it has proved difficult to establish permanent large-scale human settlements in these areas. There are a number of reasons for this, but perhaps the most important are the difficult winters and/or terrain of these areas. Nevertheless, there are parts where human-set fires and logging activities of the past markedly altered the wild vegetation landscapes.

Thus far, we have discussed how Europeans influenced the vegetation cover of the United States by introducing grazing animals, clearing forests and plowing grassland for farms, and introducing (deliberately and accidentally) plants from other parts of the world. But Europeans also influenced the vegetation cover by directly using forests and woodlands for fuel and lumber.

The timber wealth of North America was so great that in the earlier years it appeared limitless, and public reaction to the use of the forests was consistent with that misconception. The rich and varied hardwood forests of the East supplied a wide range of raw materials for a society that in its earlier phases had to treat iron as a luxury. Thus wood was used for many purposes that would later be taken over by iron and steel.

The coniferous forests were also useful in the early period, but they became more significant later—when the demand for soft woods grew in response to construction needs. The first states to be affected by this demand were Michigan, Minnesota, and Wisconsin, where there were enormous stands of softwoods. Major exploitation of these forests did not get underway until the last half of the nineteenth century. By then, logging operations in this region were well developed and were causing enormous destruction to the forest ecosystems. No care was given to forest protection because the object was to get the maximum amount of lumber to market with the minimum cost to the logging operators. The result was the destruction of many trees, shrubs, and understory plants; only vast quantities of debris (*slash*) remained after the loggers passed through an area. This slash was highly combustible

Figure 6.8(a) *Wheat field, Kansas.*

Figure 6.8(b) *Corn field, Iowa.*

during the summer months and many forests fires occurred, some of which raged over thousands of square kilometers, took hundreds of lives, and destroyed entire towns. There was absolutely no human-controlled means to stop these fires once they began, so they simply burned until wind, rain, or lack of fuel caused them to die. Aside from the immediate ecological effects, it must be assumed that extensive areas in this logged-over region are still recovering from these fires of a century ago, although most of the obvious signs are now covered by trees and understory growth. One wonders to what extent the present regenerated forests are different from those that first heard the ring of the woodman's axe.

The story was repeated in the West, for the "technique" of logging was to get all that was available and then to move on. In parts of coastal Oregon and Washington, large forested tracts were completely cut, leaving great quantities of slash, which often ignited and burned out of control for days. Unlike the area adjacent to the Great Lakes, in the Oregon-Washington region, forest often did not regenerate and the cutover land was invaded by a tight network of shrub growth. Such areas today cover many square kilometers and can be seen from the highways (Fig. 6.9).

In summary, no large forested region of the United States escaped the modifying influence of man during the past century. In Canada, large tracts of forested land in British Columbia have been only slightly modified by logging, but in many parts of the East the story was essentially the same as that for adjacent parts of the United States. One saving factor for the Canadian east is that, until recently, the taiga offered limited attractions for the lumber industry so logging was often on only a local scale. However, the world's hunger for soft-woods has altered this. These forests are now being heavily exploited, although not in the ecosystem-destroying manner of earlier years in the United States.

Figure 6.9 *Logged-off land with slash, Washington.*

So ubiquitous is the non-Amerind in North America that one is hard put to assert that the vegetation cover of any large tract is pristine. Even if an area has not been logged or cleared for farming, other exploitative acts, such as hunting and trapping, alter the previously existing ecological equilibria. In short, much of the vegetation cover of North America has been modified by human actions during the past three centuries, with the maximal changes taking place during the past century.

Not all the changes are ecologically negative and destructive. In parts of Southern California, because of a moderate winter climate, plants from virtually all parts of the world have been successfully introduced into gardens and parks. True, the pristine environment is no more in the settled parts but the resultant vegetation cover is often strikingly beautiful. Moreover, much of this exotic cover survives only because man provides water that natural climatic elements do not. Should our species disappear from Southern California, so too would much of the exotic vegetation.

6.3 Influences on Animal Life

When Europeans first arrived in North America, not a single domesticated animal was kept principally for food in the entire region. True, dogs were sometimes eaten by some tribes as a famine food and by other tribes as a delicacy, but lacking was the practice of keeping food animals so familiar to the European cultures. From the time the first humans arrived in the hemisphere, there was an almost complete dependence on wild animals for animal protein in the human diet. What had been the impact, if any, on the native wild animal species?

The first people who ventured into North America are believed to have been hunters and gatherers. We know nothing about their numbers, although we do know that they eventually dispersed to southern South America (see Chapter 7). When these people arrived and for a time thereafter there was a remarkable collection of large grazing (and browsing) mammals in North America. The animals appear to have flourished until the early post-Pleistocene, when many became extinct. One may interpret the fossil record so that it shows either that the extinctions occurred in a short period of time or that they were strung out over a long time span. If the extinctions are held to have occurred collectively in a brief time span, it is logical to search for a single common factor to account for them. This some scholars have done, and they have concluded that man-the-hunter was responsible. However, the evidence to support this hypothesis, although interesting, is not conclusive, and the counter arguments appear to carry at least as much weight. Perhaps the question will never be resolved, but many careful scholars no longer question the *ability* of early man in North America to have been a potent ecological force. Even those who do not believe that man was responsible for the extinctions base their arguments on factors other than cultural incapability of accomplishing them. This is important for it reinforces a point that has been made before in this book, namely, that so-called primitive man, for all his limited numbers, often possessed the capability of exerting powerful influence on the ecosystems in which he lived.

In any case, there were relatively profound megafaunal (large animal) changes in North America in the post-Pleistocene. But this does not mean that the region became a faunal desert. On the contrary, there is no reason to believe that there was not an abundance of mammals, birds, and fish suitable for human food. If there had not been, it is unlikely that people would have survived. When the Europeans arrived, they found what they considered an extraordinary richness of game birds, mammals, and fish. This abundance gave rise to the myth of the ecology-minded red man who lived in conscious rapport with his ecosystems and never took more game than he needed. As we stated earlier, it is true that there were ecological equilibria in

various parts of inhabited North America. But these had been achieved, not by conscious effort, but through limitations imposed by environmental elements and by the limitations of exploitative technology. Man lived more or less in balance with the ecosystems of North America because he had not yet invented techniques that permitted him to do otherwise. There is substantial evidence to indicate that Amerinds were willing to use easier techniques in hunting and fishing as soon as they gained access to them.

For example, when the Anglos arrived on the Great Plains, they found tribe after tribe of Indians possessing large numbers of horses and such equestrian skills that it was believed that the Indians had always had these animals. This, of course, was not true, for the horse was unknown in the region until it was introduced by the Spaniards. But its spread was remarkably fast and so were the necessary skills of horsemanship. Many of the Plains Indians had altered their gathering activity to concentration on the bison. Before the introduction of the horse, bison played a far less important ecological role in the lives of the Sioux, Kiowa, and other Plains tribes. The specialized hunting that greeted the Anglos was a recent adjustment to having acquired a major technological advance over hunting on foot. Furthermore, the Indians had developed a different bow from the one they used previously. The new bow was short and powerful and adapted to being shot from the back of a pony racing in and out of a herd of bison. The kill rate per hour of effort must have soared over what it had previously been. This new technology appears not to have reduced the bison herds, which were enormous, but what might have been the story of the bison had the Indians acquired good firearms before the Anglos arrived?

Another example of American man's readiness to adopt an easier method of hunting relates to the Polar Esquimo. This group of people had made an extraordinary technological and cultural adjustment to a rigorous physical environment. They subsisted almost entirely on food from animals taken from the sea, of which the walrus (*Odobenus rossmarus*) was the most important. Their existence allowed for few errors of judgment and the discipline of their lives intrigued more than one Arctic explorer. Writers have waxed rhapsodic in describing their rapport with nature. However, when these people obtained—in quite recent years—the high-powered repeating rifle, the old way was over. No longer did they spend hours at sea hoping to harpoon an animal; no longer did they spend hours carefully stalking a walrus that was resting on the ice. With their rifles, they quickly and in relative comfort shot their prey. However, often the animal was not killed when it was hit, and it was able to get to the water. Once in the water, the wounded walrus was not held to the hunter by a restraining harpoon cord. Thus, it was not long before the walrus, once more than sufficient for the needs of these people, became a rare species and required the special protective laws of the Canadian government.

The white man possessed the exploitative technology the Indians lacked and was always improving that technology. In addition, the pioneers' perception of the faunal resource led them to conclude that it was impossible to overexploit it. However, to fully understand their attitude, one must understand contemporary conditions in Europe, particularly in England. Hunting and, to a considerable extent, freshwater fishing had become chiefly the perquisite of the wealthy, and vicious laws were enacted and enforced against persons who poached (illegally hunted or fished) on a private estate. The colonists and, later, pioneers were drawn in part from the disenfranchised of Britain and Europe. Thus, one can appreciate the feeling of freedom engendered by the ability to hunt and fish as much as one liked. Indeed, the pioneers often had to hunt and fish just to stay alive

because of the time lag before the meat and products of domesticated animals became available. Thus, venison (deer meat), the meat of other mammals, and a multitude of game birds became staples. Hunting skills, particularly the ability to employ firearms, became the prerequisite for the achievement of manhood on the frontier. Most of the early folk heroes achieved their status because of alleged skilled exploits with a gun.

The wildlife of North America provided ecological stepping stones into the wilderness, allowing pioneer families to survive until food could be grown. Wildlife also early provided articles for export to Europe, of which furs were by far the most important. But as long as hunting was confined, chiefly, to the subsistence needs of the people, there was little danger of any serious reduction in animal numbers. However, as urban centers grew, a commercial market for the meat of wild birds and mammals grew. Moreover, a fad developed in the eastern United States and in Europe for beaver hide hats. They became a "necessary" part of a gentleman's wardrobe in public. Then, in the middle of the nineteenth century, a market developed, also in the eastern United States, for bison hides tanned with the hair on. These *buffalo robes,* as they were known in the trade, were used in horse-drawn carriages as protection against the winter cold.

These and other marketplace pressures increased the impact of the white man on the wildlife of North America. The *commercial* assault on the faunal resource resulted in dramatic reductions, near extinctions, and the extinction of a few species. The bison (buffalo) story has been told many times. Herds that numbered in the millions as late as 1850 (probably closer to 12 million than the 60 to 70 million one often reads about) were reduced in a few years almost to extinction (Fig. 6.10). Fortunately, the bison was saved from extinction and now occurs in sufficient numbers to more or less guarantee its continued ex-

Figure 6.10 *American bison* (Bison bison).

istence. The story of the passenger pigeon (*Ectopistes migratorius*) is not so well known. Early in the colonial period, this bird occurred in astonishing numbers (Fig. 6.11(a)); there are accounts that pigeon flocks passing an area darkened the sky for hours. But they were hunted for eastern markets and were killed by the hundreds of thousands—even millions, by some accounts. Moreover, the hunting usually took place during the nesting season, so many more pigeons died than just those shot. Within a few years, only remnant populations survived. However, even these were not given protection, and the last known passenger pigeon died in the Cincinnati zoo on September 1, 1914 (Fig. 6.11(b)).

This appears to be a clear-cut case of human-caused extinction, but recent studies cast doubt on this belief. The chief problem is the enormous numbers in which these birds occurred, which is not characteristic of bird populations under normal ecological conditions. Thus, some scholars have raised the possibility that the hunting pressure, great as it was, was only one element among several contributing to the sharp decline of the pigeon's numbers and that disease may have been an important factor.

There seems to be a general notion that

European man exterminated many animals in North America, but such is not the case. Given the casual way in which the ecosystems of North America were used, it is remarkable that less than a dozen species of larger vertebrates have been eliminated. These include the passenger pigeon, carolina parakeet, heath hen, and California grizzly bear. However, many species that were once common have become exceedingly rare. This is a result of their being hunted for their hides or as predators or because of habitat destruction. It is a measure of the durability of the wildlife of the region that so few animals have been exterminated—thus far.

We will now consider the animals introduced by man into North America. The first hunting tribes may well have introduced domesticated dogs, for certainly these animals were introduced long before the advent of the Europeans. No other vertebrates are definitely known to have been introduced into the Western Hemisphere until Columbus' first voyage. The dogs and humans undoubtedly carried with them a collection of arthropods such as fleas and lice, and they may also have been the unwilling hosts to certain microscopic animals inhabiting the interior of their bodies. The material goods carried by these people may also have contained

Figure 6.11(a) *View of Fort Erie with migration of wild pigeons, 1804. A contemporary watercolor by Edward Walsh.*

Figure 6.11(b) *One of the last of the now extinct passenger pigeons* (Ectopistes migratorius).

some small animal travelers that later became established in the New World. But taken together, aside from man's self-introduction and introduction of the dog, little was added to the fauna of the region. The arrival of the Europeans changed the picture dramatically.

Before Europeans came to the Western Hemisphere, there were only a few important domesticated animals. In North America there was only the dog. But with the Europeans came a large array of domesticated animals plus a large number of vertebrate and invertebrate nondomes-ticated animals, which they introduced both accidentally and deliberately. Principal among the earliest introductions were work animals and animals that provided food.

In the section on vegetation we mentioned the ecological impact of the introduced grazing animals on the natural grass cover. However, there was also an indirect impact because the farms had to be laid out to allow for pastures, feeding areas, and areas where grain for animal feed might be grown. These areas, of course, had an effect on the wild vegetation cover.

In addition, the spread of domesticated farm animals led to the persecution of all native animals suspected of being predators. This included large and small carnivorous mammals and various birds. The eradication of predators is such a deeply ingrained philosophy in rural America that it had a profound effect on the nature and functioning of many of the ecosystems of North America.

Moreover, the misunderstanding of the roles of larger carnivores in the ecosystems of North America lasted until recent years. It was finally determined that there is a place for such animals in those ecosystems that man has not entirely subjugated to his will.

The discussion of predators leads directly to a discussion of crop pests or agricultural pests. Most states in the United States and provinces in Canada have long designated certain native animals as pests. As such, these animals are "fair game" all year. Bird and mammal species principally are on the pest lists, but occasionally other vertebrate classes are also represented. The lists also include many invertebrates, principally insects, of which many are nonnative (they will be discussed below). The point is that the designation of "good" and "bad" animals leads to impacts on the ecosystems, because people try diligently to eliminate the bad ones. This not only alters the faunal picture directly, but it may affect the entire ecosystem in ways that are often incompletely understood.

The use of chemicals to control or destroy vertebrate and, especially, invertebrate pests has a long history in agriculture and other areas of human activities. However, until recent decades, the main chemicals used were arsenic and sulfur. Sulfur was fairly effective against some arthropod pests in the fields and caused few problems, but arsenic was always a threat to human consumers. Thus, the farmers' problem was that there were no really effective insecticides and crop losses due to insect attack was a regular and economically devastating feature of agriculture.

This problem seemed to be solved just after World War II, when DDT, a chlorinated hydrocarbon, began to be used in agriculture. The effectiveness of this chemical for insect control was phenomenal. This stimulated the research and development of a variety of chemicals designed to control the hundreds of species of arthropods that attack crops. The early success soon became clouded, however, by the discovery that (1) DDT is extremely persistent in the ecosystems where it is placed or to which it strays, (2) in addition to the target insects, many useful arthropods were destroyed, (3) the persistence of DDT in animal tissue—principally fat—raised questions about the wisdom of its use, and (4) the target species were developing DDT-resistant populations, thus presenting a new and more difficult situation than existed before. At this time, DDT is no longer used in most of North America because of legal restrictions. However, there is repeated pressure to allow its use in "special" situations, such as in combating certain insects in forests.

The trend at the moment is toward the use of a group of chemical insecticides called organophosphates. Because these chemicals quickly break down in ecosystems, they are thought by some persons to be safe. However, they are extremely toxic, not only to insects, but to man; they contain substances that were first discovered in nerve gas research.

However, we should not lash out against the idea of using insecticides in agriculture, for there exists a constant struggle between man and arthropod on the farm. And man has not even come close to winning—in an effective and ecologically acceptable manner.

Whatever the outcome of the use of organophosphates or any other insecticides, the chemical contest with the insects may be fruitless. That is, no matter what chemicals are thrown at the insects, resistant populations will probably develop.

Although the deliberate introductions of domesticated animals into North America were events of ecological significance, perhaps of even greater importance were the deliberate introductions of nondomesticated animals. It appears somewhat incredible that given the faunal richness of North America, man has made efforts to introduce exotic animals, but such is the case. Some of the introductions failed because the animals did not become established; and some animals became established locally but never spread. But other animals not only became established but also spread widely and became permanent elements in the North American fauna.

Because they are of the greatest significance to this discussion, we will begin with the deliberately introduced wild birds (class Aves). Five of these introductions have altered the avifauna of much of the North American continent, whereas others have had only local effect. Of these, the introductions of sparrows and starlings had unfortunate results.

The European sparrow (*Passer domesticus*) appears to have been liberated around the middle of the nineteenth century in Brooklyn, New York. This now heavily urbanized area was then farmland, and it is alleged that the sparrow was brought in to help combat insect pests. (If this was the reason, it was a poor one, because the bird is principally a seed eater.) For a time, the sparrow population remained small and the distribution was local. However, additional introductions were made, and eventually there was an explosion of numbers and a rapid geographical dispersal. There is some evidence that the sparrows reached the West Coast by following the railroad route— where food was available along the right-of-way and around cattle pens and loading areas. The date for their arrival on the West Coast is 1886—about thirty years after their initial introduction in the East. For decades, the sparrows existed in swarms, and they were still numerous in the 1930s. Since then, their numbers have declined. At the present time, although they are common in most urban and human-settled parts of North America, they are no longer seen in great flocks. Their decline may be due in part to the disappearance of horses from the cities, because the grain-containing excretions of the horses were a source of sparrow food. Also, it is now uncommon for people in urban areas to keep chickens in their backyards and this further reduced the available grain food. One interesting feature of the bird is that its feather colors and

shades are changing much more rapidly than those of its European relatives. Some scientists believe this is evidence of a rapid rate of evolution.

Although the European sparrow has now found a place in the ecosystems of North America and no longer appears to be an economic or ecological threat, this cannot be said of the European starling (*Sturnus vulgaris*). The first attempt to introduce this bird is not clearly recorded, but it is known that one group of starlings introduced into Central Park, New York City, began to breed in 1891. Only a few years later, they were seen far beyond the park and indeed had begun to demonstrate that they could be serious pests. Starlings tend to gather in large flocks and are extremely harmful to orchard crops. They also nest and roost in large flocks, and their excrement can quickly become an esthetic as well as a medical problem. Within a few decades starlings had established themselves over most of the eastern half of the United States and in parts of southeastern Canada. Starlings were first noted on the West Coast in the early 1940s and since then they have established breeding colonies in Washington, Oregon, and California. Moreover, it is expected that a sharp increase in their numbers will soon occur. If this comes about, western agriculture will sustain a great economic loss.

The three other birds that were major introductions—Hungarian partridge (*Perdix perdix*), chukar (*Alectoris graeca*), and ring-necked pheasant (*Phasianus colchicus*)—were "happy" introductions in terms of their economic value and did not cause any significant ecological damage. All three are native to the Palearctic region (as are the sparrow and starling), and were introduced to provide sport for hunters in areas where native birds suitable for such recreation were either lacking or rare. The first introduction began during the latter half of the nineteenth century. In the case of the ring-necked pheasant, the results have been spectacular, for this bird is well adapted to several agricultural ecosystems, particularly those in which grain is grown. It now ranges widely in the West and Middle West; in the principal wheat-growing regions, the annual kill by hunters numbers in the millions.

Native birds have also been moved around to some degree. Most notable is the bobwhite quail (*Colinus virginianus*), which has been successfully introduced into parts of Oregon, Washington, Idaho, and British Columbia. In addition, the California quail (*Lophortyx californicus*) has been introduced into other parts of the West, and the wild turkey (*Meleagris gallopavo*) has been transported and liberated into parts of the Far West where it is not native.

Although no one recorded the event, the first wild mammal introduced by man into North America was probably the house mouse (*Mus musculus*). This small rodent frequently travels in the goods and gear of man so it is likely that it arrived with the first Europeans. The black rat (*Rattus rattus*) probably arrived just as early, because this rodent was accidentally moved about by man in almost all parts of the world. The brown, or Norway, rat (*Rattus norvegicus*) was slower to reach the shores of North America, being first noted

on the East Coast in 1775. The three animals eventually dispersed over most of the continent, with the brown rat preferring coastal cities and harbor environments. The economic cost of these accidental introductions cannot be accurately computed, but it has been and continues to be enormous.

As one might expect, there were relatively few attempts to introduce large wild mammals, because the region offered a variety of native species suitable for hunting. However, on some of the large cattle ranches of the Southwest, Asian and African species of herbivores were introduced, and recently wild sheep from the Mediterranean region have been liberated in parts of Arizona. It is hoped that these sheep will establish populations large enough to permit an annual hunt.

In addition, some of the introduced domesticated animals escaped human controls and established themselves as wild (feral) animals. Probably the most well-known of these is the horse. Feral herds of horses were once common in the West (Fig. 6.12). Besides the horse, there are feral sheep, goats, cattle, burros, and pigs. The ecological impact of these animals on the ecosystems they inhabit has been only little studied, but there is sometimes a con-

Figure 6.12 *Feral horses, western United States.*

flict between native and feral stock for available food and water. It has been shown, for example, that the feral burro competes directly with the desert bighorn sheep (to the detriment of the bighorn) for scarce water supplies. But in many instances, the populations of feral animals are so modest that their ecological importance is slight.

There have been few instances of successful ecological establishment, outside of captivity, of nonnative small mammals (except rats and mice). One exception, however, is the coypu (*Myocastor coypus*), an aquatic rodent native to southern South America. It has a superficial resemblance to the beaver (*Castor canadensis*) and its fur can be used for garments that are inexpensive compared to those made with beaver skins. The animals are easy to raise in pens and coypu farms were developed in several parts of the United States several decades ago. In one such area in Louisiana, a hurricane destroyed some cages; the inhabitants escaped into adjacent swamps and quickly became established there. Soon, muskrat trappers were catching more coypu (called *nutria* in the United States) than muskrats and receiving a better price for the pelts. The coypus found the ecological conditions excellent, multiplied rapidly, and dispersed widely from their point of liberation. They appear to have contested with the native muskrat (*Ondatra zibethica*) for the available food and territory, with the result that muskrat numbers declined. Later, when coypu pelt prices fell, it was too late to eradicate this animal. By then, it had spread throughout most of the lower Mississippi drainage.

The coypu has continued to disperse and no one is certain as to what the ultimate limits of its geographical range will be. Moreover, the animal was also liberated (or escaped) later in parts of Oregon and Washington, although thus far its numbers have been modest in those areas. The animal has proved to be ecologically disruptive everywhere it became established in North America. It is also destructive to dikes and levees because of its burrowing activity, and it has the potential of becoming an agriculture crop pest as well.

There have been few introductions of exotic fish into North America, although of those that have been brought in, some are of great ecological significance. The most important aspect of the introductions is the interregional movement of fish (as well as amphibian and crustacean) species. The geological history of North America had a profound effect on fish geography. The Rocky Mountains and adjacent ranges are barriers to fish dispersal, so the freshwater fish fauna of the East is markedly different from that of the West. Most notable is the almost complete absence of native perch in California and the total absence of native freshwater catfish in the most western states. Of course, there is a native freshwater fish fauna, including trout and a host of large cyprinodonts (top minnows). Some of these attain considerable size but lack the fighting qualities prized by fishermen. Nonnative aquatic animals from the eastern states were introduced into the West beginning in the second half of the nineteenth century. The introductions included various species of sunfish, the two species of black bass, several species of catfish, bullfrogs (*Rana catesbeiana*), green frogs (*Rana clamitans*) and nonnative (to the West) crawfish.

The fish introductions were made chiefly to provide recreational fishing. The frogs were brought into the region first as laboratory animals and then liberated. The crustaceans appear to have been dispersed, at least in part, by fishermen who purchased the live animals from bait dealers and, after a day of fishing, dumped the remaining crawfish into the water. They are now almost ubiquitously distributed in California and are often a pest because they burrow in the banks of irrigation ditches.

In addition to the native United States aquatic animals introduced into the Pacific Coast area, exotic fish have also been liberated there and elsewhere in North America. Ecologically, the most significant of these was the European carp (*Cyprinus carpio*)—a fish that should never have been brought in because it is extremely destructive of the ecosystems it inhabits and can quickly turn an excellent bass pond into a muddy hole.

In recent years, there has been great interest in aquariums where tropical freshwater fish are kept for their bright colors and interesting habits. For the most part, this had no ecological effect on the North American ecosystems. The principal exception is Florida, which is the first stop for many of the fish after they leave the tropics. Because the southern part of Florida enjoys a mild climate and freshwater areas are common, it is not surprising that some of the fish have been accidentally or deliberately introduced into the streams and lakes. Thus far, there do not seem to be significant changes in the freshwater fish fauna, but some of the exotic species have become established locally. However, natural ecological barriers such as cold winters may prevent the dispersal of most of these fish beyond the waters of southern Florida.

Although most of the fish species that have been moved about in North America are freshwater forms, the marine environment has not been overlooked by man. The most outstanding example of this was the successful transporting of striped bass (*Saxatilis roca*) from the East Coast to the West Coast. This was accomplished in the latter part of the nineteenth century—a considerable feat, given the transportation facilities of that day. This fish, which frequents the shore area of the sea as well as freshwater streams that enter the marine environment, encountered optimal ecological conditions in the shore areas of California and parts of Oregon. Within a few years of its introduction, it had grown so numerous as to form the base for a commercial fishery. Later, in the twentieth century, it was decided that the fish could be

taken only by sport fishermen and its commercial sale was banned. It now supports one of the most important sport fisheries in California and parts of Oregon. Nothing is known about its effect on the aquatic ecosystems into which it was introduced, but whatever dislocations resulted apparently have passed. The striped bass is now a part of the West Coast fauna.

Although early man may have introduced a few insects by accident, human-caused introductions could not have been of much importance until the arrival of the Europeans. Although not well documented, we may assume that from the beginning of European settlement, invertebrates, mostly insects, have been introduced and this has continued to the present day. So numerous are the species it would require many pages just to *list* them. Most of the introduced species are native to the Palearctic region largely because of the intense trade and travel that has long occurred between North America (the Nearctic) and Eurasia (Palearctic). Ecologically and economically, the accidental introduction of arthropods has caused more difficulty than all the other animal introductions.

Because there were so many introductions, our selection of examples is somewhat random. We shall begin with the cabbage white (*Pieris rapae*) butterfly (order *Lepidoptera*) because it is known to almost everyone—even the urban dweller. This little white butterfly is found in an extraordinary range of habitats. Because the food of its larvae is principally plants such as cabbages, the insect is one of the more significant pests in North American agriculture. The first introduction(s) took place early in the colonial period and subsequent introductions have probably been made many times since.

Of much greater ecological and economic significance was the introduction of certain beetles (order *Coleoptera*), and perhaps the most serious beetle pest introduced is the Japanese beetle (*Popillia japonica*). The introduction, which was accidental, took place early in the twentieth century on the East Coast. Since then, it has spread widely. It has come to be one of the most serious pests in North American agriculture, particularly for orchard crops.

Turning to another order of insects, the bees, ants, and wasps (order *Hymenoptera*), we find several troublesome accidental introductions and one beneficial (for man, at any rate) introduction. The European honeybee (*Apis mellifera*), which was first introduced during colonial times, is the beneficial example. Not only is its honey highly valued, but it is important as a crop plant pollinator. Over the years, it has repeatedly escaped, and feral European honeybees are now found over much of North America.

The most troublesome members of this insect order accidentally introduced into North America are ants. Of these, the Argentine ant (*Iridomyrmex humilis*), a native of Argentina, Uruguay, and possibly southern Brazil, is an outstanding example. The first introductions occurred in or around New Orleans sometime during the second half of the nineteenth century. There was a later introduction(s) to the West Coast. This ant pest spread widely in the southern United States and is also common in many parts of California. Moreover, because it is an exceedingly aggressive species, it eliminates almost all the native ants. This may appear to have limited significance, but it might have been important to some of the invaded ecosystems. In any event, the pest and nuisance aspects of the accidentally introduced Argentine ant are well known to those who live in the regions of North America where they occur.

Another economically important order of insects is the flies (order *Diptera*). It is likely that the first member of this populous order to have been introduced by man was the house fly (*Musca domestica*). And this may have occurred thousands of years before the arrival of Europeans, having been introduced by the human invaders who came via the Bering land bridge. So intimately associated with our species is this insect (as the name *domestica* suggests) that it is difficult, if not impossible, to describe its past travels with man.

The introduction of animal-caused and/or carried disease is an important element in the story of man's influence on the animal life of North America. The group of animals most important to this discussion is the class *Hexapoda* (insects), for this enormously successful class includes many of the organisms that transmit disease. And within this class, the order *Diptera* has most of the identified culprits. The so-called yellow fever mosquito (*Aedes egypti*), (*so-called* because there are many other mosquitoes in the tropics that are also yellow fever vectors) is the principal urban vector of yellow fever. It may have been introduced into North America in the late seventeenth or early eighteenth century. A question still remains as to the origin of the virus causing yellow fever, but there is no question that the disease was unknown in North America until the urban vector was introduced. Yellow fever epidemics once constituted a major medical problem in the Southeast United States, but the disease is now unknown there because of vaccinations and vector control.

Human malaria, caused by several species of single-cell plasmodia (genus *Plasmodium*) may have been unknown in North America (or the Western Hemisphere) until after the Europeans arrived in Latin America. There were, however, various native vectors, all of which are mosquitoes and belong to the genus *Anopheles*. Malaria became a medical problem in many parts of the United States—wherever suitable vectors occurred. The greatest problem was in the South and Southeast, but the disease was also reported from time to time in California. Malaria has also been largely eradicated from North America through vector control and other measures.

Little is known about human disease in North America before the arrival of Europeans, and there is not a great deal of information for some time after their arrival. However, the available data suggest that infectious disease was not commonplace among the Amerinds. In contrast, the Europeans had an almost encyclopedic list of infectious ailments—which they transported with them to the New World. Among these infectious diseases were smallpox, diptheria, whooping cough, measles, chickenpox, typhoid, streptococcus infections, and staphylococcus infections. The Amerinds, not having had any experience with these diseases, were highly prone to infection, with sometimes dramatic results. There has not yet emerged a definitive study of the impact of exotic infectious disease on the Amerind populations of North America, but there is reason to believe that it was great. Through human-induced transport, therefore, the medical geography of North America underwent a major transformation, and the implications reached deeply into the ecological structure of the region.

6.4 Influences on Surface and Subsurface Waters

We cannot accurately determine man's influence on the waters of North America in pre-European times. However, we do know there was little attempt to divert surface waters and the few attempts that were made were of a local nature. One example occurred in the Southwest, where some irrigated agriculture was practiced. There is some slight indication that water diversion was practiced in parts of the Great Basin region by gatherers who had learned that there is a relationship between the amount of water wild plants receive and the harvest, they produce. Their technique was to dam small natural gullies to force the water up and out of the channel during periods of runoff. The water was directed out onto the land by crude, low stone dikes. (There is some dispute among scholars as to whether or not such construction ever occurred, but the supporting evidence is strong enough to be mentioned.)

As we have seen, fire was used in various ways in hunting and gathering and, on occasion, escaped control. This not only had impact on vegetation cover, but also increased the normal rate of runoff. Such effects, however, may well have been local and only rarely, if ever, achieved great magnitudes.

Little is known of the general sanitary habits of the Amerinds, so one can only speculate about water pollution. However, human numbers were usually modest in any area, so it is not likely that the water sources were significantly polluted by human wastes or other inputs of human origin. Thus, at present, we can assume that the human influences on the waters of North America were slight before the arrival of the Europeans.

There were no significant changes in the hydrology of North America in the period immediately following colonial settlement. On the East Coast, the settlers, having come from a humid environment, knew little of water control for irrigation, and their efforts were limited to digging wells and damming small water courses in order to supply power for grist mills to grind grain. Increased vegetation removal slowly led to greater flash runoffs during storms or when snow melted, so flooding became more frequent, at least locally, with the passage of years.

In the West, in those areas settled by Spaniards, irrigation techniques were introduced. However, these were rudimentary for many years because both capital and the technical ability to produce other than local water diversion were lacking. There is some evidence that overgrazing by introduced cattle and sheep led, in some cases, to severe flash floods in parts of the Southwest, but there are also alternative hypotheses to account for observed landform changes.

In the East, it became increasingly apparent over the years following colonization that surface transport, unless greatly improved, would hamper people's desires to communicate and market their products. Not until the nineteenth century were significant efforts made to upgrade transport routes. Thus, because it had long been understood that water transport is one of the most economical means of moving goods, the Eastern United States was marked by a great interest in canal construction. The most famous canal, the Erie, was completed in 1825; it linked New York with the Great Lakes. During the next few decades, a virtual mania for canal building appears to have seized investors, and hundreds of miles of canals were constructed. Many of these returned little profit to their investors but did result in marked changes in the hydrography of the region. The Welland Ship Canal, opened in 1887, which permitted shipping between Lake Erie and Lake Ontario, was of great ecological significance because it allowed the sea lamprey (*Petromyzon marinus*) to cross the previous barrier. This resulted in great ecological change and economic loss in the Great Lakes.

By the end of the nineteenth century, many changes had taken place in the surface and subsurface waters of the eastern parts of North America. By that time, industry had become established, large urban settlements had grown along the waterways, and the habit of dumping wastes into the streams was well established. Foul-smelling water had become an accepted price for progress and almost no thought was given to the ecological consequences. There is evidence that flooding was more frequent and severe as a result of changes made on the watersheds. And because most of the population depended on private wells for domestic water supplies, there was a medical problem associated with polluted water. There was little or no understanding that human excretions from outdoor privies might find their way into the local ground water and thus in the water drawn from adjacent wells. Gastrointestinal infections were common, with typhoid particularly prevalent during the summer months. The high infant mortality of much of rural America in the nineteenth century was due in part to polluted water. This problem persisted well into the twentieth century; indeed, some rural areas of the United States are still afflicted.

Twentieth Century. If hydrographic changes in the nineteenth century were principally an Eastern phenomenon, then the twentieth century was the era of principal hydrographic change in the West.

West of approximately the 100th meridian, atmospheric precipitation is often so unreliable that no advanced agricultural-industrial society can exist there unless it controls and modifies the surface and subsurface waters. However, to do so requires advanced engineering technology and an abundance of capital. When both became available in the twentieth century, a period of modification of the earth's waters began that has seldom been equaled elsewhere in the world.

West of the Mississippi there are several great rivers, such as the Platte and Missouri, that are tributaries of the "father of waters" and drain a great region. The next great river westward is the Colorado, whose name is derived from the heavy load of red silt it carries. For much of its course, the Colorado is an *exotic* river in that it flows through regions much too dry to support or augment it. Also, until recent years, it was given to wild fluctuations of

Figure 6.13(a) *Irrigated field in Arizona.*

Figure 6.13(b) *Salt damage (white areas) in an irrigated field, Imperial Valley, California.*

flow from season to season and year to year. Moving northwest, there is the mighty Columbia River, with some impressively large tributaries. And in California there is the Sacramento-San Joaquin system, with its numerous tributaries draining mostly from the Sierra Nevada. Vast stretches of land between the rivers lie parched under the sun, even though water is only a few kilometers away, flowing between rows of green cottonwoods (*Populus* sp.), or a few meters below the surface.

The people gained skill at water diver-

sion before the twentieth century—much of this came from their efforts at mining gold from alluvial deposits, at sites where sufficient water was not available locally. Water had to be carried in ditches and wooden flumes from the mountains to where it was needed. Wells were common, although pumping was often a problem when large quantities of water were required. But the problem of getting economical power for pumping water in agricultural areas was not really solved until well into the twentieth century, when the development of specialty agriculture for Eastern markets justified investment in pumping equipment. As the agricultural value of some of the land in the West became apparent, more effort was directed toward developing means by which water could be lifted or transported in order to get these lands into agricultural production (Fig. 6.13(a), Fig. 6.13(b)).

It was seen quickly that a major problem was the extreme variability of precipitation, and water impoundment was to be the means of solving this problem. Dam construction increased in sophistication and, before four decades had passed, many of the rivers of the region were under at least partial control. The dams often served more than the purpose of water storage. They were used to generate electricity and were sites for recreation.

Figure 6.14(a) *Trinity dam and lake, northern California.*

Figure 6.14(b) *Hoover dam and Lake Mead on the Colorado river.*

ways to areas often far away from their sites. For example, water is conveyed by tunnel to Los Angeles and San Francisco from dammed sources far distant (Fig. 6.14(a)). Water from the Colorado River irrigates citrus groves hundreds of kilometers away after it travels through a network of lesser dams and ditches (Fig. 6.14(b)). The Western United States is now largely a hydraulic culture dependent on complex modifications of the surface and subsurface waters. The hydrographic modification of the West is the most striking aspect of human modification of Western ecosystems, except for the urbanized areas in Southern California and those adjacent to the San Francisco Bay.

In addition, looking westward shows us that the days of canal building are not over. In recent years, ship canals have been constructed along the Texas Gulf coast, and a large project is currently underway to connect Arkansas to the Gulf of Mexico by a ship canal.

But hydrographic changes in the twentieth century have not been restricted to the area west of the 100th meridian. Perhaps the best known major hydrographic change is the Tennessee Valley Authority (TVA) project. It created a necklace of manmade lakes extending over 900 kilo-

However, almost no attention was paid to the ecological aspects of such hydrographic modification, and even today these aspects have not been fully researched. It was discovered rather soon, however, that water temperatures down stream from a dam are often greatly altered—usually lowered—and that the ecology of the water trapped above a dam is markedly different from before and supports different biota than had previously existed.

But the dams were not isolated items of construction. They were tied in various

meters and controlling the runoff of over 100,000 square kilometers. By such engineering feats, the hydrography of a large part of the East was greatly modified. South of the Tennessee River, we may witness another major modification of surface waters. Here, instead of impounding water, the engineers have set about *draining* some of the vast swamps of Florida. The state of Florida lies mostly on limestone of coral origin and most of it is only a little above sea level. With an eye to increasing the available land area, governmental agencies have dug canals to drain freshwaters to the ocean—often causing severe ecological damage (Fig. 6.15).

Not only has the face of the land been greatly modified by the dams and canals that have been constructed, but in every case, there have been far-reaching ecological changes as well. Some of these result from the changes in the chemical content and thermal properties of the water and some result from the practice of stocking the man-made lakes caused by damming, with fish.

We have focused attention on some of the large projects and major changes, but there are also thousands of small dams and ponds that collectively comprise an astounding amount of hydrographic manip-

Figure 6.15 *Drainage ditch, southern Florida.*

ulation. The result of the large and small projects is that "wild rivers" are now scarce in the conterminous United States (although they are still numerous in Canada and Alaska).

It may appear that water lifted from wells is no longer important in this region, but this is far from true. Many large urban areas augment their imported water supplies with water pumped from wells, and a significant percentage of the irrigated agriculture in the West depends on water pumped from subsurface sources. The demand for such water often exceeds the supply, and the result is that for decades, much of the United States—East and West—has faced the phenomenon of lowering water tables. The drying up of wells is a serious problem in all parts of the United States where there is heavy dependence on ground water.

Such a change in the hydrography is almost invisible because it occurs beneath the surface; however, on occasion the lowered water tables do become visible. The most frequent way this occurs is called *subsidence*—the surface slumps when the supporting water below has been removed. (Subsidence is also associated with the removal of petroleum in some cases.) Another way lowered water tables may become apparent, chiefly at coastal sites, is through the intrusion of salt water into strata where only fresh water occurred before. Perhaps the most dramatic phenomenon associated with lowered water tables is found with artesian wells. An artesian well is one from which water flows to the surface without being pumped. The water was trapped below impermeable rock strata and thus was not able to rise to its natural level. If the natural water level is above the ground level, a free-flowing artesian well results when the impermeable rock strata is broken through. However, such wells will flow only as long as the water takeout does not exceed input. If the takeout exceeds input, the water level will drop below ground level and pumping will be required.

As the northeastern United States and southeastern Canada became industrialized, the abundant rivers and streams provided convenient means to carry away from factories the wastes that were produced there. The rivers also provided abundant quantities of water for manufacturing processes, and this water, once used, could also be conveniently discharged into a nearby stream. It was known early that these discharges often produced foul odors and might kill some fish, but the general feeling was that this was a natural part of industry and need not cause major concern.

In the early years of industrialization, the alteration of water quality must have been only local. Moreover, for some time there was little input of sewage into the streams because cesspools and similar devices were able to take care of human excreta. However, as the intensity of industry increased and the size of urban populations grew, the water quality changed. More and more cities adopted sewer systems, and sewage was often discharged into streams with no treatment whatever or with so little treatment as to allow the discharge of pathogens and toxic materials. Industry demands enormous quantities of water, and waste water is generated in similar quantities. These waste waters were (and are) discharged into convenient nearby streams bearing with them thousands of possibly toxic substances.

Sometimes the toxic nature of a discharged substance is discovered long after it was regularly discharged as a safe substance. One recent example of this is mercury (Hg). Mercury was long believed to be safe when discharged into freshwater ecosystems, for people thought it sank to the bottom, worked its way by gravity into the mud, and therefore would not get into the food web. However, mercury undergoes some interesting changes given certain circumstances. It can combine with other chemical elements to form methyl mercury. In one form, methyl mercury poses no ecological problem, for it is volatile and passes out of the water as gas. However, in another form, which is created in the presence of water of a low pH, it becomes available to certain anaerobic bacteria, who take it up in their metabolic activities. From there, it may move through a succession of living organisma and end up in a fish or in man.

Methyl mercury is not new, however. The chief ore of mercury, cinnabar, occurs in many parts of the world, and the mercury that weathers naturally from cinnabar must have been entering fresh water for millions of years. However, the widespread industrial use of mercury has now greatly extended the geographic areas where the compound is found and has undoubtedly increased the *levels* of this compound in the fresh waters near factories. Like many other toxic substances, methyl mercury may start out in a food web at concentrations so low as to be difficult to detect, but as it moves through the food web, it may attain high concentration and become a serious biological danger.

It would be reassuring if we could list all the compounds that have been introduced into the fresh waters of North America through human actions and then described the ecological effect(s) of each one. However, at present our knowledge is sorely lacking here, for only recently have we turned our attention to this. In any case, the task is herculean, for *each year* thousands of new compounds appear in industrial use, and as yet there has been only a modest effort to determine the ecological dangers, if any, when they are discharged into surface waters.

However, we *are* aware that industrial pollution has had a gross impact on the freshwater ecosystems of North America. There is scarcely a stream of any size in the East that has not had its ecology altered as a result of contamination from factories. In some streams, nothing survives except anaerobic bacteria.

Of equal importance is the pollution from sewer systems. A large share of the human population of North America lives away from the ocean, and the daily quantity of excreta is monumental. So too is the task of moving it away from its urban sources. Again, the nearest streams have long served as the most convenient places to get rid of the unwanted material. The result is that some rivers and lesser streams have become sewers. The stench and sight of these rivers are shocking enough, but the ecological aspects are even more serious. Most important is the threat of disease, because all too often one city's sewage becomes the drinking water of the next town downstream. The water supply is made safe only by careful and expensive treatment. Heavy quantities of sewage in a stream greatly alter the chemical nature of the water, particularly the free oxygen available. Thus organisms with a high oxygen requirement disappear and are replaced by those with lower oxygen requirements. There is often an increase in water temperature—which also results in biotic changes. In short, many rivers in the East now possess ecologies markedly different from those prevailing before the development of large human population clusters.

Lakes also are affected by pollution, and in parts of North America, lake pollution is a major ecological problem. There are some fundamental differences between a polluted stream and a polluted lake. In a stream, the pollutants are transported away, so, given a rest from added inputs, a stream may rapidly heal itself. A lake, on the other hand, contains the pollutants for some time. In large lakes, centuries may be required to totally replace polluted water. Thus, lake pollution often has a greater ecological significance then that of streams and rivers.

The most outstanding example of lake pollution in North America is the Great Lakes. All these lakes are polluted to some extent as a result of human action but the degree of pollution varies—from low in Lake Superior to very high in Lake Erie. The principal problem is *rapid eutrophication*—rapid biological aging of freshwater lakes as a result of inputs made by man

which is most pronounced in Lake Erie. For many decades, Lake Erie has been the recipient of monumental quantities of industrial wastes and sewage; in recent years, as a result of heavy inputs of nutrient materials, there have been explosions of algae populations (algal blooms). These, in turn, greatly lowered the oxygen content of the deeper water, so the animals that require more oxygen than is now present were eliminated. The ecosystem is dramatically altered, and the lake supports fewer highly *desirable* fish species than before—although *total* processed fish meat has increased. Nutrients for algae growth is only one aspect of Great Lakes pollution, however. Thousands of chemical compounds are discharged into these waters, and only limited information is available regarding the possible ecological effects. Moreover, the water near the shores of some of the lakes has been made unsafe for swimming due to the large quantity of untreated human sewage.

In addition to industrial and sewage pollution, there are other important sources of contaminants. The insecticides used in agriculture are often transported into freshwater sources, where they may have an impact on the aquatic insect fauna and thus on the fish that feed on the insects. Any freshwater body in an agricultural area contains measurable quantities of the chemicals applied to the fields. The short-term effects are still in doubt. In time, we will be able to fully determine their ecological significance.

The other major agricultural source of stream pollution is the *feedlots* of the Middle West (and elsewhere) because enormous quantities of animal excretions accumulate as a by-product of feedlot operations. At present, there is little commercial market for this material, so it accumulates and is later washed or drained away during periods of water runoff. Eventually, it reaches nearby rivers and lesser streams, causing them to become foul-smelling sewers (Fig. 6.16).

In recent years, there has been increas-

Figure 6.16 *Hog feedlot, Middle West.*

ing attention paid to thermal pollution, the raising of the normal temperature of a stream or body of water by human actions. This phenomenon is now most closely associated with power plants that use large quantities of water for cooling. As the number of nuclear power plants increases, so too will the amount of heated water being returned to streams and lakes. Thermal pollution has not been greatly studied as yet and scientific opinion of its ecological impact varies, but we can be certain that oxygen levels are lowered and faunal and floral changes occur in water so affected.

The exact changes are determined by local factors. If nuclear power plants become common away from the oceans, the thermal properties of a significant proportion of the fresh waters of North America will someday be altered.

In summary, there have been major modifications in the hydrography and chemistry of the water in the conterminous United States and parts of Canada. Only those areas of sparse human settlement in the higher latitudes appear not to have been modified to any important degree—as yet.

6.5 Influences on Soils and Geomorphology

As we have seen, manipulation of the plant cover and hydrography of an area almost assures that the crustal features will also be influenced. Thus, because the plant cover

and hydrography of many of the North American ecosystems have been modified to a marked degree by human action, there have been significant changes in the earth's crust as well. The most pronounced change relates to the soil—accelerated soil erosion and soil impoverishment have been significant in the recent history of North America.

Virtually nothing is known of the man-soil relationship in North America before the arrival of Europeans. However, from our general understanding of Amerind agricultural practices, it can be assumed that some accelerated soil erosion occurred, but given the small size of the farming populations, it is likely that such disturbances were local and of limited ecological significance. In the nonagricultural areas, burning may have produced accelerated soil erosion, especially in those parts of the West where rainfall occurs in violent but isolated episodes. However, there has been no significant attempt thus far to determine what effect the hunters and gatherers had on the soil cover in North America.

The early European settlers possessed few advanced tools and practiced a level of agriculture that was little different from that practiced by the Amerinds. The plow was rare or lacking altogether during the earliest years, so a form of hoe culture prevailed. However, this necessitated the removal of trees to make way for crops. Thus, although the European populations were small in the beginning, accelerated soil erosion began almost at once. The soils, which had developed under different conditions from those the settlers were familiar with in England, were light and friable. Once cleared and exposed to heavy rainfall, these soils washed away with a rapidity that dismayed the farmers. Even level land was not immune. Moreover, in many instances, the farmers were hampered by difficult and stony soils, inadequate drainage, and short growing seasons. Thus, large areas were simply abandoned, and the farmers moved on to uncleared land and repeated the process. This was the pattern in New England for the first century or more, and the result was that the soils were greatly altered.

To the south, another pattern evolved. A plantation style of agriculture developed that emphasized only a few crops. This led to the intense use of land which was cleared and kept clear in order to accommodate the crop plants. The result was

Figure 6.17 *Accelerated soil erosion, Mississippi.*

that before the end of the eighteenth century, large tracts were almost denuded of their soil cover and farming was difficult, if not impossible. Farmers often sold or abandoned their land and moved west—to occupy new lands suited to cotton, which by 1800 had become king in the agricultural south. For much of the East and Southeast the major ecological story was soil destruction and land abandonment.

In defense of these early settlers, however, one must remember that they found themselves in ecosystems about which they knew little. Their earlier knowledge of farming did not help them in ecosystems that were only superficially similar to those of Western Europe. Nevertheless, the story of soil loss in the Southeast does not end with the beginning of the nineteenth century. Land continued to be misunderstood and hence misused even into recent decades—and often by persons who were not in an economic position to alter their practices even if they were motivated to do so.

In 1862, the United States government passed the Homestead Act—a piece of legislation whose ecological consequences are still being felt. This act provided an escape for the thousands of farmers whose earlier efforts in the East and Southeast had been defeated and provided an all but irresistible lure for the landless. The law provided a quarter section (160 acres) to anyone willing to "prove up" the land, that is, develop the land into a farm. If the conditions were met, the land became the private property of the homesteader. The result was that vast areas that had not previously been cultivated were now turned over by plow and planted to crops.

One feature of these new farms is important to note. Unlike the eastern region, the lands subject to homesteading were surveyed by the Federal government so the farms were rectangular rather than the irregular shapes that had resulted from using natural features ("metes and bounds") to delimit property boundaries. Although the rectangular farms led to a more ordered landscape and one much easier to describe for legal purposes, it also encouraged practices that led to ecological degradation of the soil. The farmers tended to plow furrows parallel to the field boundaries and not take note of and adapt to slope conditions. Thus the furrows were likely to cross the contour at sharp angles, which led to accelerated soil loss when it rained. Soil loss reached high levels in the 1930s, at which time major steps were begun to correct the situation. Although many parts of the country were affected, the area most greatly influenced was between the Rocky Mountains and the Mississippi River (Fig. 6.17).

Figure 6.18 *Dust storm, Colorado.*

While plowing practices were eroding the soils of the Middle West, in the dryer land farther west accelerated erosion also occurred—not only because of plowing practices but also because of the overgrazing of livestock. Great herds of sheep and cattle were thrown onto natural pastures unable to support such populations for protracted periods, with the result that the soil washed away on many parts of the range. Moreover, the practice of clearing forests, which often bared the exceedingly friable soils to the rain, also added to the soil loss.

Another aspect of accelerated soil erosion is that caused by wind rather than water. Wind is a significant problem in parts of western Texas, Oklahoma, and eastern Colorado—a region of great variability in precipitation and one in which man has not learned to live safely, in an

ecological sense, during the past century. The region is attractive for wheat cultivation, so when wheat prices are high, farmers have extended the cultivated lands into areas that are known to be ecologically marginal for that grain. In the 1930s, a depressed economy and drought combined to bring about disaster. Winds picked up the topsoil by the millions of tons and blew some of it out over the Atlantic Ocean. Unfortunately, however, memory is short in the region and another dust bowl condition developed in the early 1950s but with fewer social repercussions than the previous event (Fig. 6.18).

When soil is removed, it must come to rest someplace else. In North America this often has been lakes and harbors, where the sediments clog channels, change the general ecological conditions, and cause great economic loss through the "silting up" of areas behind dams.

Thus, one of the greatest dramas of man-induced ecological change in North America during the past two centuries has been accelerated soil erosion. No large part of the region under human settlement has competely escaped and some areas have been so altered that they now have extremely impoverished ecosystems.

In addition to accelerated soil erosion, man has altered the soils of North America in many other ways. Chief among these are the chemical changes that occurred by accident or design and that derive mainly from agricultural practices. The United

Figure 6.19(a)
Landscape after being strip-mined for coal, Missouri.

Figure 6.19(b)
Hydraulic mining in the Klondike before the method was made illegal.

States long ago passed the point where agriculture was sufficiently productive through the use of animal or green manures alone. These generally possess a low available nitrogen (N) content and thus are not suited to the high-yielding crop varieties now cultivated. As a result, farmers increase, at least seasonally, the normal amounts of nitrogen, phosphorus (P), and potassium (K) in their soils by chemical fertilizers. (The long-term effects of these practices on the soil are not completely known.) Moreover, in addition to fertilizers, a number of chemicals are employed in agriculture to control crop pests and unwanted vegetation. Many of these chemicals have a long life in the soil and thus later alter the microflora and microfauna of the soil. The ecological implications are great, but again, it appears too soon to make any estimates of the long-term ef-

fects. Recovery time might be much slower in regions with low precipitation that require irrigation.

One of the newest techniques of altering the soil is practiced in parts of the West where the desert soils often possess too high a mineral content to permit irrigated agriculture. In some instances, fresh water has been spread over the land to leach out large quantities of minerals and make the land suitable for irrigated farming of cotton or other crops. The flushed water, however, has often proved to be a problem, because it is often not suitable for further use in irrigation. This was illustrated when highly mineralized Colorado River water was later used in Mexico for irrigation.

Next in importance to soil changes in discussing man's influence on the land is mining. For our purposes, we shall examine five types of mining: (1) strip mining, (2) hydraulic mining, (3) placer mining, (4) hardrock mining, and (5) open-pit mining. In some instances, types may merge but the five are generally fairly distinct (Fig. 6.19(a), (b), (c), (d)).

Strip mining is the stripping of minerals from the surface of the land or, to be more exact, the removal of unwanted soil and rock (*overburden*) lying on top of the desired minerals and then the stripping of the minerals. This type of mining, which is

Figure 6.19(c)
Hardrock gold mine, South Dakota.

Figure 6.19(d) *Open pit copper mine, Arizona.*

Figure 6.20 *Tailings left by mechanical gold dredges, Canada.*

most used for coal, has so altered large tracts of land, mainly in the eastern United States, that they are visible on geological survey maps. The technique is efficient from an economic point of view, but land mined in this way will show the effects, not just for decades, but for thousands of years. There is an almost total alteration and major destruction of the affected ecosystems, and the ecological succession in an area after it has been mined is only partially understood at present. Long-range effects include major changes in the water quality, less food for aquatic and nonaquatic organisms, and major vegetation changes. Esthetically, an area so exploited leaves almost everything to be desired. However, the destruction is in response to soaring power demands made by most of the human residents of the ecosystems of North America.

Hydraulic mining, fortunately, has become of only historical interest, because it has been outlawed almost everywhere. In the last century it was a favorite technique for extracting gold from alluvial deposits in California, but the *detritus* (soil and stones) washed down from the Sierran foothills clogged rivers and led to serious flooding. This finally resulted in legislation forbidding the technique.

Placer mining is the extraction of minerals from stream-deposited materials, and in North America, the most important of such minerals is gold. A wide variety of techniques have been used in placer mining—hydraulic mining might be considered one of these—and each technique affects the shape of the earth's crust in different ways. The most modest placer mining operation involves one man using a gold pan to separate the heavy metal from the unwanted gravel and sand. One man makes little impression on the landscape, but hundreds of thousands of men using gold pans, can, in a remarkably short time, make their presence obvious by the gravel piles they leave behind. Adding even simple equipment, such as a sluice box, makes the geomorphic changes great and of long duration, because the characteristic result is many mounds of gravel debris. These mounds, called tailings or waste piles, persist in old, abandoned gold diggings for many years. As equipment becomes more sophisticated, more gravel is worked each day and the tailing piles become correspondingly greater in size and durability.

Figure 6.21 *Tailings from a lead-zinc-copper mine and mill.*

Many stream and river beds have been completely altered as a result of such tailings being deposited by large, mechanized gold dredges (Fig. 6.20).

Hardrock mining is an underground form of mining, so many of its effects are not visible. However, the ore must be taken to the surface and tailings are a visible feature. One of the most conspicuous signs of human presence past and present in much of the West is hardrock mine dumps. They scar thousands of hillsides and will persist in the dryer areas for many centuries (Fig. 6.21).

Open-pit mining is self-descriptive. It is the extraction of minerals by simply following the ore from the surface down into the ground. Iron and copper are the chief minerals exploited by this method. The local geomorphic changes can be dramatic and are readily seen on United States Geological Survey maps.

In varying degrees, *all* the ways man has affected the earth that we have already discussed have sculpted the landscapes of North America. We are using the term *sculpting the land* as a convenient catchall for the activities of man other than agriculture and mining that have reshaped the earth's crust.

Road construction is one of the most notable of these activities. In the earlier years of European occupance, the road net was minor, of exceedingly poor quality, and of limited ecological significance. The advent of the railroad began to change the land, however. Low grades were required for tracks, and the requirement led to a degree of cutting and filling not previously known. But the railroad was a minor influence when compared with the impact of automobiles on the physical landscape. Starting modestly in the early part of the twentieth century, the automobile has emerged as a monster swallowing entire landscapes in its quest for traction and parking space. The most recent chapter of the road-building story is the Interstate Highway System. The construction of this network of highways has resulted in what may well be the greatest amount of earth moved by man in any single engineering effort in human history. As the pyramids of Egypt symbolize the engineering effort of the Pharaohs, the major highway construction in the United States and Canada stand as the monument to engineering effort and energy consumption in North America in the twentieth century.

The automobile's influence on geomor-

Figure 6.22 *The effects of the automobile on the American landscape.*

phology goes far beyond super highways, however, for it also includes paved city streets and paved areas for parking. It is no exaggeration to state that in the United States and parts of Canada, urban areas have been made subservient to the automobile and the most obvious geomorphic alterations are directly due to this conveyance (Fig. 6.22).

Last to be mentioned here are the changes brought about through the use of machinery to carve building sites from otherwise unsuitable terrain. To some degree, this occurs in most of the major regions in North America where there are high population densities and limited level terrain. However, it probably occurs most frequently in the larger cities along the West Coast, especially in the hilly areas of Los Angeles.

6.6 Influences on the Atmosphere

The one aspect of man's influence on the ecosystems of North America that commands the most public interest and attention is air pollution. There are several reasons for this but chief among them must be the fact that, unlike many other aspects of environmental modification, contaminated air is often painfully evident. Moreover, because almost everyone is a contributor to air contamination, the topic is of great personal, political, and economic interest.

As we have seen, not every major region of the world is as yet a major source of air pollution. Western Europe and Japan are plagued with this problem, but elsewhere air pollution tends to be localized, such as in São Paulo, Brazil. Most of the sources of atmospheric inputs due to human agency are in the Northern Hemisphere, and the major contributor in that hemisphere is North America.

As in previous chapters, it is useful to distinguish between macro- and microclimatic (bioclimatic), or atmospheric, phenomena. The distinctions between the two classes are not sharp for there is no general agreement on the lower limits of a macroclimate or the upper limits of a microclimate. In general, we shall use *macroclimates* to refer to areas greater than one-fourth of the land mass of North America and *microclimates* to refer to smaller units—down to units as small as a backyard garden.

Before the arrival of Europeans in North America, man was not in a technological position to introduce many contaminants into the atmosphere. Burning did contaminate the air to some extent, because ash, other particulate material, and carbon dioxide were released, but the effects must usually have been ephemeral. In 1540, the first European to sail along the California coast indicated he saw a smoke haze. This was probably caused by fires set by the Amerinds and may be taken as the first report of air pollution in the Los Angeles basin. If, as some believe, Amerinds played a major role in creating and maintaining the grassland over much of the Great Plains, we might wonder about the climatic implications of changed surface albedos (reflection) and the possible associated changes in local precipitation patterns.

We must turn to more recent times, however, if we wish to find incontrovertible evidence that man has modified the microclimates of North America. And the most significant of such modification relates to the input of contaminants. Not until the latter part of the nineteenth century, and then only under local circumstances, did air pollution receive public attention. As in England and the Continent (see Chapter 4), the first major source of air contamination was the burning of coal for industry and heating. And this occurred principally in the northeastern United States and adjacent Canada. The steel mills were a major source of such contamination, and the large urban centers in the East burned coal almost to the exclusion of any other fuel. This often resulted in a black pall of suspended particulates over the cities in the winter, when the air was stagnant.

The twentieth century has witnessed the explosion of automobile numbers and thus the consumption of astronomic quantities of petroleum-based fuels. Chief of these is gasoline, which contains a variety of added chemicals to facilitate combustion. In the early decades of the automobile, there was little problem from the combustion of hydrocarbon fuel. However, in the 1930s, the automobile was becoming a significant source of air pollution in some parts of the continent.

After World War II, the populace engaged in an automobile buying spree. By that time, the automobile had ceased to be principally a means of transportation. It had become one of the most tightly interwoven of all the elements that go to make up the complex cultural fabric of North Americans. For example, it had become a status symbol, a means to escape parental supervision for courting couples, and psychological compensation for people whose feelings of inferiority were lessened when they commanded an automobile engine. Moreover, earning enough money to buy a car became the means by which millions of adolescent males demonstrated their adult status. Among urban youths, male and female, the automobile has become one of the major means of personal expression, and this is often true of older persons as well. In short, any attempt to establish an acceptable substitute for the automobile must take into consideration the cultural attributes this device has acquired.

The automobile is also directly and indirectly one of the most potent forces for environmental modification in the hands of man. And the environmental element that appears to be the most immediately threatened in terms of life processes is the atmosphere. The internal combustion engine of the automobile utilizes hydrocarbons, which are usually sold in the form of gasoline. Gasoline also includes a number of added substances, such as lead, to improve the quality of engine performance. The combustion of gasoline is almost always incomplete, and the exhaust, unless specially treated, contains gaseous and particulate materials that are introduced into the atmosphere. Chief among these materials are nitrates of oxygen, hydrocarbons that have not been combusted, lead, carbon monoxide, and carbon dioxide. Each of these substances contributes to air pollution, often to a remarkable degree.

Thus, the automobile has become a major source of atmospheric contamination, but there are a number of other sources as well. Chief among these are electricity generating plants that use coal or oil for fuel. They discharge particulate matter and varying quantities of sulfur dioxide into the air. A wide variety of chemicals and particulate matter are discharged from factories. And these contaminants are not confined to the more heavily industrialized parts of the continent, for mine smelters in the West may produce profound impacts on local ecosystems.

As we have noted in earlier chapters, the input of contaminants into the earth's atmosphere is only part of the story; there are important meteorological and physiographic aspects as well.

The term *smog* has come to be applied to air pollution or contamination regardless of the nature of the pollution. Actually, the term should be restricted to situations where fog combines with smoke, such as in parts of Europe where coal is or was the chief heating fuel. In North America, serious air pollution is often related to temperature inversions and/or landform configurations that tend to trap air and its collected pollutants in a given area for varying periods of time. Sunlight acts upon the nitrates of oxygen to turn the air a brownish color, resulting in *photochemical smog* that often contains high percentages of biologically destructive ozone (Fig. 6.23).

Photochemical smog is a conspicuous feature of air pollution in much of California, and ozone levels frequently become so high that the physical activities of schoolchildren must be reduced so as not to impair their health. Damage to vegetation may be pronounced, and some crops formerly grown in air contaminated areas of southern California are no longer present because of the toxic atmospheric conditions. Tree damage has been repeatedly noted in publications, and there is general agreement that photochemical smog has a disturbing effect on many other living things. The total area in California subject to this contamination has been growing in

size during the past few decades, and now only the extreme northern part of the state is free of the problem. Although industry is a major contributing source of contamination, the automobile is the most important source. Moreover, the problem is worsened in much of California by the fact that the terrain favors the development of air traps that allow for lengthy buildups of contaminants. Much of the region lies near the eastern periphery of a high pressure cell (in the summer months), in which the air descends from higher levels. Such air tends to be stable and becomes warmer as it moves downward. There also tend to be regular inflows of relatively cool marine air that is trapped below the warmer air mass above. The latter puts a "ceiling" on the marine air layer (an air temperature inversion), and the contaminants contained therein are trapped. Flushing of the contaminated air is impeded by the mountains to the east, which act as dikes to prevent dispersal of the toxic atmosphere.

Elsewhere in the United States and Canada, serious air pollution is most often associated with high pressure cells that *stagnate,* and trap within them many of the contaminants introduced by man. In the areas where this occurs, which include much of the eastern region, the automobile makes an important contribution. However, contributions of electricity generating plants and factories are much greater than in the Far West. The damage to living things in the ecosystems, however, is probably as great as anyplace else.

These are the more dramatic aspects of microclimate alteration, but there are other kinds, such as alterations in air temperatures and humidity. In agricultural areas, for example, any changes in plant cover result in changes in the microclimates. Although this is widely recognized, it has been little studied, except for some research dealing with the microclimates of specific crops. In the Far West where irrigation is widely practiced, the changes in air temperature and humidity can be pro-

Figure 6.23 *Smog in the Los Angeles area.*

nounced. The Coachella Valley, for example, is an irrigated desert. During the summer months, the air suggests the humid tropics because of the heat, and humidity caused by evaporation from irrigated land and the transpiration of vegetation. One can feel a similar effect in large cornfields in the Middle West in the summer, although there, the effect is not all caused by human manipulation.

The microclimatic changes we have discussed thus far have been accidentally influenced by man. Another and increasingly important category is the deliberate modification of weather. Not long ago anyone claiming to be able to make rain was considered a quack or a charlatan. Today, rainmaking and other attempts to modify local weather are respectable areas of scientific investigation.

Rainmaking is of the greatest general interest where atmospheric precipitation is less than required for agriculture or other water needs. The technology of rainmaking is still in an early stage, but the usual procedure is to seed cumulus or other highly unstable clouds with silver iodide crystals, which act as hygroscopic nuclei. Many problems have yet to be solved, and a major one is seeding in such a way as to assure that the precipitation occurs where it is wanted. There are also legal problems that must be resolved before artificially induced precipitation can become a widely used technique.

Attempts have also been made to modify undesirable types of weather, such as hurricanes. It is felt by some that the way to avoid the loss of life and property associated with hurricanes is to develop techniques to suppress hurricanes. Thus far, experiments in this direction have not been particularly successful, and there remains the problem of what happens to the energy that must be dissipated when the natural processes of the storms are interfered with.

In recent years, increasing attention has been focused on the possible influence man may be having on global climatic patterns. There is an unfortunate lack of reliable long-term weather data for most of the world, which greatly hampers research on this topic. It is now generally accepted that there are natural climatic fluctuations over time, but the mechanism(s) controlling them have not yet been clearly identified. Recognizing the existence of natural climatic variations makes the task of identifying human influences all the more difficult—indeed, it makes it mostly conjecture. However, we know we are doing things to the atmosphere that *ought* to alter its temperature, and from this base, we may guess the end results. Three major categories of contaminants in the global circulation may be recognized: particulate matter, gaseous materials, and radioactive materials.

Particulate matter has been increasing rapidly in recent decades, and North America is most likely the major source of such materials. It comes largely from industrial processes, although agriculture is increasingly making a contribution through the use of materials that may be transported long distances from where they were applied to a field. DDT, for example, is transported over great distances and is now a part of the global air circulation. The amount of dust and other particulate matter in the air has been increasing sharply, and there is concern that the result may be changes in the mean temperature of the earth's atmosphere. However, there is little agreement yet as to the magnitude of the change or its direction. Good arguments can be advanced to indicate a cooling of the atmosphere due to more solar radiation being reflected into space. On the other hand, the temperature of the atmosphere may rise because long-wave terrestrial radiation would not escape into space so fast. Perhaps the most significant point is that there is now general recognition that inputs of contaminants are great enough to raise the possibility of global climate change.

The possible implications for temperature change caused by particulate matter are similar to those caused by the increase of carbon dioxide in the air (which is a direct result of the combustion of fossil fuels, coal and petroleum). In this case, however, the argument that the average air temperature would increase appears to be sounder, because carbon dioxide allows the passage of short-wave solar radiation but acts as a fairly efficient trap to long-wave terrestrial radiation. If carbon dioxide buildup is causing the air to warm, some of the implications are interesting. One important implication is the possible melting back of the polar ice caps, with an attendant rise in sea level all over the world.

Radioactive contamination of the air might be included under particulate contamination but is best considered separately. The increase of radioactive materials in the atmosphere began when the first man-caused atomic explosion took place in New Mexico in 1945. Shortly after the New Mexico explosion, atomic bombs were dropped on the Japanese islands. The next explosions in the atmosphere came a few years later, when the United States detonated various atomic devices on Pacific Islands. Other testing has been conducted by Russia, China, France and India. One fact emerges from these tests. Wherever a test is made in the atmosphere, it is only a matter of time before radioactive materials will be rained down on the entire hemisphere in which the test occurred.

Before the testing of atomic weapons in the atmosphere began, some physicists insisted that the *average* fallout per unit of earth area (per square meter) would be so modest as to pose no threat to the earth's ecosystems. These calculations were basically accurate. The overlooked aspect, now fully appreciated, is the *concentration factor*. This refers to the phenomenon that living organisms tend to concentrate certain chemicals as the chemicals move through a food web. Without prior careful experimentation, it is difficult to predict where, how rapidly, and in what organisms the significant concentrations will occur. In general, however, concentration tends to increase as the particular element in question is moved through a food web. Thus, an element that may start out at nonlethal levels may reach concentrations that can produce highly undesirable reactions. Another aspect of concentration is that it may occur almost immediately in simple ecosystems. Thus, it was found in the tundra that lichens were concentrating radionuclides at a very high level. These were then being moved directly to the caribou (*Rangifer*), which depend on lichens for food in winter (Fig. 6.24).

A further problem is that some of the radionuclides travel in ecosystems with other elements. A notable example is ce-

Figure 6.24 *Caribou (Rangifer) on Alaska's north slope.*

sium 137, which moves with manganese and is deposited in human muscle tissue. Even more serious is strontium 90, which moves with calcium and thus is easily cycled, through dairy herds, to milk and thus into human beings. And operating all the time is the concentration factor.

Fortunately, most atmospheric testing has been halted. However, the People's Republic of China and France continue to test atomic devices in the atmosphere.

The last aspect of macroclimate change that we shall note is the possible changes of altering the albedo (reflectivity) of the earth's surface. In other chapters, we concluded that albedo changes for the region under discussion are probably minor and hence of limited importance. In North America, however, the albedos have been so greatly altered over such large areas that we must ask if this has resulted in any major changes in the climates. It is possible to design experiments that might cast light on the question, for it is a matter of concern—if not for the present, at least for the future, when albedo changes may be much greater than they have been.

Chapter Seven

Human Influences on the Ecosystems of Latin America

7.1 Cultural Background

Carbon dates for early man in Latin America (see Chapter 1 for a discussion of carbon dating) do not seem to show a geographical or temporal pattern, but this probably reflects the accident of site discovery. For example, there are older known dates for parts of South America that for Central America. In particular, a carbon date for northern South America indicates the possibility of human presence at about 16,000 years BP, but no evidence of similar antiquity has been found for any part of Central America although some recent finds in Mexico are of comparable antiquity. In the extreme southern part of the South American continent, human sites dated by various means indicate that man was there at least 10,000 years BP. Thus, the available evidence suggests that a conservative estimate of man's arrival in South America would be 16,000 to 20,000 years BP.

The material culture of these first people is little known, because they did not leave many traces of their presence in the ecosystems of the region. However, there is general agreement that these people were hunters and gatherers who had developed an impressive stone weapon technology. They hunted large mammals (and probably many small ones as well) over much of the mainland of Latin America. Since it is likely that the physical geography of this region was different from what it is now—because of extensive glaciation in North America and much mountain glaciation in South America—we cannot be certain what sort of beasts were available to these hunters. However, the faunal array undoubtedly differed in many ways from that of the present time.

The emphasis on hunting gave way to a more generalized type of gathering economy. However, hunting remained of primary significance in special faunal situations, such as the grasslands of Argentina and Uruguay, which still supported a hunting society as late as the beginning of the sixteenth century. During the latter part of the hunting and gathering stage, people must have begun to experiment with plant cultivation and keeping animals in captivity. Systems of slash and burn cultivation were developed that were similar to those in other parts of the world. The times and places of these efforts are known only in a general way, except for the domestication of maize (*Zea mays*) in Mexico. It now seems that the faint beginnings of seed crop production occurred as early as 7000 years BP in the area just southeast of present-day Mexico City. But this was not a fully developed system. At this early date, food obtained from cultivated plants was of only limited significance in the human diet. It was not until perhaps 3000 years BP that agriculture assumed major dietary significance, and even then, not necessarily over the entire region in which agriculture was to occur later.

A separate tradition of agriculture may have developed in some coastal valleys of Peru at a somewhat later date, but this is not yet conclusive. There is growing evidence of significant culture exchanges between Mexico and Central America and parts of South America long before the first Europeans arrived on the scene.

In addition to slash and burn cultivation, sedentary farming systems developed what might be characterized as sedentary hoe cultivation. At the time of first recorded European contact, hoe cultivation was developed chiefly in the central Andes and south central Mexico.

There is one area of controversy that shows no sign of being satisfactorily resolved for many years (if ever), that is, whether there were cultural contacts across the Atlantic and Pacific Oceans in pre-Columbian times and, if there were, what influence, if any, they had on the cultural development of this vast region. The available evidence indicates that there were contacts across the Pacific prior to the European entry into Latin America, but it appears that pre-Columbian New World cultures developed without significant influence from transoceanic sources.

In 1492, when Columbus arrived in the Americas, a vast array of cultures and technologies existed in the region. Societies ranged from the simplest (in a material sense) gatherers to complex societies depending on elaborate technologies and social structures for their functioning. Principal examples of the latter, from north to south, were the Aztecs, Mayas, and Incas. The Aztec and Maya civilizations were the principal cultural units forming what has come to be known as Mesoamerica, and the Inca civilization was spreading widely beyond its Andean heartland (see Maps 7.1a, on page 213 and 7.1b on page 214, which show the approximate geographical distributions of the human economies of Latin America around 1492).

The first generally accepted date of European contact with the Americas was 1492, when Cristobal Colón (Christopher Columbus), in command of three ships and experienced Spanish seamen, dropped anchor in a bay on the north shore of the mountainous island that became known as Hispaniola. This began a century of conquest during which a few men from Spain politically subdued most of the region we know today as Latin America. Early in the sixteenth century, Africans, most of which appear to have been from West Africa, were forcibly added to the list of intruders.

Although Spain and Portugal sought to maintain a monopoly on this vast domain and discouraged the entry of other European nations, efforts were soon underway by nationals of England, Netherlands, and France to find a place under the American tropical sun. These efforts first took the form of piracy (as often as not sanctioned by the pirate's government) but later they became more regularized and took the form of establishing settlements in what had previously been Spanish or Portuguese areas of control. England was most successful in the Caribbean region,

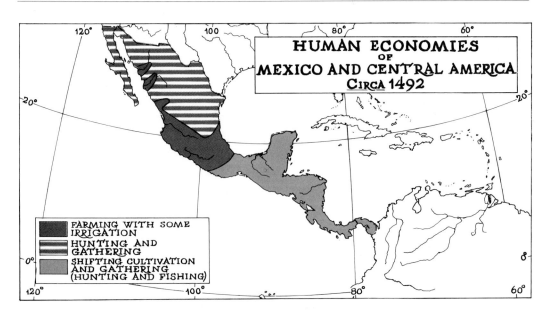

HUMAN ECONOMIES OF MEXICO AND CENTRAL AMERICA CIRCA 1492

FARMING WITH SOME IRRIGATION

HUNTING AND GATHERING

SHIFTING CULTIVATION AND GATHERING (HUNTING AND FISHING)

Map 7.1a

but the other powers plus England ultimately occupied small parts of South America.

United States influence in Latin America began in the nineteenth century. By the middle of that century, the United States had begun a process of territorial acquisition that culminated in 1903 with rights to a canal route in Panama. Of much greater significance, however, was the economic power of America. For many decades, most Latin American nations depended on the United States as a principal market for agricultural, mineral, forest, and marine products, and in turn, a large percentage of Latin American imports originated in the United States.

The Amerinds were a pivotal element in the post-fifteenth century history and development of Latin America. During the colonial period, they provided the labor that was the critical factor in economic development, and they made a major cultural contribution by teaching the colonists about domesticated plants. Moreover, they often produced children with the Europeans, thus melding the genes of both peoples.

At the time of the first recorded European contact, Amerinds lived in all but a few parts of this vast region. There were more than 20 million inhabitants in Mexico, for example, and the tiny isthmus of Panama contained more than 1 million. Parts of the central Andes had major population clusters, and dense clusters also occurred elsewhere on the continent. The Amazon Basin had a comparatively sparse population but at earlier times may have supported dense local clusters of people. The islands of the Caribbean were almost all well populated. The island of Hispaniola is thought to have had not fewer than 600,000 and perhaps as many as 1 million inhabitants when Columbus first landed there.

By the end of the sixteenth century, Amerind populations had declined sharply in almost all parts of Latin America. The factors most responsible for this decline were culture shock resulting from conquest, mistreatment by the invaders, and most important, diseases brought by the Europeans for which the Amerinds had no natural immunity. The lowest point for Amerind populations came, for most of the region, in the middle of the seventeenth century.

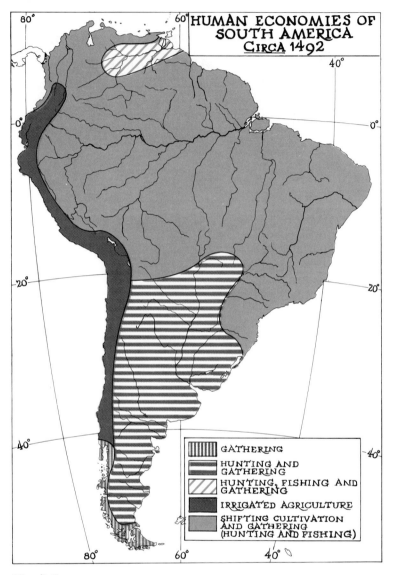

Map 7.1b

No demographic disaster of greater geographic extent has occurred in historic times. The sudden removal of millions of people within little more than a century gave the ecology of much of Latin America a sharp wrench. These people had been the ecological dominants in their ecosystems.

The geographic distribution of the remaining Amerinds was irregular. In some areas, especially several Caribbean islands, the decline amounted to complete or nearly complete disappearance. However, in other areas, particularly in parts of the Andes in Peru and Bolivia, isolated parts of the Amazon basin, Guatemala, and parts of Mexico, large numbers of Amerinds survived. One wonders if the highland habitats had certain features that favored these groups. But perhaps the pattern was purely accidental.

The Indians also survived biologically, as mentioned previously, through the mixing of their genes with those of the invading Europeans. This mixture resulted in the predominant racial type in Latin America today—the *mestizo*. There has also been a genetic exchange between Amerind and African and between African and European (Caucasian). Thus, in the centuries since conquest, there has been a continuing exchange of genes among the people of this region. Even Amerind groups that today are geographically isolated may well have had sexual contacts with non-Amerinds in the past four centuries.

In speaking of Amerind survival, we are referring here to genetic, that is, biological, survival. If there are any genetically pure Indians remaining, they have probably adopted elements of alien cultures.

7.2 Influences on Vegetation

Almost every major biome is represented in Latin America. Thus, the vegetation cover was always extremely complex. The largest remaining area of tropical rainforest occurs here (mostly in Brazil, Fig. 7.1), and in contrast, there are deserts with local areas where rainfall occurs only once in several years. Grasslands, such as the Pampa of Argentina, cover millions of square kilometers, and cloud forests, shrouded in almost perpetual mists, hug slopes high on the windward side of mountains from Mexico to Chile. Some of this complexity and variety is conveyed by Maps 7.2a to 7.2d on pages 216–218, but they can do little more than suggest the ecological complexity of the region. However, wild though much of this vegetation may appear, little of it has escaped some degree of modification by humans. Some landscapes, for as far as the eye can see, were almost wholly created by man.

We know very little about the culture of the first persons who entered this region except that they were hunters and gatherers. However, we may assume that they used fire for various purposes, including hunting (fire drives). And as we noted in earlier chapters, fire applied to a parcel of land, even at intervals of several years, may retard normal succession or inaugurate a new succession.

We know much more about the development of crop plants and agricultural systems. Some of the important plants, now cultivated worldwide, were first domesticated in Latin America. A complete list of these domesticated plants would be quite extensive, but the most important are maize (*Zea mays*), peanut (*Arachis*), manioc (*Manihot*), potato (*Solanum tuberosum*), sweet potato (*Ipomoea batatas*), cacao (*Theobroma*), cotton (*Gossypium*), tomato (*Lycopersicon*), beans (*Phaseolus*), squash (*Cucurbita*), and peppers (*Capsicum*). And of all of

Figure 7.1 *Amazon rainforest seen from the air.*

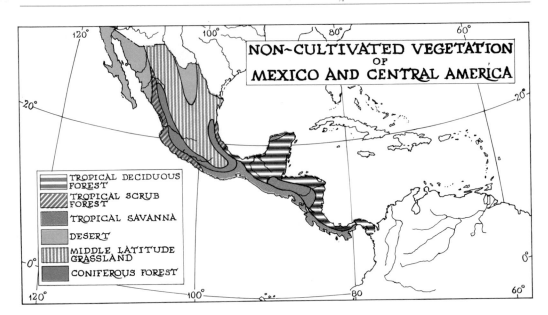

Map 7.2a

these, maize was the most important in terms of the reliance placed on it by farmers around 1492. Some crops important at the time of the Spanish conquest partly gave way to introduced crops. This was the case with certain grains in the Andean highlands, although some Amerind crops still survive there today.

It is useful to divide our discussion of the ecological effects of agriculture into two parts—slash and burn cultivation and sedentary cultivation. Unless indicated otherwise, remarks related to distribution refer to the time when first European contact (and recording) took place, that is, near the beginning of the sixteenth century.

Regardless where the major crop plants

Map 7.2b

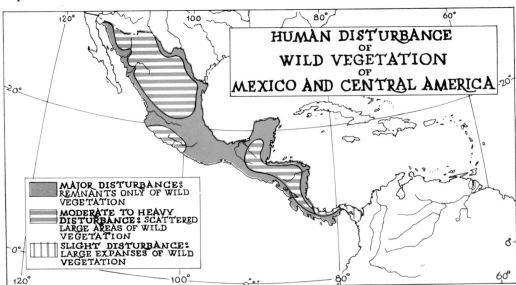

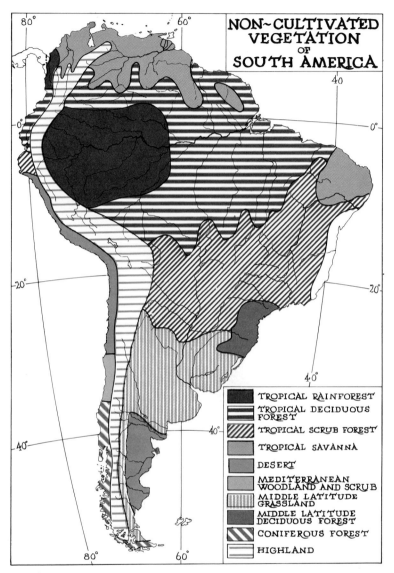

NON-CULTIVATED VEGETATION OF SOUTH AMERICA

■	TROPICAL RAINFOREST
▤	TROPICAL DECIDUOUS FOREST
▨	TROPICAL SCRUB FOREST
▨	TROPICAL SAVANNA
▨	DESERT
▨	MEDITERRANEAN WOODLAND AND SCRUB
▥	MIDDLE LATITUDE GRASSLAND
▨	MIDDLE LATITUDE DECIDUOUS FOREST
▨	CONIFEROUS FOREST
▨	HIGHLAND

Map 7.2c

of shifting or sedentary cultivation were domesticated, they often had a general distribution in Latin America by 1492. The most widely distributed crop, both geographically and ecologically, was maize. It was being grown from south central Chile to well beyond the northern limits of the Latin American region (see Chapter 6). It was grown from sea level to almost 4,000 meters above sea level in the Andes, although at the highest elevation it was of no significance to the human diet. It was

grown under conditions of continuous high atmospheric humidity and under irrigation in desert oases.

Wherever maize or other crops were grown by slash and burn cultivation, the method of land preparation was essentially the same. Stone cutting tools were used to fell trees and bushes, and fires were lighted to burn the plant debris when it became sufficiently dry. By referring again to Maps 7.1a and 7.1b, you can see that slash and burn cultivation had extensive

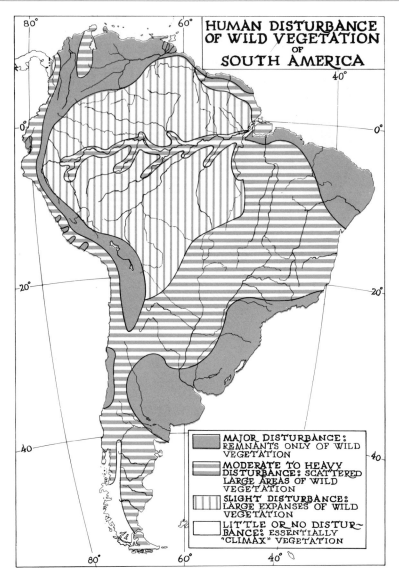

HUMAN DISTURBANCE OF WILD VEGETATION OF SOUTH AMERICA

MAJOR DISTURBANCE: REMNANTS ONLY OF WILD VEGETATION

MODERATE TO HEAVY DISTURBANCE: SCATTERED LARGE AREAS OF WILD VEGETATION

SLIGHT DISTURBANCE: LARGE EXPANSES OF WILD VEGETATION

LITTLE OR NO DISTURBANCE: ESSENTIALLY "CLIMAX" VEGETATION

Map 7.2d

distribution. The slash and burn cultivators were not little pockets of Amerinds barely hanging on in an environment they had not been able to master. They were millions of persons who obtained the bulk of their food every day through their labor on plots of land that they repeatedly cleared, burned, abandoned, and then returned to. This had been the pattern for centuries and, in some areas, for millennia. The question is: to what degree and in what ways did the actions of these cultivators modify the vegetation of the ecosystems they utilized for their crop growing activities?

To obtain answers to this question, we must turn to the documents written at the time of the first European contact and the later reports of geographers, anthropologists, and biologists. Unfortunately, the first Europeans were not trained ecologists so a balanced geographical coverage does not exist—indeed, large areas were not even mentioned ecologically. Thus, we must extrapolate from the sparse available accounts in order to make ecological reconstructions. Looking at Latin America as a whole (but excluding the islands), the region that was reported in the greatest detail was the land contiguous to the Caribbean Sea, that is, mainland Middle America (Mexico through Panama) and northern South America (Colombia and Venezuela). Thus, our discussion refers largely to sixteenth-century documents from that region.

The first permanent settlement on the mainland—the Spanish Main—was established on the Caribbean coast of Panama in 1509. This was only a few years after Columbus, on his fourth voyage to the Americas, sailed along much of that coast in pursuit of the "Indies" and its gold and spices. Columbus' son, Hernando, wrote of what he saw along this coast. It was very different from the way it is today. Along most of the coast, or at least wherever the ships stopped for exploration or supplies, were Indians (Amerinds) and fields of maize. Hernando made almost no mention of forests or trees but frequently mentioned the cultivated fields and the large number of inhabitants. Today, much of this area is forested and is sparsely inhabited.

The next useful report was provided by Vasco Nuñez de Balboa, who we have inaccurately been told, "discovered" the Pacific Ocean. However, he was probably the first European to *see* the east Pacific Ocean. Those portions of his account that relate his crossing of the isthmus of Panama describe an environment that is markedly different from conditions there now. He crossed the isthmus from north to south, but at no place in his account does he draw attention to forests or to any difficulty of passing through the vegetation. When he arrived at the principal river that drains this part of Panama, the Río Chucunaque (which flows into the Pacific) he did not attempt to embark in rafts or canoes to avoid the overland journey. Rather, he kept on his way, overland, until he sighted the South Sea, which he claimed for Spain. He then journeyed westward, again overland, to another major river (Río Bayano, today). Using broad paths (*caminos*, in his account). he moved with ease up the river valley from one Indian village to another, passing the intervening cultivated fields. Again, he made no mention of forest. Throughout his journey, Balboa passed through a vegetation landscape, opened and kept open by thousands of shifting cultivators. At some undetermined time in the past these cultivators had removed the climax forest and produced a new set of ecosystems consisting of cultivated fields (cleared by annual fires) and second-growth vegetation in various seral stages. Today, much of this area is so forested as to make an overland journey extremely difficult and is now regarded as a wilderness.

One more account is pertinent to our investigation. In 1516 a soldier was given orders to make a reconnaissance of the western part of the isthmus. Fortunately, he was not only a good observer but a literate one. He left an account of his observations and in some respects his remarks are even more instructive than are those of Balboa. This man and a group of other soldiers rode horses down the Río Bayano valley. Today, no one would attempt such a feat because the heavy forest would make it nearly impossible and there would be almost no grass to feed the horses. But the soldier and his troops traveled through this area and out to the west, passing over land covered mostly by grasses. The pastures struck him as being so fine that he compared them favorably to those of Castile. Even making allowance for homesickness, one must conclude that this was a grassy nonforested area, not even covered by second-growth brush to any important extent. The soldier continued onward into western Panama after which he and his group turned around and returned.

Gonzalo Fernandez de Oviedo y Valdez,

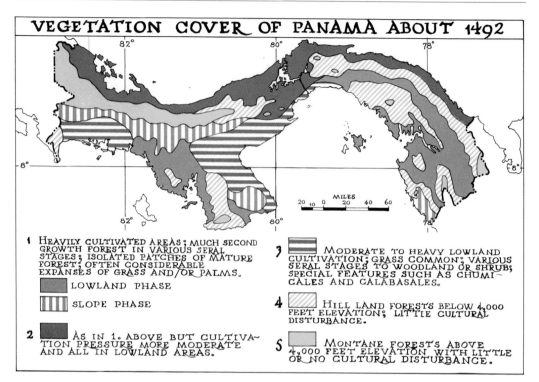

1 HEAVILY CULTIVATED AREAS; MUCH SECOND GROWTH FOREST IN VARIOUS SERAL STAGES; ISOLATED PATCHES OF MATURE FOREST; OFTEN CONSIDERABLE EXPANSES OF GRASS AND/OR PALMS.

LOWLAND PHASE

SLOPE PHASE

2 AS IN 1. ABOVE BUT CULTIVATION PRESSURE MORE MODERATE AND ALL IN LOWLAND AREAS.

3 MODERATE TO HEAVY LOWLAND CULTIVATION; GRASS COMMON; VARIOUS SERAL STAGES TO WOODLAND OR SHRUB; SPECIAL FEATURES SUCH AS CHUMICALES AND CALABASALES.

4 HILL LAND FORESTS BELOW 4,000 FEET ELEVATION; LITTLE CULTURAL DISTURBANCE.

5 MONTANE FORESTS ABOVE 4,000 FEET ELEVATION WITH LITTLE OR NO CULTURAL DISTURBANCE.

Map 7.3

a Crown agent, was the foremost chronicler of the sixteenth century, and to him we owe much of what we know about the population of this and other parts of Latin America. Oviedo, in speaking of the Indian population of Panama, said he thought that there were 2 million or that "they were innumerable." Obviously, he did not take a head count, but he was a careful observer and a restrained reporter.

A reconstruction of how the vegetation may have looked in Panama at that time is presented in Map 7.3, above which will become more interesting if you compare it with Map 7.4 on page 221 which shows how the vegetation was distributed in 1963.

Wherever the Spaniards went in Central America and Mexico, they encountered large numbers of Indians, although some regions were less densely occupied than others. It appears that Panama may have been one of the more densely settled areas of mainland Middle America, excluding the southern Mexican highlands.

The reports of the first Spaniards in areas other than Panama often included comments about grasslands and the Indians' practice of setting fire to them every year. One study, which will be discussed in greater detail later in the chapter, showed that Indian-set fires were a major factor controlling the extensive grasslands in the center of Honduras.

As Map 7.2a, which shows wild vegetation, indicates, there are extensive pine-covered areas in Central America and Mexico. Note that pines do not extend south of southern Nicaragua and that some of the pine distribution occurs within humid tropical climates (Fig. 7.2). Recent studies indicate that part of this distribution was influenced by man through the practice of setting fires in the area. Fire often plays a significant role in the ecology of the genus *Pinus* by favoring pine seedlings over those of broadleaf tree species. An interesting feature of some of the hardwood forest lands adjacent to the pine

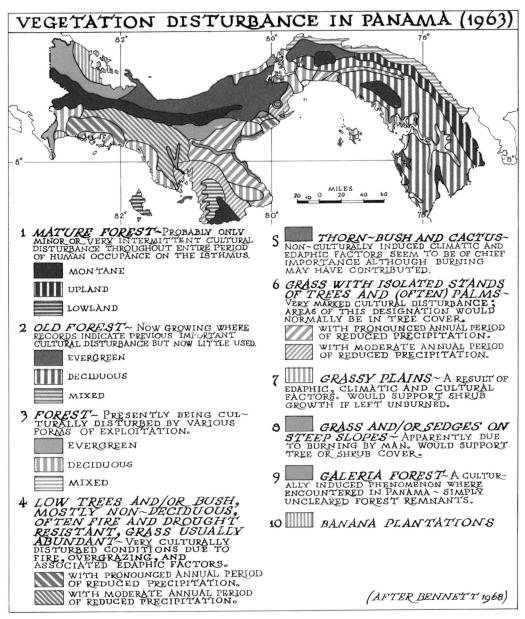

VEGETATION DISTURBANCE IN PANAMA (1963)

1 MATURE FOREST~PROBABLY ONLY MINOR OR VERY INTERMITTENT CULTURAL DISTURBANCE THROUGHOUT ENTIRE PERIOD OF HUMAN OCCUPANCE ON THE ISTHMUS.

- MONTANE
- UPLAND
- LOWLAND

2 OLD FOREST~ NOW GROWING WHERE RECORDS INDICATE PREVIOUS IMPORTANT CULTURAL DISTURBANCE BUT NOW LITTLE USED.

- EVERGREEN
- DECIDUOUS
- MIXED

3 FOREST~ PRESENTLY BEING CULTURALLY DISTURBED BY VARIOUS FORMS OF EXPLOITATION.

- EVERGREEN
- DECIDUOUS
- MIXED

4 LOW TREES AND/OR BUSH, MOSTLY NON~DECIDUOUS, OFTEN FIRE AND DROUGHT RESISTANT, GRASS USUALLY ABUNDANT~VERY CULTURALLY DISTURBED CONDITIONS DUE TO FIRE, OVERGRAZING, AND ASSOCIATED EDAPHIC FACTORS.

- WITH PRONOUNCED ANNUAL PERIOD OF REDUCED PRECIPITATION.
- WITH MODERATE ANNUAL PERIOD OF REDUCED PRECIPITATION.

5 THORN~BUSH AND CACTUS~NON~CULTURALLY INDUCED CLIMATIC AND EDAPHIC FACTORS SEEM TO BE OF CHIEF IMPORTANCE ALTHOUGH BURNING MAY HAVE CONTRIBUTED.

6 GRASS WITH ISOLATED STANDS OF TREES AND (OFTEN) PALMS~VERY MARKED CULTURAL DISTURBANCE; AREAS OF THIS DESIGNATION WOULD NORMALLY BE IN TREE COVER.

- WITH PRONOUNCED ANNUAL PERIOD OF REDUCED PRECIPITATION.
- WITH MODERATE ANNUAL PERIOD OF REDUCED PRECIPITATION.

7 GRASSY PLAINS~ A RESULT OF EDAPHIC, CLIMATIC AND CULTURAL FACTORS. WOULD SUPPORT SHRUB GROWTH IF LEFT UNBURNED.

8 GRASS AND/OR SEDGES ON STEEP SLOPES~ APPARENTLY DUE TO BURNING BY MAN. WOULD SUPPORT TREE OR SHRUB COVER.

9 GALERIA FOREST~A CULTURALLY INDUCED PHENOMENON WHERE ENCOUNTERED IN PANAMA ~ SIMPLY UNCLEARED FOREST REMNANTS.

10 BANANA PLANTATIONS

(AFTER BENNETT 1968)

Map 7.4

lands of Central America is that they contain partly buried pine tree stumps that have resisted rotting and termites because of their resin content. These are sometimes sought for fuel by the Indians living in the region. The stumps are evidence that shows that areas in the pine region that have had a reduction in human population have less fire pressure and tend to return to hardwood forest. To some extent, then, the pine forests of Central America and parts of Mexico are climaxes maintained by human-set fire.

Pines are not the only trees that appear to be favored by human-caused disturbances in this region. There are unusual concentrations of certain tree species in places that have had a long history of

Figure 7.2 *Pine forest, Honduras.*

human occupance and burning for cultivation. For example, in parts of the Yucatán peninsula, there are well recognized plant associations associated with former land use by shifting cultivators. Among these is the *guarumal,* which is a concentration of the guarumo, *Cecropia* spp., a softwood tree that is an early successional element in abandoned fields all over tropical America. Ordinarily, the guarumo occurs in limited numbers in any given place and is found mainly on riverbanks and in the forest where a tree was blown down and sunlight was able to penetrate to the ground. But a guarumal may cover many hectares and persist for a couple of decades. Of particular interest, because of their commercial significance, are mahogany (*Swietenia macrophylla*) and tropical cedar (*Cedrela mexicana*). Usually, both these trees are widely scattered elements in the tropical broadleaf deciduous forest of Central America, but in some areas that have a history of slash and burn cultivation and later abandonment, these trees attain such densities as to be easily recognized as unique vegetation associations and attractive for timber exploitation.

We generally know less about the ecology of shifting cultivators in South America than we do of those in Middle America. However, as the studies of South America accumulate, the same general picture is emerging. It is clear that shifting cultivators have created or extended savannas and continue to modify the vegetation of the continent. One recent study showed that shifting cultivators were able to open up and keep open large tracts of formerly forested lands in northern Colombia. Moreover, the accounts of some travelers tell of an indiscriminate use of fire on the part of shifting cultivators— they ignited grasslands just to see them burn even though they were not able to farm them.

It is often assumed that shifting cultivators had no effect on the vegetation of the Amazon basin, but this assumption rests almost entirely on the untested hypothesis that this region was not a suitable habitat for man. There are great areas of savanna and islands of grass and bush in the basin of the Amazon that await investigation. Assumptions about the past human demography of the basin seem to be contradicted by reports of "black earth" found in sites that appear to have had dense human populations. There is no reason to believe that much of the basin was any more difficult for shifting cultivators to exploit than were similar ecosystems in Asia that have long histories of human settlement. As climatic data for the basin accumulate, it becomes increasingly evident that large parts of the region fall into the tropical monsoon climate (Map 1.2 on pgs 22–23) which, in many respects, is ideal for shifting cultivation. The Amazon basin may contain some interesting cultural surprises.

On the southern periphery of the basin lies a great area of mixed trees, shrubs, and grasslands which Brazilian geographers have divided into many categories such as cerrado, caatinga, and pantanal. The abundance of names reflects the complex physical appearance of the vegetation of this region. Therefore, to make generalizations about the ecological factors that control their occurrence is difficult, if not impossible. There has been considerable controversy over the years about which factors are chiefly responsible for producing and maintaining these plant formations. It is clear that soils and climate play important roles. It is also clear that fire set by man played a role in the *total* ecosystems and continues to do so.

In the northeast of Brazil, extensive areas are dominated by palms (some of which are economically significant) and it

has been shown recently that these areas are influenced by fire and tend to expand in area under a fire regime. Thus, the unusual concentrations of palms may have begun when shifting cultivation was the principal land-use of the region.

Palm savannas are found over almost all of tropical Latin America and are also partly the result of fire, because palms survive fire much more easily than other trees and shrubs. In addition, it was a widespread practice among the shifting cultivators not to cut down palms when they cleared a plot because the trees provided food, shade, construction material, and sap for wine making.

In parts of southern Brazil and a little beyond, are found large stands of the Brazilian pine (*Araucaria;* Fig. 7.3). This is not a true pine but has been given its name because of a superficial resemblance to trees belonging to the genus *Pinus.* The ecology of these valuable timber areas is not understood, but fire appears to be important. Thus, when this area was under shifting cultivation, the *Araucaria* may have been one of the important successional tree genera after a cultivated plot was abandoned.

The detailed effects of pre-Columbian sedentary farmers on the wild vegetation is so far little known. In the Mexican region, a picture emerges from the literature of

Figure 7.3 Araucaria *forest and lumber mill in southern Brazil, about 1925.*

greatly modified landscapes in the areas where people were settled (as one would expect), but it appears that much of the forest that was located at some distance from settlements remained. It should be kept in mind that the dense human population reported in the early sixteenth century for the basin of Mexico and adjacent regions of sedentary farming required a large supply of fuel for cooking and heating—probably thousands of metric tons of wood each year.

In the Andean region, the human population was large, although there is still no agreement about the number of persons living there at the time of conquest. One recent estimate places the human population of the central Andes at the time of conquest at about 4.6 million persons. Since the populations were clustered in Andean basins, the ecological densities were high. In addition to being sedentary hoe cultivators, the people kept thousands of llamas and alpacas. There is reason to believe that high human population densities existed in parts of the Andes for many centuries before the Spanish conquest of the area. What changes did these people impose on the vegetation?

Unfortunately, we can only answer tentatively. It has long been accepted that the treeless expanses of the inhabited parts of the Andes (Fig. 7.4) were natural in that they were simply responses to the climate and soil. However, there is evidence to contradict this belief. It has been reported that trees were growing, at least in the early part of this century in spots too steep to farm; that trees imported into much of the area grow without difficulty; that there has recently been progress in introducing exotic trees (*Eucalyptus*) that grow very well without special care; and that fruit trees introduced since the conquest are grown without special care. In addition, there are repeated allusions in the literature to the use of animal dung for fuel. The fuel needs of those dense human populations must have been enormous and probably led to the widespread use of animal dung for heating and cooking after the wood fuel source was all but destroyed. Therefore, although the picture is fragmentary at present, it appears that long before the Spaniards arrived in the Andes, man had greatly modified the trees and shrub vegetation.

Before proceeding to the European period it will be useful to summarize. All available evidence indicates that shifting cultivators were able to alter the wild vegetation over immense parts of Latin

Figure 7.4 *Indians farming at approximately 4000 meters elevation in the Andes.*

America in pre-Columbian times. The alterations took many forms, of which the more important were the creation and/or extension of grasslands; the creation of special plant associations; the creation of complex vegetational mosaics composed of mature forests, shrubs, grasses, and second-growth forests and woodlands. The total areal extent of these changes is still unknown, but few regions of the tropical lowlands and agriculturally suitable uplands did not feel the ecological brunt of this economy. The Spaniards, for the most part, entered a world that man had been modifying for thousands of years. In the more limited areas of sedentary hoe cultivation, the local vegetation changes were undoubtedly profound and long-lasting. This is especially true in the Andes, where the ecological impact of herding must be added to that of agriculture. In short, by 1492 the vegetation cover had become, to greater or lesser degrees, artifacts of human use and misuse.

As noted earlier in this chapter, there was a drastic decline in Amerind numbers in the 150 years following their first contact with Europeans. This resulted in vast tracts of land being almost completely vacated by humans and the attendant relaxation of ecological pressures that had been exerted for many centuries. Decade by decade, forest reclaimed the lands that once knew the ring of the stone ax and the roar of human-set flames. All but the most durable evidence of human occupancy passed from the scene.

When the Spaniards arrived, they quickly introduced a great host of domesticated plants and animals and different agricultural systems that involved a new complex of tools and technologies. Most important to the wild vegetation cover of Latin America and to the general ecology of this region was the introduction of cattle, sheep, horses, burros, and goats and the introduction of plow cultivation.

We still do not have adequate knowledge of the ecological shock caused by the introduction and proliferation of livestock into much of Latin America. By all accounts, the numbers of these animals grew so fast that great herds were reported on the Caribbean islands and in Mexico, Argentina, and elsewhere within a few years. The cultural consequences were great, and major ecological dislocations occurred, particularly in parts of central Mexico in the sixteenth century. The most immediate effect on the vegetation was caused by burning to improve pasturage. But in addition, heavy grazing removed the grass cover from many areas, which induced accelerated soil erosion. Goats did not become economically important everywhere, but where they did, they had the same effect on the land that they had in the Mediterranean region (see Chapter 4). Nowhere was this more pronounced than in Venezuela, where damage to the shrub cover caused great areas of hill slopes to be laid bare. Sheep became a problem in parts of Mexico and in Southern South America because their numbers were often too great for the carrying capacity of the native pastures. The result was destruction of the wild vegetation. Moreover, as we mentioned earlier, grazing and browsing mammals tend to be selective feeders. Thus, those plants that were not favored increased while the exploited species decreased in relative abundance. This has been occurring on thousands of square kilometers in Latin America for the centuries since conquest, but the details are as yet little known.

The principal effect of the plow on vegetation was to extend the area under sedentary cultivation and thus to extend the areas over which wild vegetation had been almost completely suppressed.

The Europeans also introduced charcoal manufacture (Fig. 7.5). There is no evidence that Indians in pre-Columbian America deliberately made charcoal, but it was a major fuel in sixteenth-century

Figure 7.5 *Hardwood cut and stacked before being partially covered with earth and ignited to produce charcoal.*

Europe because it was the only significant improvement over firewood then available. Thus, the Spaniards in Latin America cut large quantities of hardwood and turned them into charcoal—a practice that is still followed in many parts of the region.

In the sixteenth century, the discovery of mines in Mexico and the Andean region rapidly created an enormous demand for timbers for mine shafts and charcoal fuel for smelters. A widespread removal of trees quickly followed in those areas where mines were located. Today, much of the treeless conditions of the older silver mining regions of Mexico, for example, may be attributed to the miners.

Curiously, logging for construction timber, dye wood, and tannin was a very localized activity for many years and did not have much impact on the vegetation in the sixteenth and seventeenth centuries. However, exceptions that might be noted are logwood (*Haematoxylum campechianum*), a tree found in the Yucatán peninsula that was once greatly desired as a source of black dye material, and brazilwood (*Caesalpinia*), which, appropriately, was exploited in Brazil and was desired for the red dye it produced. In addition, many trees were cut for their bark, which was used for tannin, and many trees provided construction materials, but this logging was not on a scale of major ecological significance.

Up to the latter part of the nineteenth

century there was a growing tendency to establish plantation-type agriculture which focused on whatever native or exotic crop promised high economic returns. In the early years, the areal extent of the plantations was usually limited, and most of the important plantations were developed on the Caribbean islands. On the mainland, the plantations were concentrated along the coast of Brazil, south of Recife. There, cotton, tobacco, indigo, sugar cane, and other cash crops were grown. The great inland coffee plantings came later. In Mexico, a variety of plants were grown commercially, but they seldom reached a level that would have importantly affected the wild vegetation.

With independence from Spain in the early nineteenth century, new economic and hence new ecological forces were loosed on the region. These had and are still having effects on the vegetation cover. For example, the increased attention given to the production of livestock, particularly cattle, has had two principal effects on the vegetation. It extended the pasture areas at the expense of tree-covered land, and it fostered the introduction and wide diffusion of exotic pasture grasses drawn chiefly from Africa.

Deforestation associated with cattle is found chiefly in the tropical monsoon and tropical savanna climatic regions (Map 1.2 on pages 22–23). Although accurate figures of the areal extent are not available, it is known that large areas were cleared of trees in the past three decades to make way for the raising of cattle. Nowhere is this more evident than in parts of Central America and Panama, but similar changes have been taking place in Colombia and Venezuela as well. After the trees are removed, they are frequently replaced by grasses of African origin belonging to the genus *Panicum,* which are usually introduced to Latin America by way of Brazil. Another grass, known variously as *faragua* or *jaraguá (Hyparrhenia rufa;* Fig. 7.6) has been so widely dispersed by man in the last thirty years that it now is one of the most important pasture grasses in tropical savanna regions from Mexico to Argentina. There is disagreement as to its geographic origin. Some authors assume a Brazilian origin and others, an African origin. Whatever the case, faragua is now widely and firmly established far beyond the borders of Brazil. It is so aggressive that, once established, it is all but impossible for tree or bush species to reinvade the pastureland. In the last few decades, with the aid of man, it has taken over much of the grasslands of Central America. From a commercial point of view, this is advantageous because the native grasses tend to be low in nutrition for livestock and tend to die back during the dry season. Faragua

Figure 7.6 *Jaraguá* (Hyparrhenia rufa) *pasture, Venezuela.*

thrives under conditions of annual drought and human-set fire. The *Panicum* grasses tend to require more moisture than faragua but offer better nutrition for livestock. Some of them have spread over and now dominate, at least visually, vast areas in the tropical monsoon climates of Latin America.

This does not exhaust the list of forage crops introduced by man into Latin America. For example, in the southern part of South America, the most dramatic and economically significant introduction was alfalfa, also known as lucerne (*Medicago sativa* and other species), a native of Europe (Fig. 7.7). This excellent pasture plant may have been introduced into Argentina as early as the eighteenth century. However, it did not achieve significance until the nineteenth century, when political independence opened the way for the development of a major livestock industry. Since 1890, alfalfa has been the best of the Pampa pastures. Its spread was encouraged by the holders of large landed estates by their insistence that after tenant farmers used the land for a stipulated period they must plant it to alfalfa. This legume was so successful that during its peak, in 1918–1919, it covered approximately 8.7 million hectares. (The present area is smaller, possibly around 5 million hectares.) Alfalfa remains the single most important pasture plant in the country, although others have been introduced and there are many native species as well.

One result of grass introduction is that today the pasture areas of almost all of Latin America are complex collections of native and exotic species. And these grasses have formed new ecosystems whose interactions are only partly understood. So great has the African contribution been to the pastoral scene that a geographer, James Parsons, has recently described the process as the Africanization of the grasslands of Latin America.

Nonnative crop plants were also introduced into Latin America. These in-

Figure 7.7 *Alfalfa in Argentina.*

troductions began as early as the second voyage of Columbus and have continued into the present time. Among the earliest were bananas, rice, sugar cane, and the major European grain crops—wheat and barley. Later, plants carried directly from Asia including the mango tree (*Mangifera indica*), certain citrus, and perhaps some further varieties of bananas and rice were introduced. Coffee was introduced into Haiti in the eighteenth century. Only a few crop plants from subsaharan Africa came in, but among them were watermelon (*Citrullus*) and okra (*Abelmoschus*). One tree crop, the breadfruit (*Artocarpus*) was introduced by the British into the West Indies to provide an inexpensive staple food for slaves. Breadfruit never became important in this region but the tree was later carried widely about the mainland of Latin America. Today it and the mango are widely established and are often found growing without human aid.

The establishment of political independence of many Latin American nations in the nineteenth century stimulated the livestock industry and greatly influenced the development of large-scale commercial ag-

ricultural enterprises—the plantations. The crops associated with plantations have changed from time to time during the past 175 years—often in response to changing world-market opportunities. Those that emerged as being the most significant are bananas, sugar cane, coffee, and cotton. Nations based their entire economy on one or another of these crops and thus became enmeshed in the difficulties of crop monoculture. Our concern here is the influence these activities had on the vegetation cover.

The principal and most obvious effect was the substitution of the commercial crop for the wild vegetation, usually woodlands or forests. The replacement is especially complete where sugar cane, bananas and cotton are grown. To cultivate these crops, it is the practice to remove virtually all the existing tree cover and then to maintain highly artificial ecosystems. In the earlier days, before the use of fertilizer and other modern techniques, soils were often soon worn out and fields were abandoned to grow back to a dense second growth of trees and shrubs. Today, with land more expensive and scarce, the tendency is to manage it in a way that precludes the reinvasion of woody vegetation.

Coffee culture may promote a variety of vegetation changes, depending on what region is involved. In the highland coffee areas of Colombia and Costa Rica, for example, it has long been the practice to retain a certain number of wild trees in the plantation areas in order to provide shade for the coffee trees. However, in Brazil, where a large quantity of coffee is produced, the practice is to grow the coffee trees without shade so the wild vegetation is almost completely eliminated. Coffee plantations cover large tracts and dominate the landscape of most of El Salvador, the central part of Costa Rica, parts of the highlands of central Colombia, and large areas in Brazil, particularly in the states of Paraná and São Paulo.

Banana plantation ecosystems are generally distributed in the Caribbean slope and parts of the Pacific slope of Central America and large areas in Mexico. In the past few decades, the Guayas lowlands of Ecuador emerged as the largest banana-producing area in the American tropics, although recently its production has declined because of disease.

Cotton (*Gossypium*) was domesticated in both the Old World and the New in pre-Columbian time. When the first Euro-peans arrived in America, they found that cotton was grown in widely scattered areas of Latin America and was woven into textiles. Until the invention of the cotton gin, however, cotton was of little commercial importance, and even afterward, most Latin American countries found it difficult to compete with the United States. For a time, during the United States Civil War, cotton was grown in some quantity in Latin America, particularly in coastal Brazil where there were established plantations that had been planted to tobacco and indigo (*Indigofera*). However, with the close of hostilities, United States production soon relegated the Brazilian effort to a position of little significance.

After World War II, it became profitable for some Latin American nations to cultivate cotton for export and to develop textile markets within the region. The largest areas of cotton growing are in Mexico and Central America (Fig. 7.8) although several South American countries also have significant cotton plantations. In Peru, for example, a fine grade of cotton is grown in irrigated fields, where the predecessors of the Incas also farmed this crop.

One of the most obvious changes in vegetation caused by cotton cultivation is the creation of a simplified ecosystem in which no other plant is tolerated. But another

Figure 7.8 *Cotton field at harvest time, Nicaragua.*

ecological effect is caused by the use of insecticides, especially the chlorinated hydrocarbons, although they do not act on wild vegetation. Cotton is particularly subject to the depredations of insects and it has become a general practice in Latin America to use heavy dosages of DDT and similar persistent insecticides. (However, because almost no official attention has been given to this practice, there are few data available to indicate the quantitative ecological effects.) The input of such insecticides into the cotton fields of Central America has been so great that the ecology of the cotton field soils has been seriously dislocated and probably the ecology of adjacent areas as well. There is no question that the threat to the health of some agroecosystems (agricultural ecosystems) is great and will become more apparent in the years ahead.

Cotton is not the only crop to which insecticides or other chemical agents are applied. Every plantation crop in the region is treated this way, but as yet we have little idea of the effects this has on the short- and long-term functioning of the affected ecosystems.

Thus, one economic activity or another has often produced a negative effect on the wild tree cover. But in addition to economic activities, forest fires and deforestation to allow for the spread of farming families have also destroyed the wild tree cover.

Forest fire is a problem chiefly in the parts of Latin America where long seasonal droughts occur. In recent years, some of the most serious fires occurred in the state of Paraná, Brazil. In 1963, after a prolonged drought that was followed by a late frost, the vegetation became extremely dry. When the farmers in the region set their annual spring fires to clear off unwanted vegetation, the result was a holocaust that destroyed much of the vegetation cover on approximately 2 million hectares of land.

From the earliest days of the colonial era to the present, many parts of the region lacked adequate roads. Since World War II, however, most Latin American nations have been building roads to stimulate economic development and to foster the well-being of their inhabitants. The benefits are evident almost everywhere, but the construction has frequently stimulated ecological changes that reduced the economic benefits.

Chief among these changes is vegetation destruction caused by formerly landless farmers. All too often the pioneers in areas newly opened by roads are poor folk who lack the capital and technical knowledge to exploit the land correctly. For example, for hundreds of years following the Spanish settlement in Costa Rica, the area between the capital city and Panama was not traversed by an all-season road. Population clusters that developed in this region were isolated from the capital and had difficulty in getting access to markets. A road (Fig. 7.9) was recently completed through this area, and the economic and social improvements are significant. However, the road also allowed poor farmers and their families to settle near it on small plots of land. These farmers proceeded to clear their land of wild vegetation and to plant it to subsistence crops. But much of this land is in steep slopes and receives torrential rainfall during parts of the year. At the middle elevations (1500 to 2000 meters), fine oak forests were cut, and the wood that was not used for cooking and heating fuel was burned in the fields. Without supporting deep roots, the exposed surfaces quickly gullied and soil loss became general. Even if such lands were abandoned, it would be a long time before a protective tree cover could become established.

In other cases, roads are built to open new regions for settlement. The ecological effects of these endeavors are little understood, but such construction is always accompanied by changes in the wild vegetation. A notable recent example is the TransAmazon Highway in Brazil, which is

Figure 7.9 *Land cleared of forest adjacent to the Pan American highway in southern Costa Rica.*

being pushed into a region where few people live. Although it is still too early to identify all the implications of this project, many ecologists believe that it threatens the ecological stability of a large part of the region.

Three further aspects of man's impact on the vegetation of Latin America must be mentioned: (1) the commercial exploitation of trees for lumber, (2) the continued major exploitation of wood for fuel, including charcoal, and (3) the introduction of exotic trees for the production of raw materials.

Major timber exploitation for lumber is a relatively recent phenomenon in Latin America, although small-scale exploitation has existed from the early colonial period. Except for some special situations, the forests of Latin America are mixed hardwoods, that is, many species occur within a given unit area; this makes commercial exploitation difficult. Only a few of the hardwood species can be marketed outside local areas, so exploitation is usually concentrated on those few species, principal among which are mahogany (*Swietenia*) and tropical cedar (*Cedrela*). These trees and a limited number of other valuable species have been "mined" out of many areas where they formerly were abundant. With the development of the plywood industry in recent years, more hardwood species have been exploited for veneers,

but as yet this has had only minimal effect on the forests.

Softwoods are rare over most of Latin America. It has been noted that pines (*Pinus*) do not occur naturally south of Nicaragua, and only a few softwood species found south of Central America have useful properties. Probably the most important of those few is the so-called Paraná pine (*Araucaria*), which is found chiefly in parts of southern Brazil. The wood of this tree is extremely useful, and therefore, the tree has been overexploited. In addition, many fires occur in the region where Paraná pines grow, so the tree's range is being markedly reduced (although fire is a part of the tree's ecology).

Pine (*Pinus*) exploitation in Central America and Mexico is frequently done in such a manner as to seriously injure the ecosystems. In Mexico, in spite of enlightened forestry laws, there is still a serious problem with logging practices that result in soil destruction and general impoverishment of the logged-over ecosystems. In addition, in some instances after an area is cut and the loggers leave, poverty-stricken squatter farmers take advantage of the cleared land. They plant a few crops and then, having added to the ecological destruction, move on to another area. The human pressure on Mexico's highland forests cannot be overstated. It is only with the greatest difficulty that the Mexican government is able to prevent the total destruction of the vegetation in many of these ecosystems.

In an age when fossil fuels are burned so much that the atmosphere of the entire globe is affected, the fact that man still burns enormous quantities of wood for fuel may be overlooked. But the intensive use of fossil fuel is still highly regionalized. In many parts of the world, including Latin America, large numbers of people find it economically necessary to depend chiefly on wood and its derivative, charcoal. The result is that millions of metric tons of wood are consumed each year and the drawdown on the wood reserves adjacent to population centers is obvious.

In addition to the people's use of wood for domestic purposes, the mining industry is a major user of charcoal, particularly in parts of Brazil. For example, it was recently reported that in the state of Minas Gerais, the smelters needed approximately 900,000 tons of charcoal annually. This hunger for charcoal in Brazil's iron industry (coal is scarce in Brazil) has resulted in the removal of the woody vegetation from enormous areas. But it is probably the *domestic* demand for charcoal that is presently having the greater effect on vegetation over most of the Latin American region. Charcoal vending is such a humble feature of so many Latin American cultures that it is usually overlooked even by those who collect economic data. In charcoal making, hardwood trees are used almost exclusively, and near the coasts, these include certain trees belonging to mangrove ecosystems. In fact, it is said that some mangrove species, particularly the red mangrove tree (*Rhizophora mangle*) yield the best charcoal. Thus, these ecosystems have often been much reduced or, in some areas, virtually destroyed.

The lack of trees in parts of Latin America has led to the importing of exotic species to obtain the wood required for lumber, fuel, and other purposes. *Eucalyptus* from Australia has been most impor-

tant in this practice. Many landscapes in South America, not so long ago completely devoid of tree cover, are now shaded by these trees, which have been carried far and wide by man in the last century. Argentina probably has the largest area set out in *Eucalyptus,* but extensive growing stands are also found in Uruguay, Chile, and the highlands of Peru, Ecuador, and Colombia. Pines (*Pinus*) have also been widely introduced into South America. Again, it is Argentina that has the largest area planted. One of the most interesting aspects of this introduction and establishment of trees is that they are often found growing in areas, such as the pampa and middle elevations in the Andes, where it was once believed trees could not grow. This is relevant to questions about pre-Columbian human influences on the wild vegetation of areas that were treeless at the time of European conquest.

In addition, there has been much experimentation over the years by governments and private individuals to test the wood-producing potential of exotic trees from Africa and Asia. It is hoped to find trees that produce wood in economically useful quantities. Most of these efforts, however, have not yet resulted in significant alterations of the Latin American landscapes, even though the list of plants involved is fairly long.

This brief review of human influences on the vegetation of Latin America has shown that man has been influencing the plant cover at least since the establishment of agriculture several thousand years before the arrival of Europeans—perhaps even before agriculture. The arrival of the Europeans marked the opening of a new chapter in the plant geography of Latin America, and the influence of man since the early sixteenth century has become ever more complex. In the past 100 years, particularly, that influence has increased because of the growth of economic activity and the rapidly increasing human population. The picture that emerges is of a vast region whose vegetation cover has been modified by human actions over a long period of time and in which one cannot describe any vegetation complex as pristine, in the sense that man has not exerted some influence on it.

7.3 Influences on Animal Life

With the exception of parts of Mexico, Latin America comprises one of the major zoogeographical units of the world, the Neotropical region. The richness and diversity of the animal life in this area is unsurpassed anywhere. Moreover, the region contains many unique elements, such as the only living primates with prehensile tails. The Amazonian part of the region is justifiably famous for the magnificent collection of birds it contains, a large portion of which is not found outside the region. Although lacking the great herds of mammalian herbivores such as in Africa, the Neotropical region contains a collection of mammals that for species diversity is among the most unusual in the world.

Zoogeographers recognize various subregions, some of which we will use in our discussion. Map 1.4 on page 28 shows the location of the Neotropical region.

The first people we have knowledge of in Latin America were hunters and gatherers who appeared on the scene at least 20,000 years ago. As in the case of North America, we can only conjecture about the effects these people had on the wild animals they took for food and raw materials. Moreover, we do not even have a clear picture of what the larger mammals were that man hunted in Latin America. We are certain that he hunted now-extinct mammoths in Mexico and parts of northern South America, but we do not know to what degree, if any, he specialized on this animal. We also do not know if the hunting pressure contributed to the disappearance of these great proboscideans in Latin America. Until we have more precise knowledge of human numbers, hunting techniques, food requirements, and the size of the exploited animal species population, we cannot construct useful ecological models. We can only observe that early man exerted some influence on the number and distribution of the animals he hunted most and that, through the use of fire, he may have created or extended ecological conditions favorable to some species and unfavorable to others.

Our discussion can best begin, then, with the advent of agriculture. Associated with the development of agriculture in some places in Latin America was the domestication of animals, but animal domestication never achieved the level of cultural and genetic complexity as it did in Eurasia. The principal known domestications and the often assumed region of origin are shown in Table 7.1. Two regions stand out—the Andes and Mesoamerica (southern Mexico to Costa Rica). Alpacas, llamas, and guinea pigs (*Cavia*) were domesticated in the Andes; turkeys (*Meleagris gallopavo*) and certain stingless bees (family Meliponinae—the more important genera being *Melipona* and *Trigona*) were domesticated in Mesoamerica. To this modest assemblage may be added the muscovy duck (*Cairina*) which may have been domesticated in northern South America. Not shown in Table 7.1 are the many animals which were kept in early stages of animal domestications. There are many records of Amerinds keeping deer (*Odocoileus* and *Mazama,* principally), tapirs (*Tapirus* spp.), peccaries (*Tajacu* spp.), quail (several genera), ducks (order Anseriformes), and jungle fowl such as chachalacas (*Ortalis*) and guans (*Penelope*). It appears that most

Table 7.1
**Some Animals Domesticated in Pre-Columbian
Latin America**

Common Name	Zoological Name	Region of Origin
Turkey	*Meleagris gallopavo*	Mesoamerica
Alpaca	*Lama pacos*	Andes
Llama	*Lama peruana*	Andes
Muscovy duck	*Cairina moschata*	Northern South America
Guinea pig	*Cavia porcellus*	Andes
Stingless bee	Fam. Meliponinae	Tropical America

of these animals were kept for food, but this is not always clear. In addition, a much larger collection of animals were kept as pets. Pet-keeping was prevalent among shifting cultivators in the tropical forest regions, particularly in the Brazilian-Guianan subregion.

The agricultural practices probably played important roles in the animal life of Latin America. The extensive areas of second-growth vegetation would have encouraged the increase of some animal populations, such as the whitetail deer (*Odocoileus virginanus*), grey fox (*Urocyon*), and coyote (*Canis latrans*). The fox and coyote were common in the Central American subregion but now are rare in many parts of that area. (They were never found south of it.) Among the birds that may have profited from some of the changes wrought by farmers are the finches (family Fringillidae). They are seed eaters, so grasslands tend to encourage their numbers. The animals that might have been negatively influenced were those of strictly arboreal (tree) habitats. Among these are many of the primates and other mammals such as the sloths. Those birds that require old forest would also have found their habitats much reduced in some areas, such as in many parts of Central America.

Although we cannot be certain, it appears that shifting cultivators did not usually overexploit the animal species available to them. However, in the sedentary hoe culture in parts of the Andes and Mexico, the opposite may have been the case. By the time the Spaniards arrived in the central Andes (1530s) the wildlife in the region of human habitation had been brought under the close control of man. Hunting was strictly managed by the ruling Incas and it is recorded that meat eating was highly controlled. Annual state-controlled hunts were staged during which all large predators were destroyed, if possible; certain animals were retained for food; and the wild vicuña was captured, shorn of its fleece (which went to the ruling caste), and then liberated (Fig. 7.10). There is no way to determine the faunal picture before the rise of the dense human populations associated with the Inca Empire. However, whatever it was it must have become greatly altered and perhaps impoverished before the arrival of the Europeans.

In Mesoamerica there is no evidence of state-controlled hunting, but there is evidence that meat from game birds and mammals was sometimes not abundant locally. The fact that insect eating (both the adult insect and eggs or larvae) was so pronounced in central Mexico at the time of Cortez's arrival makes one wonder if there was a paucity of available animal protein from other sources such as birds and mammals.

The advent of the Europeans initiated complex events that had major influences on the animal life of Latin America. There is little evidence that the Europeans wished

Figure 7.10 *Llamas, alpacas and vicuñas grazing at approximately 5000 meters elevation in the Andes.*

to exterminate the Amerinds, as they did in North America and Tasmania, but the end result was almost the same. The strong social control in much of North America regarding Caucasian sexual contact with Amerind women resulted in only a modest amount of genetic exchange and a major reduction in the size of Amerind populations. By contrast, the exchange of genes between the Caucasians and Amerinds in Latin America began almost at once. This sexual acceptance played a significant role in the partial genetic preservation of the Amerind in what is sometimes referred to today as *la raza nueva* (the new race). (There was also an exchange of genes between Africans, Caucasians, and Amerinds, but this tended to be more localized.) Thus, the biogeography of *Homo sapiens* was importantly influenced by the arrival of Caucasians and Africans. It is still being influenced by representatives of other human gene pools such as mongoloids (Chinese and Japanese) in Mexico and parts of South America.

On the second voyage of Columbus, horses, cattle, sheep, and some fowl were taken to the New World, and on later voyages major efforts were made by the Spanish and Portuguese governments to introduce more domesticated animals. From a numerical point of view, many of these introductions were spectacularly successful. Cattle, for example, when given the opportunity, often multiplied into huge, semiwild herds. (The impact on the vegetation was noted above.) The herds grew most numerous in Mexico and Southern South America, particularly in the Argentine area where cattle soon escaped human control and established great feral herds that roamed at will over the pampa. These were frequently pursued by Indians who, like their North American brethren, acquired horses from the Spaniards and became superb horsemen.

To the deliberate introductions of animals in the century of conquest must be added the animals that arrived by accident. Obviously, there are few reliable data as to place and dates of introduction. It may be assumed, however, that the first shipload of Spaniards brought with it a collection of human ectoparasites and endoparasites that probably included fleas (*Pulex* spp.) and intestinal worms (helminthic animals). Moreover, the plants and animals that were deliberately introduced undoubtedly carried with them an exotic microflora and fauna. Some elements of the latter may have emerged later as "pests." It is during this earliest period that we might assume that the black rat (*Rattus rattus*) and the house mouse (*Mus musculus*) were accidentally introduced.

The accidental introduction probably having the most serious consequences in terms of human health—the introduction of the mosquito *Aedes egypti,* the urban vector of yellow fever—may have occurred in the seventeenth century. The impact of this disease on human populations in the Americas cannot be overstated. Where the virus causing yellow fever came from— Africa or the Americas—still is not determined, but there is no question that before the arrival of the urban-dwelling *Aedes* there were no urban yellow fever epidemics.

The question of the geographic origin of yellow fever turns principally on the fact that yellow fever is a zoonosis and affects primates other than man. In the Neotropical region, most of the primate species appear to be sensitive to the disease, some especially so. Treetop mosquitoes native to the American tropics are the vectors in treetop ecosystems. The question is, did yellow fever occur in these animals (mosquitoes and primates) before the arrival of the urban vector? Although this question may never be resolved, it is clear that the introduction of *Aedes* greatly altered the habitats of the warmer parts of the Americas as far as man is concerned.

Another major influence on human disease in Latin America was the accidental introduction of the organisms that cause human malaria. The insect vectors are mosquitoes of the genus *Anopheles,* of which there are many native species in Latin America. However, the agent causing the disease, unicellular animals of the genus *Plasmodium,* may have been carried into Latin America in the blood of infected conquistadores and Africans. Human ma-

laria has an ancient history in Spain, particularly in the southern part, which is where many of these men came from. Because several species of *Plasmodium*—*P. vivax, P. ovale, P. malariae,* and *P. falciparum*—cause human malaria, more than one introduction is indicated.

Another mosquito, *Anopheles gambiae,* was accidentally brought into eastern Brazil from West Africa in the early twentieth century. But a major campaign was launched to erradicate it and it appears to have been successful. Also in recent time, a fluke, *Schistosoma mansoni,* which causes schistosomiasis, was introduced into northern South America.

After political independence, a number of animal introductions were made, mainly in Argentina and Chile, for recreational purposes, perhaps chiefly because of the immigration of Western Europeans into this region. The introduced animals that became established in Argentina are listed in Table 7.2. (Accidental introductions and livestock are not included.) Three of the major zoogeographical regions were drawn upon—the Nearctic, Palearctic, and Oriental regions. It is not surprising that the northern middle latitudes are so well represented because, in many respects Argentina is their ecological analog in the Southern Hemisphere. Unfortunately, there is no available published information regarding the ecological effects these introductions may have had on the Argentinean ecosystems. However, the population explosions of exotic animals often reported in other parts of the world have not yet occurred here.

In recent decades, an important aspect of animal introductions has been exotic fish, particularly trout and, to a lesser degree, salmon. Although widely distributed in the northern part of the world, the family Salmonidae, which includes trout and salmon, did not occur in the Neotropical region until man carried some members of the family into it. Trout subsequently became important in Chile and Argentina. In

Table 7.2
Some Animals Introduced into Argentina

Common Name	Zoological Name	Source Region
	Mammals	
Red deer	*Cervus elaphus*	Palearctic
Fallow deer	*Dama dama*	Palearctic
Axis deer	*Axis axis*	Oriental
Caribou	*Rangifer tarandus*	Palearctic
Mule deer	*Odocoileus hemionus*	Nearctic
Blackbuck	*Antilope cervicapra*	Nearctic
Wild pig	*Sus scrofa*	Palearctic
Muskrat	*Ondatra zibethica*	Nearctic
European rabbit	*Oryctolagus cuniculus*	Palearctic
European hare	*Lepus europeus*	Palearctic
	Birds	
Pigeon	*Columba livia*	Palearctic
California quail	*Lophortyx californicus*	Nearctic
European sparrow	*Passer domesticus*	Palearctic

Chile, where the introductions were extraordinarily successful, it is said that there is some of the most outstanding trout fishing in the world.

The successful fish introductions into Chile are listed in Table 7.3. Although the introduction of trout was quite advantageous, the introductions of carp and goldfish were of questionable value. These fish are notorious for their ability to degrade aquatic habitats through their feeding habits, which greatly disturb the substrate.

Trout have also been introduced into Andean streams and lakes, into western Panama (rainbow trout), on the slopes of an extinct (?) volcano, and into rivers and lakes of Mexico. In recent years, there has been growing interest in fish culture in artificial ponds in Latin America, and this has led to the introduction of *Tilapia* spp. from the Ethiopian region. The effort is still a modest one so it is not known if any of these introductions have as yet found their way into lakes and rivers. However, this is to be expected eventually.

There is no evidence that any animal species on the mainland of Latin America was threatened with destruction during the colonial period. This was mainly because there was relatively little commercial interest in the native fauna at that time. However, in more recent years, several factors have altered this picture: markets developed for the animals (meat, hides, feathers, live animals); the human population increased, so exploitation of the fauna for subsistence increased; and animal habitats were altered or destroyed by the altered or intensified land-use practices associated with the growing human population.

Commercial exploitation of the Neotropical fauna became important in the late nineteenth and early twentieth centuries because women's clothing of the period required large quantities of bird feathers. Although there are few reliable numerical data, there is general agreement that large numbers of birds possessing the kind of plumage demanded were slaughtered until laws were passed to end the

Table 7.3
Exotic Freshwater Fishes in Chile

Common Name	Zoological Name	Source Region
	Salmonidae	
Rainbow Trout	*Salmo gairdneri irideus*	Nearctic (California)
Eastern Brook Trout	*Salvelinus fontinalis*	Nearctic (USA)
	Cyprinidae	
Carp	*Cyprinus carpio*	Palearctic
Goldfish	*Carassius carassius*	Palearctic
Tench	*Tinca tinca*	Palearctic
	Ictaluridae	
Catfish	*Ictalurus melas*	Nearctic
	Poecilidae	
Mosquito Fish	*Gambusia affinis*	Nearctic
	Atherinidae	
Matungo	*Odontethes bonariensis*	Neotropical
	Cichlidae	
Chanchito	*Cichlasoma facetum*	Neotropical

Source: Buen, Fernando de, "Los Peces Exóticos en Las Aguas Dulces de Chile,"
Investigaciones Zoológicas Chilenas, Vol. 5, pp. 103–137.

practice. At present, most of the exploited bird species have more than recovered from the excessive hunting pressure.

The hide and pelt trade has had a different result, however. In recent decades, a large market has grown in the United States and Europe for the hides of caimans (alligator-like reptiles that occur from Mexico to Argentina in fresh water), crocodiles, and the larger cats, particularly ocelots (*Felis pardalis*) and jaguars (*Felis onca*). Snake and lizard skins have also been important; they are used for shoes, women's handbags, and similar articles. The cat pelts are used chiefly for coats and decorations on women's garments.

Because the statistics on this trade kept by most countries are so incomplete and inaccurate, we can only make general statements about the effects of this exploitation on the animals. However, even the most cursory on-the-scene survey forces one to conclude that the animals are rapidly being reduced in number and their natural geographic distribution of Latin America is rapidly shrinking. For example, in the parts of Mexico and Central America where caimans were common and often abundant only thirty years ago, now they are either rare or have been eliminated. Similar conditions are encountered in Panama and many parts of South America where these reptiles were formerly reported in great numbers. Aside from faunal alteration, one wonders how else the ecosystems are affected when a once abundant predator is largely or completely removed. It must be assumed that some basic changes have taken place in the aquatic ecosystems from which these reptiles were removed, but thus far there has been little attempt to study these changes.

The larger cats have seldom been abundant, but jaguars have been completely hunted out of large parts of Latin America. Even where laws were passed to eliminate commercial hide hunting, such as in Mexico, the illegal practice continues. With but few areal exceptions, the jaguar has become rare in Mexico, Central America, and Panama. Moreover, it is no longer reported in much of its former range in South America and is not even safe in the Amazon basin.

The ocelot has had approximately the same history, although this smaller cat appears better able to persist in regions disturbed by man. It is a threatened species in mainland Middle America and in the more densely settled parts of South America, where it was formerly common. However, as with the jaguar, almost nothing is known about the total population size of the ocelot, so we must rely largely on hunters' reports of species numbers.

Meat hunting is exerting growing pressure on the game animals in this region, and the resource is not able to stand up to the pressure. Although some countries, such as Mexico, Argentina, Brazil, and Chile, have hunting laws on the books, subsistence hunting continues, often clandestinely. The chief source of animal protein in the diets of millions of Latin Americans is the local wild animal resource. This is a continuation into modern times of the condition that prevailed when the Europeans arrived. The major change is that land-use practices have altered so there is less land for shifting cultivation and an ever-growing amount of land for pasture, and the widespread use of firearms enables the people to easily overexploit the

diminished animal resource. Fish are also an important part of the picture. The alteration of stream flows and the widespread (illegal) use of dynamite for fishing has frequently destroyed or greatly reduced the fish resource in streams.

Meat hunting has two economic aspects: hunting for subsistence and hunting to sell the meat. In many parts of Latin America, the meat of wild animals is sold openly in markets. Thus, although subsistence hunting for food exerts extreme pressure on the resource, hunting to sell the meat is responsible for the outright destruction of it in many places. The most favored animals are tapirs (*Tapirus* spp.), deer (*Odocoileus, Mazama,* and other genera), paca (*Cuniculus paca,* a Neotropical rodent that has excellent meat), and agouti (*Dasyprocta aguti,* another Neotropical rodent whose meat, although good, is inferior to that of the paca). In addition, many other mammals and birds and reptiles are hunted for meat. Among the reptiles, the most important are lizards of the genus *Iguana,* which occur from southern Mexico to Argentina and are much sought after for their mild meat. They are often assiduously hunted during the part of the year when the females are gravid (contain eggs) because the eggs are considered a great delicacy. It is still common to see piles of living but trussed-up iguanas in the market stalls of Latin America. This trade, often uncontrolled, has resulted in a sharp decline in the animal's numbers during the past decade, and iguanas have all but disappeared from large areas where they were formerly abundant. However, the ecology of these animals is little known, so there is no way to appraise the ecological consequences of their being eliminated.

During the past twenty-five years, an active market has grown in the United States and Europe for live wild animals from the Neotropical region. This trade is directed to three principal outlets: pet stores, zoological gardens, and medical research. The pet trade alone is responsible for the removal of thousands of mammals, birds, and reptiles from the Neotropical region each year. Chief interest is in primates, particularly ringtail monkeys (*Cebus* spp.), spider monkeys (*Ateles* spp.), and squirrel monkeys (*Saimiri* spp.). The birds that are most important are members of the parrot family, which have often been imported illegally into the United States, and toucans, which are purchased because of their huge, colored bills and bright plumage. Juvenile caymans, sold as alligators, are imported in large numbers, although most cannot survive under the conditions in which they will be kept. There is also a trade in snakes, turtles, and lizards. Fish have also been an important part of the live animal trade, with the bulk coming from the Amazon basin. There is no evidence that the populations of the exploited fish species have been affected in a significant way by this activity, however.

Zoological gardens purchase a wide array of Neotropical animals, but they are probably not a threat to the fauna, except possibly such rare forms as the Andean bear (*Tremarctos ornatus*) and the Andean tapir (*Tapirus pinchaque*). It is sometimes argued that the animals are safer in a zoo, but it is better to give a threatened species complete protection in its home ecosystem.

The importance of primates in medical and biological research has greatly increased in recent years. Not only do non-human primates resemble *Homo* in many ways useful to medical research, they are also sometimes wild reservoirs for human disease, including, in the New World, yellow fever. There has also been great interest in primates by students of animal behavior (ethologists) and a number of primate study centers have been established in the United States and elsewhere. The primate species cannot stand up to this heavy exploitation and they are diminishing rapidly in many parts of Latin America. However, our ignorance of the ecology of most of these animals is so great that we cannot even guess at the effects on the ecosystems from which they have been or are being eliminated.

The growth of human population in Latin America and the changes in land-use are having the greatest effect by reducing

Table 7.4
Some Changes in the Mammalian Fauna in El Salvador, 1800 to 1968–1969

Common Name	Zoological Name	1880	1968–1969
Opossum	*Didelphis marsupialis*	C	D
Spider monkey	*Ateles geoffroyi*	A	NP
Howler monkey	*Alouatta villosa*	R	NP
Giant anteater	*Myrmecophaga tridactyla*	D	NP
Cottontail	*Sylvilagus floridanus*	C	D
Squirrels	*Sciurus* spp.	C	R
Coyote	*Canis latrans*	D	NP
Jaguar	*Felis onca*	D	NP
Puma	*Felis concolor*	D	NP
Tapir	*Tapirus bairdii*	R	NP
Collared peccary	*Tayassu tajacu*	C	NP
White-tail deer	*Odocoileus virginianus*	C	NP

Key: A = Abundant C = Common D = Diminished from earlier years R = Rare NP = No longer present

Source: Daugherty, Unpub. Dissertation, Univ. Calif. 1969

Table 7.5

**Some Recent Changes in the Mammalian Fauna in Panama
(Southern Veraguas Province)**

Common Name	Zoological Name	1910	1963
Opossum	*Didelphis marsupialis*	C	C
Capucin monkey	*Cebus capucinus*	C	NP
Howler monkey	*Alouatta villosa*	C	NP
Tree squirrel	*Sciurus* spp.	C	R
Hog-nose skunk	*Conepatus semistriatus*	C	C
Coati mundi	*Nasua narica*	A	R
Ocelot	*Felis pardalis*	C	R
Puma	*Felis concolor*	P	NP
Jaguar	*Felis onca*	P	NP
White-tail deer	*Odocoileus virginianus*	A	R

Key: A = Abundant   C = Common   P = Present   R = Rare
     NP = Not present

Source: Bennett, Ibero-Americana: 51, 1968

the available habitats of many animal species. The most massive and obvious change is the removal of tree cover to make way for pastures, which we discussed earlier. As the forested areas shrink, so do the ranges of the vertebrates and invertebrates that require trees in order to live. On mainland Middle America, the forest removal has eliminated forest animals from many areas. Occasionally "islands" of forest were left, and they contain small populations of animals that are completely cut off from the rest of their species. The microevolutionary implications would be interesting to investigate were it not that the larger mammals and birds are soon hunted out of these tiny refuges. Even the nongame animals last only until the islands of vegetation are cut and planted to grass.

Alteration in the faunal picture resulting from vegetation clearing is most easily seen on the Pacific slope of Central America and Panama. Here, the recent growth of pasture areas, commercial agriculture, and human populations have dealt the wildlife a major blow. Two studies were made of these phenomena and the basic results are presented in Tables 7.4 and 7.5. Table 7.4 tells the recent story for El Salvador, and Table 7.5 shows the current picture for parts of Panama. In South America, the major recent impact of land-use on animals is in Argentina, Chile, Uruguay, southern Brazil, and northern Colombia and Venezuela. The process is an ongoing one, and each year the effects become more intense and enlarged in area. Given the present conditions, one may expect widespread exterminations within the next fifty years. No large region in Latin America is exempt from these influences by man, not even the Amazon basin.

7.4 Influences on Surface and Subsurface Waters

Human influence on the surface and subsurface waters of Latin America began with the development and spread of agriculture. In the beginning, agriculture had only local, short-lived effects on the water, but as human populations increased and more intensive forms of agriculture were developed, man increasingly manipulated or influenced the hydrographic features of the region.

Although the local stream flows in areas where slash and burn cultivators were numerous must have been modified to varying degrees, it was in the areas of sedentary hoe culture that the most striking changes occurred. These areas were principally in central Mexico, the central Andes, and the coastal valleys of Peru.

When Cortez and his followers entered the valley of Mexico in October, 1519, they beheld one of the most extraordinary water control systems devised anywhere up to that time. The basin's floor had a number of lakes—actually, mostly one big lake—in which dikes, dams, canals, and causeways had been constructed. An intensive agriculture was carried on in parts of the lakes on land that had been built by human effort. These islands, called *chinampas* (Fig. 7.11), supported many crops, the chief of which was maize. Because the area was a basin of interior drainage from

which there was no surface outlet and some of the lake waters were therefore saline, parts of the waters had been walled off. Behind these walls, fresh water from hillside runoffs was admitted, thus keeping the more saline waters away. In addition, away from this controlled lake environment within the basin of Mexico, and extending westward were various irrigation works that guided the seasonal flow of water onto the fields.

The development of irrigation systems in the Andean region is still the object of intensive study. The first known efforts of water control occurred in the oasis-like coastal valleys of Peru. They were later extended to the highland basins and valley slopes, where they underwent further development and modification. The irrigation works near the coast included great canals, sometimes extending for many kilometers, aqueducts, which carried water across valleys, and elaborate terrace systems (Fig. 7.12), which may have been built partly to conserve water as well as soil. The earthworks associated with this water control may still be seen over thousands of square kilometers. They are evidence of the huge input of human labor extending over centuries since neither machines nor the wheel were available to these people. The water in the coastal valley streams and

Figure 7.11 *Chinampa agriculture near Xochimilco, Mexico.*

Figure 7.12 *Terraced slopes in the Andes.*

the limited runoff in parts of the central Andes were brought under the almost complete control of man.

Irrigation was also practiced north of this region, especially in limited parts of Ecuador and Colombia. However, these were comparatively modest systems and did not achieve the levels of hydrographic modification found in the central Andean region.

Data on water pollution among pre-Columbian farmers is not available. However, we can make some assumptions about their problems based on current conditions where modern public health standards have not been adopted or are of limited significance.

The principal problem was most likely the disposal of human body wastes, and the physical environments of the farming groups were probably most significant in how this was managed. Today, those living in tropical forests or adjacent to rivers and streams usually defecate directly in the water. Thus, the water directly below an area of human settlement is continuously polluted with human excretions. However, large fish populations directly below villages are common, and some of the species feed on the wastes, thus acting as depolluting agents. A group of armoured catfish, including such genera as *Plecostomus*, which possess mouths adapted to feeding on the algae adhering to rocks, may also be extremely abundant below a lowland tropical village. It is likely that the human excretions provide an enriched nutrient base for the algae, whose abundance, in turn, provides a heightened food base for catfish.

Thus it might be argued that the input of human excretions is not pollution at all, but simply an example of a mammal using the waterways to deposit its wastes—as do the hippopotamus in Africa and the tapir in the Neotropical region. Pollution occurs when such deposition creates a *pathological* condition, which does not seem to have been the case among the Amerinds. There are few reliable data on Amerind infant mortality, but available information suggests that Amerind children were highly sensitive to bacterial infections of the intestinal tract. However, if they survived, they acquired a resistance to the effects of the bacteria and other pathogens obtained from water sources. All evidence points to human excretion-polluted water as the main cause of *present-day* infant mortality in this and other regions. Some of the pathogens may have been introduced accidentally since 1492.

The Yucatán area has had a history of serious water pollution, due in part to the limestone that underlies much of the peninsula. This limestone has created a (karst) hydrography in which much of the drainage is underground and water is often available only at sink holes (cenotes; Fig. 7.13), unless wells are driven or dug. Studies conducted several decades ago showed that water was often seriously polluted by bacteria and helminths (worms). Although the people did not defecate directly into the water sources, they often deposited excretions behind a house or hut from where it could be washed into the domestic water supply. This kind of water pollution is more rare now, but it may have been present in pre-Columbian times. The introduction of the pig (*Sus*) helped solve the problem, because pigs readily scavenge this kind of debris. There has recently been greater emphasis on the safe disposal of human wastes and protection of wells, which were Spanish introductions, from surface contamination.

Human waste disposal must have been an extremely difficult problem in the basin of Mexico, but we have little information on conditions there when the Spaniards arrived. However, the casual way in which Amerinds usually deposited their body wastes suggests that water pollution may have been an important problem everywhere people lived in Latin America.

Figure 7.13 *The sacred cenote (sink hole) of Chichén Itzá, Mexico.*

Although reliable statistics are difficult to obtain, it is evident that the chief cause of childhood mortality in Latin America today is intestinal infections, and probably the principal share of these are caused by contaminated (polluted) water. There may be a density dependent factor operative here, that is, infant mortality may increase exponentially with increased local population densities.

The Europeans frequently found that they could teach some Amerinds little about water control that they did not already know. The biggest contribution the Spaniards could make was variants on the already established water control technology. These variants were chiefly important where Amerinds had been most successful in modifying the natural hydrography, such as in central Mexico. There, the Spaniards introduced brickwork aqueducts, dug wells, and other engineering features previously unknown in the region. The Spanish impact on water control in the central Andean region was far less significant, however.

The basin of Mexico appears to have stimulated the engineering imagination of generations of men after the conquest years of 1519–1521. Because of the problems associated with floods that occurred there with varying frequency, schemes to *drain* the basin were devised early. As noted earlier, the basin of Mexico (Fig. 7.14) has interior drainage, that is, it does not have a natural surface outlet—a feature common to the arid and semiarid parts of the world. During Amerind time, specifically, right before the Spanish conquest, the people in the city of Tenochitlán lived a semiaquatic existence, embarking in their canoes when the floods came. But the Spaniards were unwilling to live like this.

Even before the arrival of the Spaniards, the Aztecs tried to control the floodwaters by building walls and dikes, but these were inadequate when the floods were severe. The first major effort to drain the basin—

Figure 7.14 *The basin of Mexico and Mexico City.*

by the construction of a tunnel—was begun in 1607, but it was unsuccessful in spite of the great effort expended. It was not until 1856 that a project was begun that would lead ultimately (in 1900) to the effective drainage of the basin. This drastically altered the hydrography of what had become a densely settled area and created a host of unforeseen problems. Chief among these was the exposure of soft lake bed sediments, which are picked up and transported by the wind. As a result, there is a high incidence of bronchial disorders among the inhabitants of Mexico City, not to mention the discomfort they experience in breathing air filled with this fine dust. Moreover, the city rapidly expanded onto the land bared by the receding water and the heavier buildings built there settled into the soft sediments. This problem appears to have been compensated for in the newer construction, but the ground floor of the old Palacio de Bellas Artes, for example, has sunk well below its former level.

Floods remain a problem in the basin, and in years of heavy precipitation, parts of Mexico City are underwater for days. This basin has become one of the largest urban centers in the Western Hemisphere in terms of human population (estimated to be more than 6 million in greater Mex-

ico City), but the site was extremely ill chosen.

We shall pass over the colonial period and discuss hydrographic changes in the modern era, beginning with Mexico. Then we shall proceed southward, drawing attention to those modifications that are either largest in scope or of great ecological significance.

Mexico has a troublesome physical environment for anyone wishing to control its wild ecosystems. For example, it was no accident that one of the foremost Aztec gods was Tlaloc, the fierce-looking god of rain, for no problem is greater in Mexico than its scarcity of water. It was not until fairly recent decades that man was successful in controlling the fresh water of large parts of the country—even though water availability has always been the main farming problem.

Most of the water-control projects are dams and many have been constructed in the northwestern part of the country. In this region, atmospheric precipitation is extremely variable and unreliable. Stream flows often alternate between raging floods and low, almost still water. In pre-Columbian times, some of the Amerinds used the floodwaters to irrigate floodplains for agriculture, but little or no diversion of water was involved. A growing population added stimulus to water-control efforts, and dams were constructed on the larger rivers of Sonora and Sinaloa. Dam construction extended southward, and a particularly large water impoundment was recently constructed on the lower Río Balsas. There are few remaining uncontrolled rivers in Mexico today.

Although the hydrographic modifications have been huge, there is little ecological information on the changes that must certainly have occurred above and below each dam. For the time being, we are limited to drawing attention to the obvious visible changes in Mexican hydrography.

Irrigation projects are, of course, associated with the dams. In fact, most of the dams in Mexico were constructed mainly for irrigation purposes. Particularly in northwest Mexico, large areas have been put under intensive irrigated cultivation. This has altered the hydrographic system, but detailed ecological ramifications are only dimly perceived because they have not yet been the object of serious attention.

Hydrographic modifications in Central America tend to be modest, chiefly because of a lack of suitable rivers, a lack of capital, and generally good rainfall reliability. However, there have been some major projects—of these the most important is the Panama Canal. The narrowness of the isthmus of Panama was apparent to the earliest Spanish explorers. It is believed that Balboa suggested that a canal be dug—a feat quite beyond sixteenth-century engineering capability. However, the idea of a canal remained alive, and in the latter part of the nineteenth century, the Republic of Colombia (Panama was then a province of Colombia) awarded a concession to dig the canal to a French company headed by De Lesseps. After the French excavated a large quantity of earth and rock, financial and medical problems (yellow fever and malaria) brought their effort to a halt. In 1903, the United States obtained the canal-building rights from the newly born Republic of Panama, and the canal opened in 1914.

The Panama Canal is a lock type, and requires that transiting ships be raised approximately 27 meters above sea level at the highest point. The canal waters, which are chiefly fresh, come mainly from the Río Chagres, which is dammed at two places. The first dam is for the control of runoff during the season of heavy rain and thus to provide a backup of water for canal operation in dry periods. Electricity is also generated. The second dam impounds the Chagres to form Gatun Lake, which is a basic part of the canal itself.

Because the waters of the canal are largely fresh, the canal has not been a significant route between the Caribbean and the Pacific for marine organisms. However, the canal's effectiveness as a barrier may not be so great as formerly believed. Gatun Lake has a fairly large number of tarpon (*Tarpon atlanticus*), a fish native to Caribbean waters that enters rivers freely. There is also an anchovy of Caribbean (?) origin and jack (*Caranx* sp.), a "marine" game fish.

Nevertheless, all available evidence indicates that the freshwater Gatun Lake has been an effective barrier to the dispersal of vertebrates between the two oceans. On the other hand, it may not have been effective against the inter-ocean dispersal of small invertebrates. For many decades, it has been the practice for ships to fill their ballast tanks with marine water so they will set low enough to be manageable in the narrow confines of the canal. After the canal transit is completed, the tanks are "blown" (emptied), liberating any microorganisms that might have survived. Any biological effects resulting from these actions are presently unknown.

The present canal is growing less able to handle the demands made on it, and there is major interest in constructing a sealevel canal. This would probably cause some biogeographical and ecological changes in both oceans, but it is impossible at present to determine what these changes would be.

An intensive period of careful research is required before the construction of such a canal in order to identify where the principal dangers lie.

As large as the region is, South America has not yet been the site of major hydrographic changes. There are, of course, many small dams and irrigation projects in all the countries, but the larger schemes, which modify large landscapes, are so far limited. Curiously, two of the more outstanding examples are in two of the smallest nations on the continent—Guyana (former British Guiana) and Surinam (former Dutch Guiana), located in the northeast.

The Guyana project is adjacent to the Atlantic shore. It comprises elaborate dike and ditch systems to enable the intensive cultivation of sugar cane (Fig. 7.15). The resultant landscape is suggestive of the polder lands of the Netherlands and is similar in construction and maintenance. This intensively modified area stands in marked contrast with much of the interior of the country.

In neighboring Surinam, there is a similar complex of ditches and dikes and also the largest artificial lake in South America—Lake Brokopondo. Located on the Surinam River, this lake has a surface area of approximately 1500 square kilometers. As in similar lakes elsewhere, water plants have invaded the area and the fauna of the impounded water has undergone some changes.

Although Brazil has an enormous system of rivers, there have as yet been relatively few attempts to exploit the hydroelectric potential. Among the chief exceptions are a dam on the Rio São Francisco in the northeast, a major hydroelectric plant at Iguassú Falls, and various small impoundments in the states of Paraná, São Paulo, and Rio de Janeiro. Hydrographic modification in Argentina has been largely confined to minor irrigation schemes in the semiarid northwest of the country. Although Chile has many rivers, they are all short and do not lend themselves to major engineering schemes directed toward their modification. However, Chile does have many dams, particularly in the central part of the country, but the ecology has not been the object of much study.

Water pollution in modern-day Latin America was mentioned above. It is a problem over most of the area but is only now receiving much attention. The problem is largely the result of mismanaged human waste disposal, as we have seen, made worse by the explosive growth of human populations. Raw sewage is often allowed to enter river and lake waters in which people bathe or swim or on which they depend for their drinking water supply. However, an increasing number of governments are assuring that the domestic water supplies in the larger urban centers are not contaminated, and there is a spread of water purification plants to many smaller towns and villages. Nevertheless, millions of Latin Americans are still required to drink water that has been contaminated by human body wastes and, as stated above, this remains the principal source of lethal infections in infants.

In those areas where industry has appeared, there is usually a marked degree of water pollution. But this type of pollu-

Figure 7.15 *A typical sluice adjusted for the drainage and irrigation of the coastal lands in Guyana.*

tion has received little attention thus far. In recent decades, there has been a great increase in the use of persistent insecticides and chemical fertilizers. As in other areas of the world, some of these materials end up in surface and subsurface waters, which flow into the oceans. Once the materials reach the oceans, they get into larger circulation systems. They may ultimately cycle into ecosystems far removed from where they were applied. There is no way at present to determine the degree to which the Latin American inputs of chemicals are accounting for buildups in the marine ecosystems of the world. However, it is certain that in every nation in Latin America in which commercial agriculture plays a role, chemicals are used in the fields and some of these exert deleterious effects on local and distant ecosystems. Of particular importance in this respect are cotton growing enterprises. These tend to use large quantities of persistent insecticides and chemical fertilizers, particularly phosphates and nitrates, because the soils are often lacking in these and other critical plant nutrients.

7.5 Influences on Soils and Geomorphology

The story of human modification of the earth's crust in Latin America began when man disturbed the vegetation cover enough to induce accelerated soil erosion. Thus, the process began thousands of years ago. But with the spread of agriculture and growth of population, the ecological pressures exerted by farmers increased and accelerated soil erosion to important dimensions, particularly in the central Andes and central Mexico.

It is not surprising that these two areas, which were under intensive hoe cultivation, had a problem with accelerated soil erosion. It is almost impossible, however, to distinguish between postconquest and preconquest erosion. The most accurate evaluation seems to be that serious soil erosion was established long before the arrival of Europeans but that the Europeans added to the existing problems or substituted new ones.

It was in the Andean area that the most outstanding geomorphic changes were caused by man in preconquest times. Terraces (Fig. 7.12) were constructed by an enormous expenditure of human labor, and the resultant landscapes extend over hundreds of square kilometers. It has often been assumed that the terraces were a soil conservation measure but the evidence for this is not positive. Recent study indicates that there is considerable variation in the morphology of terrace walls so the purposes for which the terraces were constructed may have varied from place to place and through time. It is certain, however, that they do serve to reduce soil erosion on what were or had become bare slopes. Many of these terraces were so well constructed that they have survived four centuries of neglect in good condition—and some are still farmed.

Another cause for modification of the earth's crust in pre-Columbian South America was the practice of several culture groups constructing extensive systems of mounds. It is believed that these were used to cultivate crops above land that was seasonally flooded. So massive were many of these systems that they remain, centuries later, as visible geomorphic features on the landscapes.

Geomorphic changes due to mining activities were small and highly local in pre-Columbian Latin America. Most mining was of the placer type, that is, unconsolidated sedimentary materials were worked for the gold they contained. In areas of heavy rainfall, the sand and gravel wastes from these mines must have been leveled rapidly or washed away, leaving scant trace of man's disturbance. The evidences of hardrock mining are, of course, far more persistent, but hardrock mining was limited chiefly to the central Andes and there it was only very local. Andean hardrock miners usually worked alone and followed only those veins that contained extremely high metal contents. The principal metals mined were copper, silver, gold, and lead. Most mines were not deep or long. The longest thus far reported was nearly 75 meters—not great by modern standards, but impressive if one is aware that it was dug with a deer antler for a tool.

The arrival of Europeans led to new pressures on the soil cover and to greatly increased mining activity. The pattern of soil erosion must have altered sharply with the introduction and spread of grazing and browsing animals on steep slopes. During the colonial period, however, there appears to have been only local instances of accelerated soil erosion due to livestock grazing—probably because ranching was controlled in various ways by government edict in many parts of Latin America. In the post-colonial period, livestock numbers increased; and more recently, the rapid growth of human numbers has caused the animal populations to increase also.

As noted in the discussion on vegetation, overgrazing and overbrowsing have resulted in the destruction of the vegetation cover in many parts of upland Latin America. In addition, there has been direct clearing to expand the available pasture area. These practices have had a marked effect on the soils of the region. Although no detailed country-by-country study of recent soil erosion has been made, the studies that have been conducted indicate that accelerated soil erosion is one of the most serious aspects of man's recent influences on Latin American ecosystems.

No country is completely free of acceler-

ated soil erosion, but those most affected are Mexico, Colombia, Venezuela (Fig. 7.16), Brazil, and Chile. It has been estimated that Mexico has *lost* almost half its useful soil cover and that almost all the country's farming regions have suffered some damage. It is impossible to travel any distance in the highlands of Mexico without witnessing extreme soil destruction. Although the government has attempted to alter this situation, it remains and in some areas even increases each year. In some of the more settled parts of northern highland Colombia and highland Guatemala, soil destruction has elicited strongly worded reports from travelers. And the story is repeated in other countries of Latin America where improper farming practices, too many animals, and too great a human population cause a soil problem that each year reduces the ecological patrimony of the Latin American region.

Another major aspect of human influence on the soil is the chemical and physical changes that result from land-use. Even shifting cultivators in preconquest times undoubtedly influenced the physical and biotic characteristics of the soils they farmed longest or burned most frequently. The relationship of these practices to the development of grasslands has already been discussed, but it might be recalled that repeated forest cutting and burning tend to develop "clay pans" at varying shallow depths. These alter the drainage properties of the soil and thus, in a fundamental way, the biology and chemistry of the soil. Soil changes may have been dramatic in the areas where hoe cultivation was practiced. In the Andes where terracing was well developed, the soils retained by the terrace walls must, in time, have acquired properties very different from those of soils on the slopes before terracing.

In recent times, the modifications of soil quality has been rapidly increasing—because of irrigation, chemical fertilizers,

Figure 7.16 *Accelerated soil erosion in Mexico.*

insecticides, herbicides, and other aspects of modern plow agriculture. One of the more dramatic and economically troublesome changes is soil salinization resulting from irrigation. Mexico, in particular, has had serious difficulties with salinization (see Chapter 3 for a discussion of the problem in Southwest Asia). The delta region of the Colorado River (Río Colorado) and the Laguna district in northern Mexico are the most affected.

The Colorado delta is composed largely of sediments deposited by the river. The soil and the climate (and economic conditions) have made this region attractive for commercial cotton cultivation. However, the region is a desert, so irrigation is necessary and the water is obtained from the Colorado River. As was noted in Chapter 6, the water of this river is much used, abused, and modified by people north of the United States-Mexico border. Moreover, the farmers on the delta often attempt to extend the land area planted to cotton, and this leads them to engage in irrigation practices certain to result in the ruin of the soil as far as future cultivation is concerned. The destruction is caused by their spreading the water too thinly. This prevents a proper flow to transport away from the fields the salts that are carried to the surface of the soil by capillary action and the salts that are constantly being transported via water into the cotton ecosystem. The result is a buildup of minerals in the soil to the point where cultivation becomes impossible. The buildup is hastened if the water becomes more heavily charged with disolved minerals *before* field application. The end of this series of events is whitened acres covering large parts of the delta.

The problems of the Laguna district are similar, except that the water is of local origin. This area was once a showplace for *ejido* (communal) agriculture in Mexico, but soil loss due to improper irrigation attained major importance and the whitened land surfaces testify to the ecological mistake.

Although livestock ranching and farming are generally responsible for the most widespread and ecologically significant changes in the earth's crust over most of Latin America, in postconquest times, mining activities have also contributed to altering the face of the region. Minerals, particularly gold and silver, were among the principal reasons for conquest, so mining activities got underway shortly after the initial period of conflict with the resident Amerinds. The Spaniards had, for a time, an advanced mining technology. Thus, they were able prospectors for ore bodies and able miners and smelters when they located the ores. In the colonial period, both placer and hardrock mining was followed, but hardrock mining often resulted in the more obvious geomorphic changes.

Although hardrock mines were opened in many places, they rapidly became concentrated in the Andes and Mexico. The great Andean silver mine of Potosí was opened in 1545, and it, plus the surrounding mining district, supplied a large percentage of the treasure sent back to Spain annually. In the middle and end of the sixteenth century, a number of great silver mines were opened in Mexico north of the capital (the impact on the vegetation has been noted earlier in this chapter).

In the larger mining districts, the geomorphic picture was gradually altered—not only by the many shafts leading into the earth, but even more by the accumulating piles of mine wastes, or "tailings," discarded after the metal was extracted. And because many of the mines were located in arid districts, these piles of debris persist and are an often peculiar looking feature of the landscape. It seems that the most arresting sight on thousands of slopes in the drier mining districts of Latin America are mine dumps—some huge and many small—all bearing testimony to man's presence.

In place of gold and silver, present-day mines concentrated on metals demanded by modern industry. Copper, iron, and aluminum are far more useful and valuable than all the gold and silver previously taken from the earth in Latin America. The result is that the alteration of the earth's crust has been greatly increased.

Although the alterations associated with modern mining, as with those associated with the mining of the past, tend to be local, they are also longlasting. Should man disappear from the earth, the great holes he recently excavated in parts of South America would persist for hundreds, if not thousands of years. Northern Chile, where copper has been mined by the open pit method for several decades (Fig. 7.17) is a prime example.

The three main iron-mining regions in Latin America are northwest Mexico, Cerro Bolivar, Venezuela, and the Itabira region of Brazil, and the landscapes of all these areas are greatly modified by mining activities. The open-pit method, which almost always causes a major disturbance of extensive areas, is used to mine bauxite, the ore of aluminum. The pits that often result will persist for many centuries. Bauxite mining will probably increase as world demand for aluminum increases, and thus an increasing number of Latin American landscapes will be subjected to this kind of alteration. At present, the major bauxite-producing region on mainland Latin America is Guyana and Surinam, but concessions are being given by other nations. The resulting mines will have important ecological implications.

Mention should be made of small-scale

Figure 7.17 *Open pit copper mine, Chuquicamata, Chile.*

mining, chiefly placer mining which occurs over much of Latin America. Gold, platinum, and diamonds are the minerals most sought. Even these small scale operations cause a marked alteration in the geomorphic expression of local areas. This is particularly noticeable in certain diamond-mining districts where the holes and the tailings impart a distinctive and persistent aspect to the landscape.

We have seen that the overwhelmingly important aspects of man's influence on the earth's crust in Latin America has to do with the thin mantle of soil. The most serious ecological consequences of this alteration relate to human welfare.

7.6 Influences on the Atmosphere

Until the establishment of agriculture in this region, man must have exerted only minimal influences on the microclimates. Where human-set fires were sufficient to alter the wild vegetation cover, it may be assumed that changes in the microclimates occurred but they must have been on a local scale. However, with the greater alteration of vegetation associated with crop cultivation, it is certain that microclimates changed significantly. However, these changes were not even thought of until recently.

As has been pointed out, vegetation tends to create and maintain different types of local climates, which we call bioclimates or microclimates. The vertical extent of such bioclimates is generally correlated with the height and density of the vegetation cover. Thus, forests are able to maintain homeostatic bioclimates close to the ground if the cover is sufficiently dense. But when the vegetation cover is altered, the existing microclimates are also altered, sometimes considerably. This is particularly true in the tropical parts of Latin America. Moreover, the changes, if of long duration, play significant roles in plant and animal succession in the disturbed ecosystems.

Table 7.6 presents some bioclimatic differences between a forest and an adjacent cleared plot during part of the wet season in a tropical deciduous forest climate station in Panama. Note that in the forest there was little daily fluctuation in air temperature and that the range of air temperature over the man-created and maintained plot of short grass was much greater. When grass cover is removed, daily ranges

Table 7.6
Air Temperature [1] in Forested and Cleared Areas (Western Panama) [2]

Time	Forest				Clearing			
	0600	1200	1800	2400	0600	1200	1800	2400
Dry Season								
March 2	—	—	26.6	20.5	—	35.5	28.3	21.1
March 3	18.3	30.0	26.6	17.7	18.8	36.1	27.2	20.0
March 4	18.8	31.6	25.5	21.1	22.2	35.0	26.6	23.3
March 5	20.5	26.6	24.4	20.5	22.2	32.7	25.0	22.2
March 6	19.4	28.8	20.0	18.8	21.1	36.1	22.2	20.5
March 7	17.2	27.7	24.4	18.8	23.8	34.4	25.0	20.0
March 8	16.6	28.8	22.2	19.4	21.1	36.6	23.3	20.0
Wet Season								
May 13	—	32.2	31.1	23.3	—	36.6	28.8	21.6
May 14	21.1	27.7	26.6	23.8	25.5	35.0	27.2	23.3
May 15	22.7	27.7	26.6	23.8	22.2	32.2	25.5	22.7
May 16	22.7	28.3	25.5	22.7	21.1	34.4	23.3	22.2
May 17	21.6	27.7	23.8	23.3	20.0	33.8	23.3	22.2
May 18	22.7	27.7	25.0	23.8	21.1	33.3	25.5	22.2
May 19	22.7	28.3	27.2	25.5	22.2	33.3	27.7	25.5

[1] Measured within the first 25 centimeters above the ground with a shielded recording thermometer, degrees Celsius.
[2] Cattle ranch in the Province of Veraguas, 1963.
Source: Bennett, Ibero-Americana: 51, 1968

of soil temperature may increase, as is shown in Table 7.7 for a station in El Salvador. Thus, we can perceive, at least to a limited extent, how shifting cultivation or other types of agriculture involving the clearing of vegetation or the burning of grass cover can affect bioclimates. Although temperature has been the focus here, changes in absolute and relative humidity, velocity of air movement, and intensity of light all go along with changes in the vegetation cover.

To get an idea of the areal extent of these bioclimatic changes in Latin America, we can refer to any map that shows the distribution of agriculture and livestock ranching. In short, man has been modifying the bioclimates of Latin America for thousands of years, and this

modification is an important aspect of the rural and urban ecosystems today.

Air pollution is frequently considered to be chiefly related to industry. However, air pollution is common in rural regions when the farmers are burning slash in preparation for planting and when the ranchers are burning off dry pastures. Smoke affects the air over hundreds of thousands of square kilometers and significantly reduces visibility. It is a common belief among the country people in some parts of Latin America that this smoke causes rain. As yet, there are no substantiating data, but the idea is interesting and has received some attention from scholars.

There is no medical evidence that the smoky air caused by these fires has any pathological effects. It may be assumed

Table 7.7

Mean Soil Temperatures under Grass and Bare Land (at a station in El Salvador, March, 1963)

Depth (Cm.)	Bare Land (°C)	Grass Cover (°C)
2	29.4	26.7
5	29.6	27.1
10	30.6	26.8
20	30.6	27.2
50	29.8	27.2
100	28.3	26.8

Source: Daugherty, Unpub. Dissertation, Univ. Calif., 1969

that *Homo sapiens* has adapted to wood smoke, because it has been an intimate part of his bioclimate for thousands of years.

The macroclimates of Latin America are controlled by forces that have been little affected by human agency (see Map 1.2 on pages 22 and 23 for the Köppen macroclimates of Latin America). Unlike the northern middle latitudes, the region has comparatively few huge industrial concentrations that pour large quantities of particulate and gaseous materials into the atmosphere. There is industry, of course, and it is growing in size and in its potential to influence the macroclimates. Air pollution of urban areas caused by internal combustion engines and factories is becoming more frequent in Latin Amer-

ica. The basin in which Mexico City is located and the city of São Paulo, Brazil, vie with Los Angeles for the record for heavy air pollution. The burning of coal in Santiago, Chile, sometimes forms a dark cloud that makes the city appear to be on fire, and the soot-faced buildings in that city are reminiscent of similar facades in the coal-burning cities of Europe.

The effects on macroclimates resulting from forest removal is a subject of some controversy at present. It has been held that rainfall has decreased in parts of South America after extensive forest cutting, but the data on past precipitation are too limited to evaluate the current situation. Observed decreases in precipitation may only be a normal cyclic occurrence and have no relation to vegetation alteration. This question does, however, demand careful investigation because its implications are serious for a region where a rapidly growing population is increasingly affecting the vegetation cover.

Earlier in this chapter, we mentioned the use of persistent insecticides in many parts of Latin America and that some of these materials must find their way into freshwater and marine ecosystems. In addition, these materials must also find their way into the global air circulation, as they do elsewhere in the world. Perhaps a part of the DDT that is now present in the tissues of some Antarctic animals had its origin in Latin America.

Chapter Eight

Human Influences on Island Ecosystems

Islands are essentially microcosms on which ecological relationships, sometimes difficult to discern on continents, stand out in sharp detail. Islands also tend to be ecologically fragile; their ecosystems are especially sensitive to the activities of human inhabitants. Although most of the human population lives in continental locations, collectively, islands have an impressive portion of the human population living on them. Some well populated island nations are Indonesia, Japan, Republic of the Philippines, New Zealand, Sri Lanka, Malagasy Republic, United Kingdom, Iceland, Jamaica, Cuba, Haiti, Dominican Republic, Barbados, and Trinidad-Tobago. The combined human population of these islands in 1974 was approximately 366 million—almost 10 percent of the world human population.

Of course, we can discuss only a limited number of islands. Those that were selected are representative of marine islands and are illustrative of the influences of human societies on island ecosystems.

8.1 Islands of the Caribbean

Recent archeological discoveries indicate that man reached Hispaniola by 5000 years ago and perhaps even earlier. The exact route traveled is unknown, but most likely was from northern South America and thence north and west through the island arc that begins with Trinidad and ends with Cuba. Because the archeology of the islands is not completely known, there is even confusion about what Amerind groups were present in all the islands at the time of the first recorded European contact in 1492. It appears fairly certain, however, that by this date, only two principal culture groups were present (the others having disappeared)—the Arawaks and the Caribs.

The Arawaks, who preceded the Caribs

into the Caribbean islands by hundreds of years, may have been present by about 100 B.C. in all or most of the island region. They lived in villages and engaged in agriculture, fishing, and hunting—although hunting could never have been of major significance over most of the area. The agriculture was a variant of shifting cultivation—the chief variant being the construction of mounds of earth in which the plantings were made. Although maize was grown, root crops appear to have supplied more calories.

Beginning about 1000 years BP, Carib Indians, who appear to have been extremely aggressive and given to cannibalism, began to invade the Caribbean region. Pushing the Arawaks before them, they had, by 1492, reached the eastern end of Puerto Rico—a drive that came to an end with the arrival of Europeans. The Caribs seem to have possessed a subsistence technology similar to that of the Arawaks; so as far as their general use of the land and waters is concerned, we may consider the two groups essentially equal.

There is considerable disagreement about the number of Amerinds on the islands when Columbus first encountered them, mainly because the Amerinds perished so rapidly after their first contacts with the foreigners. However, at the end of the fifteenth century, Columbus tried to get a head count on Hispaniola, and the result of his census was approximately 1,000,000. We have no reason to doubt this count. In any case, the sad fact is that thirty years after Columbus dropped anchor off Hispaniola, the Amerinds on Hispaniola were almost all gone. By the end of the sixteenth century, they were extinct on Hispaniola and rare almost everywhere else in the Caribbean—although the Caribs were still numerous on some of the Lesser Antilles. Cuba, Puerto Rico, and Jamaica had generally large populations of Amerinds at the times of first contact, but their exact numbers are not known. It is clear, however, that the Arawaks were a fragile society and unable to withstand the assaults of either the Caribs or the Spaniards.

Columbus made four voyages to the New World—insisting to the end that he had reached the Indies. In any case, he "discovered" all the larger Caribbean islands and a few of the smaller ones. He was soon followed by Spaniards in search of gold and other valuables, who were successful on Hispaniola and Cuba but not on Jamaica and Puerto Rico. However, the tide of conquest quickly passed to the mainland. In 1519-1521, Cortez established New Spain; a decade or so later, Peru was also secured by Spaniards. Thus, major Spanish interests were directed to the mainland, and most of the Caribbean islands were only lightly garrisoned or settled.

Pirates, often called buccaneers, from other European nations (England, Netherlands, France) quickly moved in. They preyed on Spanish shipping—often to their and their government's considerable financial reward. Some of these pirates were extraordinary persons. The English captain William Dampier was a pillager of Spanish property on land or sea, but a naturalist by avocation. And William Morgan, perhaps the most notorious of the English pirates, at one time had a considerable price on his head in England but lived to become the first English governor of Jamaica.

However, the national groups represented by the pirates soon sought a more secure base in the Caribbean and the weakly garrisoned islands made settlement relatively simple. By the end of the eighteenth century, the Caribbean islands were a jumbled collection of colonies held by Spain, England, France and the Netherlands. France was the first of these nations to be partially dislodged—from Haiti, during the time of the French Revolution. The next European government ousted was Spain. At the end of the nineteenth century she lost all her remaining islands in the Caribbean, most notably Cuba and Puerto Rico. Cuba achieved a kind of independence and Puerto Rico came under direct United States control. (Puerto Rico now has the status of a commonwealth of the United States.) The British colonies became independent in the middle of the twentieth century, creating the nations of Trinidad-Tobago, Jamaica, Barbados, and Granada. The Dutch still retain Aruba, Curaçao, and a few others. Thus, there is great political diversity in the Caribbean area.

When the Indians disappeared, there quickly arose a demand for black slaves from Africa. Slaving was well under way before the middle of the sixteenth century and continued until the first half of the nineteenth century, when it was stopped by law. The Africans had a profound cultural and human influence on the region, but so far, this has received inadequate attention from scholars. Although the African contribution to the music of the region has been studied, the larger and more significant questions about cultural transplants have been virtually ignored. Most studies have focused on the contributions of the Europeans to the region and tacitly assume that the most important contributions of the Africans were their labor and physical presence. Numerically, the Africans were extremely successful, for today they represent the single largest racial group in the islands. Furthermore, they have intermixed with most of the other racial groups. After slavery was abolished, labor had to be sought elsewhere, and this led to the importation of contract laborers from Asia. Thus, a complex array of human races and cultures are present in this region.

Economically, the first interest of the invading Spaniards was in the acquisition of gold and other valuable metals. When major discoveries were made on the mainland and the relatively poor placer deposits on the islands had been exhausted, Spanish interest turned to the mainland. It was this lack of economic interest that appears to have been chiefly responsible for the light garrisoning of many of the islands that opened the way to British settlement which we mentioned previously. Of course, military and political events in Europe also played a role in the developments in the Caribbean. In any case, the new invaders, whether British, Dutch, or French, were mainly interested in commercial agricultural enterprises, and they were the most instrumental in developing plantation agriculture in the region. The plantation economy reached such a high level that Englishmen of the eighteenth century, when describing a wealthy man, often said he was "as rich as a West Indian planter." In recent decades, the agricultural economy has been modified by the development of industry, tourism, and mining enterprises for metals that were not particularly valuable in 1492 (for example, aluminum).

After the sixteenth-century collapse of Amerind populations, human population began a slow regrowth that reached alarming proportions on some islands by the end of the nineteenth century. In the twentieth century, the region's population has been increasing rapidly, so the number and density on some islands have reached levels clearly out of balance with the island's carrying capacity. This problem is most notable in Haiti, Jamaica, Puerto Rico, and Barbados, but no island in this region is free of it.

INFLUENCES ON VEGETATION

If the first human invaders of these islands possessed agriculture, as is believed, the history of vegetation alteration began with their arrival, because as we have mentioned many times, fire and vegetation clearing were the basic components of plant cultivation. The fact that the islands generally had a large human population by the time the Spaniards arrived suggests that most of the available land suited to cultivation was periodically used. The early Spanish accounts mention that the Indians were not disturbed by steep slopes, because their system of farming did not require the level land needed by plow farming. We can therefore assume that only the mangroves and swamps remained essentially unaltered. The effect of human influence appears to have been the creation of large expanses of second-growth woodlands.

In addition to altering the wild vegetation cover, the Amerinds were responsible for the introduction of a large number of cultivated plants. Some of these plants escaped and established themselves outside of human control, but others, being *cultigens* (dependent on human agency for their continued survival), did not become established in a wild state. Although the list of cultivated plants the Amerinds carried to the islands is impressively long, only a few reached important levels of production. Foremost among them were yautia (*Xanthasoma*), a root crop somewhat similar to the staple root crop taro (*Colocasia*) of the tropical Pacific; manioc (*Manihot*), another root crop, which often supplied the main bulk of caloric intake; maize (*Zea mays*), which appears not to have been of primary significance at the time of the Spanish arrival; sweet potato (*Impomoea batatas*), which was of great importance to the caloric contribution to human diet; and cotton (*Gossypium*), a textile plant. It is also probable that, like the Europeans who later invaded the region, the Amerinds accidentally introduced plants that were not crop plants and that some of these became part of the "native" flora of the islands. However, this is all but impossible to establish now.

As noted earlier, the Amerind population was almost totally destroyed within fifty years after Columbus landed in Hispaniola. Thus, most of the land that had been tilled over the preceding centuries was abandoned and large tracts reverted to trees. This hiatus of human influence ended a century later when the non-Spanish Europeans began to settle some of the islands. Eventually, the Spaniards also perceived economic opportunity in plantation crops, especially on Cuba, which had long been known as the "Pearl of the Antilles."

In the fifteenth and sixteenth centuries, the Spaniards introduced numerous crop plants, some from the Canary Islands and others from the Iberian Peninsula. Among these were sugar cane, bananas, and rice. There is evidence that the Amerinds were already growing a type of sugar cane, but the plant gave way completely to the crop of Asian origin and Spanish introduction. It has also been suggested that the cooking banana, or plantain, was present in the region before the arrival of Columbus. The sequence of post-Columbian plant introductions has not yet been determined, but it was a process that continued over many years and has not yet come to a stop.

The British appear not to have contributed greatly to the introductions although they did introduce the breadfruit tree (*Artocarpus altilis*). This tree, which produces a starchy fruit, was taken to the British islands in the Caribbean in hopes that it would provide an abundant and cheap source of food for the African slaves on the sugar cane plantations. Although the tree did not fulfill this function, it was carried widely about the island region. Today it grows wild in secondary forests and is used decoratively in parks and gardens.

A listing of all the alien plants taken into this region since 1492 would be extremely long. Floral elements were imported from Africa—watermelon (*Colocynthis citrullus*) and coffee (*Coffea arabica*); from Asia sweet orange (*Citrus sinensis*), sugar cane (*Saccharum officinarum*), mango (*Mangifera indica*), and indigo (*Indigofera tinctoria*). The number of grasses is extensive and they have transformed the species composition of the savannas on most of the islands. The result is a plant geography that can be understood only by knowing the history of human introductions and human alterations of the wild vegetation cover.

The development of plantation agriculture led to a new removal of tree cover, and this removal was not of the temporary nature of that associated with shifting cultivation. Plantation agriculture requires, in most instances, the removal or at least nearly complete control of the noncultivated vegetation. Although a fairly wide array of crops has figured in plantation agriculture in the Caribbean, those that emerged as the most significant economically and ecologically are sugar cane, coffee, bananas, tobacco, cacao, and indigo. In the middle of the nineteenth century, the development of chemical dyes eliminated the cultivation of indigo, and of the rest, sugar cane has easily been the most important crop in terms of money earned and area given over to it. The cultivation of sugar cane accounts for the largest share of permanent vegetation alteration in the Caribbean in post-Columbian times. It is well suited to many of the soils and climates of the region and lent itself to an economy that used slave labor. It was a commercial crop that developed with the growth of a European sweet tooth, which previously had to be content with bee's honey. Sugar not only provided great quantities of cheap sweet stuff but, when in the form of syrup, was also distilled into rum.

So great has been the demand for sugar that a number of islands in the Caribbean, Cuba being the most notable example, became economically trapped into a sugar monoculture. The result has been the removal of almost all trees over great stretches of land to make way for cane fields. The vegetation landscapes that were suitable for sugar in Cuba, Jamaica, Puerto Rico, and Trinidad, and many of the Lesser Antilles, are now nothing like they were when the Spaniards first arrived. Extending the areas of ecological change associated with plantation agriculture were coffee, tobacco, bananas, and some lesser commercial crops such as cotton.

Human clearance of vegetation was also associated with the spread of livestock, particularly in Cuba, but in many instances this was largely a local feature. However, the introduction of livestock resulted in vegetation changes induced directly by the grazing or browsing habits of the animals. These changes must date from the beginning of the sixteenth century and gradually spread from island to island as the animals were introduced.

Commercial exploitation of the forests began in a modest way in the seventeenth century with the cutting of trees useful for dyestuffs, such as logwood. The cutting (perhaps the term "mining" is more appropriate) of the tropical cedar (*Cedrela*) came later. The former local concentrations of these trees on some islands probably relate to the previous long history of shifting cultivation and abandonment of cultivated areas. This tree is usually uncommon in the forests of this region, but is favored by conditions occurring on abandoned farms.

The commercial exploitation of pines (*Pinus*) is probably not much older than this century, and began after much of the once-extensive stands had been cleared to make way for commercial crops. There has been limited commercial cutting of other tree genera, but it probably has had only minor effects on the vegetation cover.

In addition to the commercially induced changes in the plant cover are changes caused by subsistence or semisubsistence farming and noncommercial or limited commercial uses. All available information suggests that the area most seriously affected in this way is the Republic of Haiti, where an overwhelming percentage of its teeming population (estimated to be at least 4.5 million in 1971) lives on the land as peasants (Fig. 8.1). The ecological pressure is tremendous and has resulted in major modifications of the noncultivated vegetation, which is used for fuel and

Figure 8.1 *Small farm, Haiti.*

building materials and is repeatedly cleared to make way for subsistence crops.

In Puerto Rico the changes have been so great in the past century or so that most of its vegetation cover has not even been able to approach the climax stage. The effects of human crowding and intensive land-use practices are seen throughout the island. Lush though some of the Puerto Rican landscapes may appear to be (Fig. 8.2), almost without exception they are human artifacts.

And this is true, not only for Haiti and Puerto Rico, but also for many other Caribbean islands—especially most of Jamaica, Cuba, eastern parts of the Dominican Republic, and many of the Lesser Antilles. In the Lesser Antilles, one may witness the changes that take place when an essentially rural population becomes highly dense. The changes are caused by the need for more crop land, more pasture land, and more fuel. The need for fuel is often overlooked in ecological studies but a large percentage of the human population in the Caribbean area still depends almost entirely on wood for fuel. In addition, the burning of wood to manufacture charcoal is still widespread. Naturally, this results in a continuous reduction of the hardwood species suited for charcoal.

One may find other areas in the world's islands where man has altered the vegetation cover to a marked and dramatic degree, but no area exceeds the Caribbean for the total extent of such change. The plant geography of this region has become as much a reflection of cultural as of biological and physical processes.

INFLUENCES ON ANIMAL LIFE

The Amerinds who came to the islands may have brought a few animals with them

Figure 8.2 *Small farm, Puerto Rico.*

from South America—agouti, cottontail rabbits, iguanas, and armadillos. However, these introductions have not been proved, and if they did occur, they were apparently minor. The assumption that these animals were brought in by humans before Columbus is based on the fact that the biogeography of the islands where they are found indicates a long history of physical isolation from any mainland. This is especially true of the Lesser Antilles—where most of the introductions are assumed to have occurred. It should be noted that Trinidad, though an island now, is a biogeographical extension of South America. It has virtually the same biological composition as the mainland, which is only a few miles away and to which it was joined until sea levels rose in recent geological time.

Early documents also indicate the presence of guinea pigs and muscovy ducks on some islands but these seem to have been uncommon and unimportant in the ecology of the people. Also present was the dog, of which at least one breed was kept as a pet and eaten on festive or ceremonial occasions.

Bones found in middens suggest the possibility that several rodents and other animals were exterminated by man in pre-Columbian times. It would be surprising if Amerinds did not succeed in exterminating some of the native animals, because this would fit the general historical pattern of the invasion of islands by man. However, the mere disappearance of a faunal element, even if remains are found in cultural association, does not prove the case.

Whatever man's influence on animal life during the pre-Columbian period, there can be no doubt that after 1492, striking changes took place as a result of human actions. The first notable human influence was the introduction of domesticated animals of Old World origin—pigs, goats, sheep, horses, cattle, asses, fowl, cats, dogs, rabbits and ducks. The honey bee (*Apis mellifera*) was introduced surprisingly late (nineteenth century), perhaps because of the abundance of sugar.

The limitation of the areas suitable for livestock prevented the larger animals from becoming numerous except on a few islands—particularly Cuba. Some of these animals, particularly pigs, goats, cattle, horses, and cats, became feral. Feral cats appear to be of the greatest ecological significance because they prey on native birds, particularly when the birds are nesting, and on many of the smaller lizards. Thus, the cats have had a definite negative influence on the ecosystems of many of the islands. With only a few local exceptions, the other feral species appear to have exerted little impact.

In addition to domesticated animals, a number of animals were introduced for assumed economic or ecological reasons. Among these are the mongoose, whitetail deer, bobwhite quail, marine (or Surinam) toad, guinea fowl (which perhaps should be listed as part of the array of domesticated animals, although its exact source is not clear), and two species of African monkeys.

The mongoose (*Herpestes auropunctatus*), a highly prolific little Asian carnivore, was the greatest single ecological error in the deliberate animal introductions. It was first introduced into Jamaica in 1870 (soon after, into a number of other islands) to control exotic rat populations that had become troublesome. Although the animal was effective at rat control, it also preyed on native reptiles and birds, and was directly responsible for the extinction of several reptiles. As yet, no effective means have been discovered to rid the region of this animal.

Of the animals that were deliberately introduced, the most unusual must be the African monkeys. Two species, both of the genus *Cercopithecus* and both native to West Africa, were apparently introduced by planters during the eighteenth century. They are found on St. Kitts, Barbados, and Grenada. It is said that the animals escaped accidentally from time to time and thus established themselves in the wild. Although little is known of their ecology in the Caribbean, they probably have had a negative effect on some of the native fauna for they are relatively large.

The accidental introduction of animals by Europeans must have begun with the first voyage of Columbus. Certainly among the first to be introduced was the black rat (*Rattus rattus*), of plague infamy, and the house mouse (*Mus musculus*). These rodents met excellent ecological conditions and multiplied until they reached major pest proportions. (It was because of the ecological success of the rats that the later and disastrous introduction of the mongoose was made.) The brown, or Norway, rat must have arrived much later, because it did not appear in Western Europe until the latter part of the eighteenth century and only arrived in the British Colonies of North America by about 1775. Late though it may have been, however, it became a major economic pest and is undoubtedly a scourge in the local ecosystems it inhabits.

There has not yet been a detailed study of the accidental introduction of invertebrates into this region, but enough is known to sketch out the picture. Among the more serious of these introductions was that of the small, unicellular organisms that cause human malaria, *Plasmodium* spp. The islands appear to have been completely free of this disease in 1492, so perhaps the first shiploads of Europeans included some plasmodium-infected persons. Malaria was so common in many parts of the Mediterranean that its occurrence among crew members might have gone unremarked in the logs or journals. The mosquito vectors, however, were New World natives and were not introduced by man.

Of even greater consequence to human ecology and habitability of this and other New World regions was the introduction of the vector for urban yellow fever, the mosquito *Aedes egypti*. Although there is still some question about the place of origin of the yellow fever virus, there is no question that its urban vector was accidentally introduced into the Western Hemisphere sometime in the seventeenth century. However, it is still not clear where the original point of introduction was in Latin America. But by the end of the eighteenth century, yellow fever had become a scourge in the Caribbean and rendered many of the once-pleasant islands little more than pest holes poorly suited for human habitation.

Attention has been drawn to the possibility that pre-Columbian man may have exterminated some of the animals native to the islands. Whether he did or not, there is no doubt that man has exterminated some animals in postconquest times and that he is threatening the survival of many animal species. All vertebrate groups have felt this more recent human-induced pressure, and the result has been an often drastic alteration of the fauna of all the islands, especially of the smaller ones. Since 1800, at least ten mammalian forms have been completely exterminated (the actual total may be greater), and many forms will pass from the scene in the next few decades. However, because most of these mammals are undramatic in their appearance, they receive little attention except from biologists. Most are rodents, a few are bats, and two are among the most interesting members of the mammalian order *Insectivora* to be found anywhere in the world; these are two genera of the solenodons. A small number of the Haitian solenodon (*Solenodon paradoxus*) still occur in a limited area on Hispaniola; but the Cuban solenodon (*Atopogale cubana*), which once inhabited parts of eastern Cuba, has been considered extinct for several decades, although a few animals may have escaped attention. These two completely inoffensive (to man) creatures are remnants of a group that once ranged widely and successfully in many parts of the world.

The bird life in this region has been

much injured by man since 1492, and twelve forms have been completely eliminated, probably as a direct or indirect result of human actions. The effect of feral cats on nesting birds has already been noted. It was also once a common practice to capture and sell birds of the parrot family, and this may have contributed to the demise of the parrots that have become extinct within historic times—the Cuban red macaw (*Ara tricolor*), plus other macaws, but these are lacking voucher specimens; the Puerto Rican paroquet (*Aratinga chloroptera maugei*), extinct in Puerto Rico although closely related forms are still fairly numerous elsewhere in the Caribbean; and parrot(s) of the genus *Amazona*. Although some of these survive, it appears that others, known only from travelers' descriptions, are now gone. Some of the remaining forms are either rare or are becoming rare. Probably the most important factor leading to the decrease or extermination of the bird life was the major change in the vegetation cover, particularly the widespread deforestation that occurred to make way for plantation agriculture or livestock raising. It cannot be a coincidence that most of the birds that have disappeared or are threatened are those that require trees in order to live.

The native reptilian fauna has suffered greatly as a result of environmental changes, particularly because of the introduction of cats (as noted above).

Much remains to be learned of the ecological implications and costs of the changes made by man in the zoogeography of the West Indies. However, it is clear that the costs have been great and that the shock to the native fauna has been on a major scale.

INFLUENCES ON WATER, SOILS, AND MICROCLIMATES

Thy hydrography of the islands, in the main, is modest, as far as surface streams and lakes are concerned. Indeed, natural lakes are rare in this region, largely because of the humid conditions, which tend to make lakes ephemeral.

The limited land areas preclude the development of large river systems, but all the islands, except the coral islets, possess nets of surface drainage. With almost no important exceptions, these streams have been modified in various ways by human agency. Changes in runoff must have occurred when shifting cultivation became well established in pre-Columbian times, for, as noted in earlier parts of the book, removal of vegetation assures that there will be alterations in runoff. On the islands where human numbers appear to have been densest, such as Hispaniola, it may be assumed that man significantly altered the drainage characteristics of the land long before the appearance of Columbus.

Recently, man has been actively engaged in dam construction on many of the islands in the Caribbean. This has altered the hydrographic landscape and, of course, the ecological conditions. Puerto Rico is the outstanding area of dam construction largely because of its cultural contact with the United States, where the technology for this activity is so well developed.

Accelerated soil erosion is of major ecological and economic importance almost everywhere in this region. Although it probably began in preconquest time—indeed, there could have been no escaping it on some of the steeper slopes of the more densely populated islands—within the past fifty years or so, this effect of human misuse of the vegetation cover has become highly visible. It appears that Haiti has suffered the greatest soil loss, but parts of Puerto Rico and some of the steeper islands in the Lesser Antilles have also been markedly changed.

In Haiti, as mentioned earlier, the large rural population engages in near-subsistence farming, mostly on steep slopes originally covered with forest vegetation. The alteration of this vegetal cover and the prevailing farming practices has resulted in a major and continuing soil loss. This has been made painfully clear by the rapid silting-up of the only significant reservoir in the country, which was constructed recently with foreign financial assistance. Because the watershed supplying the dam is used by peasant farmers, alluvial material has washed in behind the dam and the dam is now all but dead.

Puerto Rico's soil loss is also caused by a too-large farming population trying to live on steep slopes from which the wild vegetation cover has been removed. The erosion is obvious on most of the island but is mainly so on the northern-facing (windward) slopes.

Cuba, despite its intensive sugar cane monoculture, has largely escaped major soil losses, mainly because of its combination of low slopes and soils that do not erode easily. However, soil loss has occurred locally in several parts of the country, most notably in the eastern end where the highest elevations occur.

Soil is also *mined* in the Caribbean region, because it sometimes contains a high percentage of bauxite, the ore of aluminum. The principal site of this activity, thus far, is Jamaica, where the total area devoted to it is large (Fig. 8.3), but it is also being pursued in the Dominican Republic and Haiti. Such mining greatly alters the local geomorphic expression and may, if not rigidly controlled, cause ecological damage lasting for centuries. The mining firms in Jamaica are required to return the mined areas to conditions suited to agricultural or pastoral productivity, but, of course, the ecosystems thus created differ in many ways from the ones they replace. Thus, as areas are worked out in the Caribbean (and elsewhere), new landscape complexes composed of the altered ecosystems associated with bauxite mining will increasingly come into being.

The islands in the Caribbean are not wealthy enough to be a significant source of pollutants to the atmosphere because they contain only a limited number of factories and consume only a limited amount of fossil fuels. However, insecticides are widely used in the area and, as shown earlier, some of these find their way into the general atmospheric circulation. Thus, to some extent, the region contributes to this type of pollution. It is in the area of microclimatic change that humans have been most influential, but we cannot document this statement with data from sites within the region.

Summary. The overwhelming ecological fact about the islands of the Caribbean is that so much on them is alien or at least altered from what it was before the advent of *Homo sapiens*. The graceful coconut trees, which make a visual delight of the trade-wind beaches, arrived even more recently than the Europeans. The rum is derived from an exotic grass—sugar cane—and the technical process of alcohol distillation is also exotic. The men and women who entertain are often descendents of people who only a short time ago were wrested from their African homes and transported as slaves, and the music is also partly rooted in other regions. The spreading mango trees are from tropical

Figure 8.3 *Bauxite mining, Jamaica.*

Asia, and the fruits in the markets are the produce of trees that are native to almost anywhere but the island on which they are grown and harvested. And the human scene, the ethnic and racial complexities, are further evidence that little of unaltered nature remains here.

8.2 Islands of the Pacific (Polynesia)

CULTURAL BACKGROUND

Flung across the great central reaches of the Pacific Ocean is a group of islands whose general boundary line forms a crude triangle enclosing hundreds of thousands of square kilometers. This great area which is known as Polynesia (many islands), contains a vast collection of islands of volcanic and coralline origin. When the Europeans first entered these waters, many of the islands were settled by peoples whose common cultural and biological affinities were obvious.

The Polynesian people appear to have originated somewhere in Southeast Asia. They embarked either from that mainland or closely adjacent islands to the eastern Pacific sometime during the early Christian era. Because the archeology of Polynesia is incompletely known, there is still no close agreement as to when these first people appeared in the island group. The available evidence points to Tahiti as being the first major landfall in the eastern Pacific, so from there later voyagers probably went north to find and settle the Hawaiian Islands. Others found and settled Easter Island, which is thousands of miles from any major landfall. Even later, it appears, Polynesians discovered and settled what is today known as New Zealand. The Tahiti area was probably settled by at least 800; the Hawaiian Islands by 1000; and New Zealand may have received its first Polynesian voyagers by the eleventh century. No one who has not been at sea on the great reaches of the Pacific Ocean can grasp the enormity of the accomplishment of these pioneer mariners.

At the time of first European contact, the Polynesians, although still excellent sailors, practiced both slash and burn and sedentary agriculture based on a small array of basic crop plants. The plants of importance to the eastern Polynesians were taro (*Colocasia*) (Fig. 8.4), yam (*Dioscorea*), sweet potato (*Impomoea batatas*), breadfruit (*Artocarpus altilis*), and several varieties of bananas (*Musa*). Coconuts (*Cocos nucifera*) were also extremely important, particularly on the coral atolls. Protein food was derived largely from the sea,

Figure 8.4 *A taro* (Colocasia esculenta) *garden on Tahiti.*

so fishing was highly developed.

The New Zealand region is not favorable for the cultivation of several of the Polynesian crop staples, such as bananas and breadfruit. Thus, the people there relied chiefly on yams, sweet potatos, and taro. Attempts were also made to cultivate several native plant species.

Over most of Polynesia, three domesticated animals, probably of Oriental origin (definitely not native to the Polynesian region), were found—dogs, chickens, and pigs. Of these, only the dog appears to have been introduced into New Zealand by the Polynesians.

The lives of the Polynesians have been greatly romanticized; even today, there is a widespread notion that here *Homo* has achieved a paradise on earth. This myth has hindered us in our attempts to discover how these people lived and their relationships with the ecosystems they occupied. A careful reading of the early documents indicates that life was often a bitter struggle and that the social systems were almost always of such a nature as to be shocking to any civilized person of our own time. But, these systems worked there in those times, and the systems since imposed on the area often do not stand up well under close moral scrutiny.

A curious dichotomy existed in the pre-European period of Polynesia. Children were often valued and loved greatly and never wanted for adult care—but infanticide was practiced in an almost casual fashion among many Polynesian societies, and no remorse attached to it. We may rationalize that this was done to prevent overpopulation, but there is evidence that it was often simply a means of getting rid of a child who happened to inconvenience whomever had brought it into the world.

The lives of common citizens were held in little regard by the ruling caste—a person might be killed with a sudden, unannounced blow of a club simply for getting in the way of a "great" person. These practices could lead us to conclude that there was a general lack of regard for human life in the islands. However, there was probably as high a regard for human life as is generally evinced by most human societies; only the rules for preserving and taking it differed from the rules of the whites who invaded the region.

The most careful study to date of the Hawaiian population at the time of the first reported European contact (1778) estimates the figure to be around 300,000. Estimates for New Zealand vary considerably, ranging from approximately 50,000 to 200,000 in 1769, at the time of Cook's first visit. In all cases, however, the human populations declined in the years after contact with European civilization—but not at the explosive rate as occurred in the Caribbean region. In Hawaii and New Zealand it appears that the low point for the Polynesian population occurred seventy to eighty years after the first *haole* ("white," in Hawaiian) and *pakeha* ("stranger," in the Maori language of New Zealand) dropped anchor. The Polynesians have managed to survive and have even been increasing for some decades now, although it is doubtful that many of them lack genes from non-Polynesian sailors.

There is a tacit understanding that the first significant outside cultural influence in the Polynesian region was European, but one must not overlook the possibility of casual contacts with Japanese and Chinese fishermen blown far from their waters during storms. In addition, the possibility exists that Amerinds may have occasionally reached the islands and that Polynesians not only reached the American mainland (probably in South America) but even returned to the islands. However, if these contacts occurred, we know of little possible contribution to the human ecology of Polynesia.

The first European voyage of renown into the Pacific was that of Magellan in 1520. He crossed the entire region of Polynesia but missed the islands contained within it. The Spaniards, after consolidating their hold over much of Latin America in the first half of the sixteenth century, began to explore in the Pacific. Before the end of the sixteenth century, they established the famed Manila galleon trade between the Philippine Islands (Manila being the port) and the west coast of Mexico—a trade that was to last for 250 years. Curiously, all this crossing of the ocean did not lead to the discovery of many of the Polynesian islands. It was not until 1595 that the group of islands now known as the Marquesas was "discovered"—the Spaniard responsible named them after the wife of the viceroy of Peru.

On the heels of the Spaniards came the Dutch. And then came British and French pirates, who were attracted by the Spanish shipping. Thus, pirates too became a cultural element in the Pacific in the seventeenth century.

In the second half of the seventeenth century, serious exploratory voyages were made by representatives of several European governments. Among this group were Wallis, who discovered Tahiti in 1767, and Count Louis Antoine de Bougainville, who, in 1768, discovered the Samoan Islands. The most outstanding, however, was Captain James Cook, whose voyages into the Pacific (1768–1780) on behalf of the British government established the location of the island distributions in Polynesia. Cook is also credited with the discovery of Hawaii, although there has long been a claim that he was preceded there by a Spaniard. But, as stated earlier, it is possible that a Chinese or a Japanese fisherman preceded the Europeans.

Next came the Yankee whalers. Along with some European whalers, they entered the Pacific early in the eighteenth century. Although the War of 1812 stopped United States whaling in the Pacific for a brief time, after it was resolved, the whalers returned in force. Their ships and roistering crews were seen in almost all the safe harbors in the Pacific, with Hawaii being perhaps the favorite rest spot (Fig. 8.5).

Soon after Captain Cook's detailed description of New Zealand was published, English colonists arrived. And in the late eighteenth and nineteenth centuries, a varied collection of nationals, representing the United States, England, France, and Germany arrived on other islands in Polynesia. Germany came and went rather rapidly, but France remained in the southeastern region (Tahiti), the United States remained in the Hawaiian chain, and England remained (until recently) in New Zealand. The present political arrangements in this vast area are extremely complex.

Figure 8.5 *Old illustration of the whaling fleet at Lahaina, Maui, Hawaiian Islands.*

The settlement of the Europeans ushered in new land-use practices. These included a catastrophic exploitation of the forest resources in Hawaii and New Zealand and, in Hawaii, farming methods that in some instances developed into an intensive tropical plantation agriculture based on such crops as coconuts, sugar cane, and pineapples. New Zealand, which lies partly in more middle latitudes, developed agriculture along more diversified lines.

The next traumatic event in the region was World War II. To this day, many of the effects of the war on the functioning of the ecosystems remain unstudied. After the war, parts of Polynesia were considered excellent for the atmospheric testing of atomic bombs. The first of these tests took place just at the western edge of Polynesia, in the Marshall Islands, which are part of the region called Micronesia. However, much of Polynesia was downwind from the test sites. Later, France tested atomic weapons in the atmosphere over the French Society Islands—specifically in the Low Tuomotu, which are coral atolls east of Tahiti. It should be pointed out that the entire world was "downwind" from these tests and radioactive materials released by them still circulate in the upper atmosphere. However, the greater part has come to earth and some continues to circulate in many terrestrial and aquatic ecosystems.

INFLUENCES ON VEGETATION

Hawaiian Islands. The Polynesians carried various crop plants with them on their voyages in the Pacific, and these formed the basis of agricultural production at the time of the first recorded European contact. The crops in the high islands of Hawaii, which have a volcanic origin, were essentially the same as those in all the other higher, tropical islands in this region when Wallis, Bougainville, and Cook came on the scene. The basics were taro (*Coloca-*

sia), yams (*Dioscorea*), sweet potatos (*Ipomoea batatas*), various types of bananas (*Musa*), coconut (*Cocos nucifera*), and a few others. The agricultural systems varied from essentially sedentary types to shifting cultivation. Coconuts, once in production, yielded for many years, and taro, which was frequently grown in specially constructed pits that were cultivated year after year, represented a form of sedentary cultivation. Bananas, yams, and sweet potatos appear to have been planted most frequently on sites that were prepared and used as in typical shifting cultivation.

The ecological result of plant cultivation was an obvious alteration of the vegetation cover where crops were grown on the same site year after year and a more subtle alteration of the vegetation where shifting cultivation was followed. Given the large numbers of people in Hawaii when Europeans arrived, it is safe to assume that any land that could be cultivated had been cultivated at least once. The more level lands adjacent to sites of permanent human settlement were highly humanized landscapes by the time Cook and the others came on the scene. Most of the crop plants appear to have been confined to the areas of cultivation, but it is possible that man assisted the coconut to disperse in much of the Pacific world—and perhaps to the west coast of Central America in pre-Columbian times. In addition, a number of unwanted plants (weeds) were probably inadvertently carried to Hawaii by the early human pioneers, but this has not yet been studied.

The early years of the haole invasion were not marked by significant changes in the vegetation. However, this changed abruptly when a lucrative market was discovered in China for sandalwood (*Santalum freycinetianum*). Distributed on many Polynesian islands, sandalwood was prized for its aromatic qualities and usefulness in fine cabinetry.

Sandalwood trading was organized chiefly by United States traders, who tied it

Figure 8.6 *Pineapple plantation, Maui, Hawaii.*

in with the sea otter fur trade from the west coast of North America. The cutting was carried on with scant regard for conservation practices or the ecology of the forests from which the wood was being removed. There is evidence that King Kamehameha I made some effort to conserve this valuable tree, but he had only limited success; after he passed from the scene, no restraint at all was practiced. One of the methods used to locate the trees was setting fire to a forest and then following the incense odor. Naturally, this practice injured forest tracts in general; it probably reduced permanently the never-too-abundant forest resources of Hawaii. The social aspects of the forest exploitation involved a major dislocation of Polynesian societies with the resulting rapid shattering of their cultural systems, which had already been threatened by the white invasion.

The next major vegetation changes in Hawaii were initiated by the development, in the nineteenth and twentieth centuries, of plantation agriculture. At first the plantations were based on the production of sugar cane; later, pineapples were added (Fig. 8.6). Agricultural use of the land actually involves only about 10 percent of the total area, but most of it is intensively cultivated. Thus, the former vegetation cover is completely altered, and usually completely obliterated. The landscape is one of undulating fields of sugar cane or pineapple or, to a much lesser degree, some of the frost-sensitive crops for which there is

a mainland market. Most of the intensive agriculture is on Oahu, but the other islands also have their share.

Introduction of exotic plants became important after 1778, and probably reached a peak in the nineteenth century. So many plants were introduced—said to be more than 2000 species—that the plant geography of the islands is now quite different from what it was in 1778, which in turn, was different—though to a lesser extent—from what the first human (Polynesian) pioneers encountered in an earlier day. Serious problems have resulted from some of the more recently introduced exotic plants, particularly lantana (*Lantana*), which is an extraordinary pest in Hawaii.

To summarize briefly, the vegetation of the Hawaiian islands has been altered by man since the time of the first human arrivals. Until 1778 the changes were related to agricultural practices and the introduction of a few exotics. In the post-contact period, there was serious forest destruction associated with the search for sandalwood trees and, later, the introduction and establishment of plantation agriculture. During the past 175 years, there has been a major wave of plant introductions, which has modified the ecological and floristic nature of the Hawaiian ecosystems.

New Zealand. Only a few of the crops grown by the Polynesians in the tropical Pacific were suitable for the more middle latitude conditions of New Zealand, so the list of effective crop-plant introductions in prehistoric times is limited. The sweet potato, known as the *cumara* among the Maori, was the mainstay food plant, but taro, yams, ti (*Cordyline* sp.), and gourds (*Lagenaria*) were successfully introduced and cultivated. It can be assumed that a number of exotic weeds also arrived with these people but the nature of these introductions, if they occurred, is not known at present.

In recent years, attention has been focused on the possibility of human impact on the vegetation cover before the arrival of Europeans. Part of this interest revolves around a postulated Moa hunting culture thought by some to have predated the arrival of Polynesians and the introduction of agriculture. It has been suggested that these hunters employed fire to such an extent that the vegetation over millions of hectares of land was altered into a tussocky grassland. However, although the evidence for this hypothesis seems valid, it falls short of proof.

The impact of the Maori (Polynesians in New Zealand) on the vegetation took two principal forms: clearing, associated with shifting and sedentary cultivation, and burning the forest, to encourage the spread of the New Zealand ferns (Fig. 8.7). The roots of these ferns were a major source of food, often more important than that provided by the introduced crops. If this fern was as important as the documents state and if the Maori understood that firing increased its potential yield, this is an example of plant-cover modification for an unusual reason, a reason that lies between simple plant gathering and out-and-out plant cultivation. In any case, human-set fire played an important role in the ecology of New Zealand well before the arrival of the Europeans. And, because the overwhelming majority of the Maori population lived on North Island at the time of Cook's arrival, it may be assumed that the relationship be-

Figure 8.7 *Tree ferns* (Cyathea dealbata), *New Zealand.*

tween fire and ferns was of the greatest significance there.

With the arrival of Captain Cook in 1769, attempts were begun to introduce new plants into the area. These attempts, which focused on seeds brought from England, were not successful. However, by the early part of the nineteenth century, perhaps a little before, exotic plants were being successfully introduced by accident and by design.

In the ecological history of New Zealand, the nineteenth century was marked by an almost feverish activity to introduce as many exotic plants and animals as possible. Unfortunately, the time and place of the vast majority of plant introductions went unrecorded and only appear in the record after having become an established part of the ecosystems. One writer indicated that by 1920, over 600 introduced plant species had become thoroughly established in the wild; that is, they were reproducing and maintaining themselves without human assistance. It is clear that the overwhelming majority of the introductions were accidental, their having accompanied plants or plant parts that were deliberate introductions.

During the last half of the nineteenth century, and extending for a time into the

Figure 8.8 *Small farm in New Zealand located in a once forested area.*

twentieth, a number of societies were formed in New Zealand to aid and abet the introduction of plants and animals. These *acclimatization societies* (as they were designated) were often indefatigable in their efforts to ransack the world's ecosystems for living things that might be established on what is, after all, a rather modest-sized portion of the earth's surface (approximately 268,000 square kilometers). The deliberate introductions were either for economic use or for esthetic satisfaction.

The plants introduced for economic reasons include the crop plants and tree species that today are of great economic significance to New Zealand. Among the important crop plants were wheat, white potatoes, oats, and barley (Fig. 8.8). Pasture plants, including clovers, rye grasses, and others useful for feeding livestock also formed an important part of the introductions. This is the appropriate place to mention the introduction and widespread planting of turnips (principally on South Island), although they are not a grass, because their tops and roots were grown for sheep food.

Although the native New Zealand forests contained a variety of tree species with fine timber qualities, there was a complete lack of true pines. The times and places of pine introductions are not well

documented, but they must have occurred during the early part of the last half of the nineteenth century. Many conifers (a group that includes pines and other trees) were experimentally introduced, and by the early part of the twentieth century, at least fifty species, principally from the British Isles and North America, were es-

Figure 8.9 *Monterey pine* (Pinus radiata) *forest, New Zealand.*

tablished. Only a few have had significant ecological and economic impact and, of these, the leader is the Monterey pine (*Pinus radiata*), a native of the North American west coast. This is particularly interesting because the tree is not common or widely distributed in its native area; but in New Zealand, thanks to human agency, it has encountered ecological conditions much better suited to it than those in its native habitat. So great has been the *economic* success of this introduction that the export of lumber from the Monterey pine exceeds the value of that from any other native or exotic tree in New Zealand.

The area devoted to *tree plantations,* as they are termed in New Zealand, now exceeds 500,000 hectares, and a substantial part of this area is devoted to exotic trees—of which *P. radiata* is the most important (Fig. 8.9). However, the importance of other species is increasing and thus also the area devoted to their cultivation. In Table 8.1 is a list of the more important exotic trees now present and cultivated in New Zealand.

Thus far we have discussed only the *additions* to New Zealand's flora. The opposite side is the *destruction* of the native vegetation. Beginning about 1840, livestock, particularly sheep, were introduced into New Zealand and a major economic focus on producing wool rapidly developed. However, the areas of grasslands were small, so extensive forested tracts were quickly destroyed, usually by fire, to

Table 8.1
Exotic Timber Trees in New Zealand

Common Name	Botanic Name	Native Region
Softwoods		
Eastern white pine	*Pinus strobus*	Eastern North America
Ponderosa pine	*Pinus ponderosa*	Western North America
Lodgepole pine	*Pinus contorta*	Western North America
Monterey pine	*Pinus radiata*	Western North America
Loblolly pine	*Pinus taeda*	Eastern North America
Slash pine	*Pinus elliottii*	Eastern North America
Austrian pine	*Pinus nigra*	Europe, North Africa
Maritime pine	*Pinus pinaster*	Western Mediterranean
Scots pine	*Pinus sylvestris*	Northern Eurasia
Mexican weeping pine	*Pinus patula*	Mexico
Redwood	*Sequoia sempervirens*	Western North America
Western Hemlock	*Tsuga heterophylla*	Western North America
Canoe cedar	*Thuja plicata*	Western North America
Hardwoods		
Western catalpa	*Catalpa speciosa*	United States midwest
European ash	*Fraxinus excelsior*	Europe
English oak	*Quercus robur*	Europe, North Africa
Crack willow	*Salix fragilis*	Europe
Poplar	*Populus* spp.	?
Blue gum	*Eucalyptus globulus*	Australia
Turpentine	*Syncarpia laurifolia*	Australia (?)
Silver wattle	*Acacia dealbata*	Australia
Black wattle	*Acacia decurrens*	Australia

Figure 8.10 *Clearing brush and trees with fire, New Zealand.*

provide more pasture land. This destruction was on a large scale and altered the ecology of New Zealand to an extent that only now is coming to be appreciated. Few records were kept of this land clearing, so we must compare the general descriptions made at the time of European entry with the current landscapes. This is sufficient, however, to show that the alteration was massive and ecologically destructive. The fires, which have been vividly described by eyewitnesses, call to mind similar and greater conflagrations that occurred in North America (Fig. 8.10).

Although South Island was first to receive the attention of British settlers, North Island experienced more forest removal. This was caused by a number of factors, many of which derive from the greater farming and dairying opportunities on North Island than on South Island.

Today, the plant geography of New Zealand is almost as much the result of human activity over the past 125 years as it is of the biogeographical processes that have extended over hundreds of thousands of years.

INFLUENCES ON ANIMAL LIFE

Hawaiian Islands. The impact of the first Polynesians on the zoogeography of the Hawaiian Islands was probably limited to the introduction of a few domesticated animals and a few commensals and parasites. The domesticated animals were the chicken (*Gallus gallus*), the pig (*Sus scrofa*), and the dog (*Canis familiaris*). All three of these managed to establish feral populations, although feral dogs were rare in the pre-European period.

The commensals and parasites have never been systematically listed or studied but certainly included the Polynesian rat (*Rattus exulans*). This animal may have been brought in deliberately because it is known that high-born Hawaiians engaged in rat hunts with special tiny bows and arrows. An accidental introduction is probably represented by the small lizards (family *Scincidae*), which seem to have been commensal travelers with the Polynesians in their voyages around the tropical Pacific. It may also be assumed that the Polynesians, like most of mankind, were the unwilling hosts to an array of animal endo- and ectoparasites and that these were carried to the islands.

The advent of the European period marked the beginning of sharp changes in the fauna of the Hawaiian Islands— changes due to the reduction or extinction of native forms on the one hand and the introduction of exotic forms on the other. The destruction of native animal life probably began with the destruction of the forest for sandalwood. Of all the influences on the native animals, this was probably the most significant, for most of the species of birds that disappeared forever from the Hawaiian ecosystems were forest-dwelling. Other reasons cited as the causes for these extinctions are competition from introduced exotics (but most of the introductions do not live in forested ecosystems) and the feather trade of the early twentieth century (this certainly did not help the reduced populations of some of the birds and may have been the deciding factor at the end). Some people believe that the Polynesians were chiefly responsible, because the ruling class collected large numbers of the more brightly hued species and used their feathers to fashion beautiful feather cloaks. There is some substance to this argument, at least to the extent that some species may have been overexploited. There is no evidence, however, that any species became extinct as a direct result of this activity. Fourteen bird species are now believed to be extinct, and at least six forms are considered in danger of becoming extinct.

As you must have noted, our discussion has concerned only birds. This is no oversight. The Hawaiian Islands have only one native land mammal—the Hawaiian hoary bat (*Lasiurus cinereus*). This animal is not common and seems not to have been common even 100 years ago. Taxonomically, it is closely related to New World forms and is accorded only subspecific status; this suggests that it arrived on the islands relatively recently (in a biogeographical sense). The animal requires forested areas for shelter, so it may be supposed that the reduction of forest cover has had some effect on the size of its population.

One native marine mammal has also been threatened by man. This is the Hawaiian monk seal, which is found only around a few coralline islands to the west of the high islands. During the period of whaling (and sealing) in the nineteenth century, these tame animals were brought to the verge of extinction. By chance, the species survived, although in numbers so small that their future survival is questionable.

Animal introductions during the Polynesian period were modest in number and, with the exception of the pig, of limited ecological significance. Since the establishment of non-Polynesians in the islands, however, that situation changed markedly and many animals were liberated. We shall discuss only vertebrate introductions, but many invertebrates were also introduced. Of these, the giant African snail has been most significant. Being a crop pest, it is a constant potential menace to the agriculture of the warmer parts of the American mainland.

A large number of mammal species were introduced for economic or esthetic reasons. In the last decade of the eighteenth century, Captains Cook and Vancouver introduced cattle (*Bos taurus*), goats (*Capra hircus;* Fig. 8.11), and sheep (*Ovis aries*). The animals were quickly dispersed about many of the high islands and it seems that feral populations came quickly to be established—an understandable situation because native Hawaiians had no experience with herding or herd animals. Other introductions of these three animals, representing various breeds, have occurred over the years. Today, most of the cattle and sheep are raised under human control, but feral populations still exist on some of the islands. However, these live only in extremely difficult terrain and are generally not well regarded by either ranchers or ecologists. Goats are currently of little economic value, so most of them probably exist in a feral condition; they are regarded by many as an interesting and desirable species for sport hunting. These feral herbivores have, at times, played important roles in the ecosystems of the islands through their grazing and browsing habits, but to what extent they influenced the ecology of native animals is not known.

In addition to the economically impor-

Figure 8.11 *Feral goats, island of Hawaii, Hawaii.*

tant mammals introduced, others were introduced for sport hunting. These include the axis deer, mule deer, and pronghorn (a very recent attempt to establish a game species in the islands).

Several small carnivores have been introduced, including the dog (breeds other than the Polynesian ones), cat (*Felis catus*), and mongoose (*Herpestes auropunctatus*). The mongoose was introduced in the latter part of the nineteenth century to control rodents introduced by man. It is interesting to note that the stock came from Jamaica and, as noted earlier, was not native there either. Both cats and dogs managed to establish feral populations. In the case of cats, this may be an important factor in the ecology of some native and introduced bird species. The feral dog numbers appear to be low but at times are pests to the cattle and sheep. It is also believed that feral dogs interfere with the Hawaiian goose (*Branta sandvicensis*), a rare species found only in the Hawaiian Islands.

In the time since Cook dropped anchor in Hawaii, there have been attempts to introduce more than 100 species of birds. Most of these attempts failed, but a sufficient number succeeded to result in a significant change in the avifauna of the islands. Indeed, the exotic species now outnumber the surviving land birds native to the islands. To present an idea of the geographic range of the exotics, they have

been arranged by place of origin in Table 8.2.

There is mixed opinion about the ecological effects of these bird introductions. It was long believed that the European sparrow and the myna were responsible for the decline of certain native bird species. But as noted above, most of the native birds that are in trouble are forest forms and thus remain largely outside the ecosystems occupied by these exotics. On the other hand, the future evolution of native Hawaiian birds probably will not follow the same courses they would have if such a wide assemblage of exotics not been introduced. For in many cases the exotics have occupied niches not previously occupied by (known) bird species. Thus these niches are not available for any evolutionary radiation that might otherwise have occurred among the native forms.

Of the remaining classes of vertebrates introduced to Hawaii, the most notable are freshwater fish. Hawaii is a true oceanic island; that is, it was never joined to a continental mass. Whatever life it has, had to get there by passing over the formidable physical barrier of the Pacific Ocean. Thus it is not surprising that there are no freshwater fish native to the islands. Man compensated for this by introducing several fish species that are valued for their recreational fishing attributes. Black bass (*Micropterus*) and rainbow trout (*Salmo gaird-*

Table 8.2
Exotic Birds in Hawaiian Islands

Common Name	Zoological Name	Region of Origin
California quail	*Lophortyx californicus*	Nearctic (west)
Chukar partridge	*Alectoris graeca*	Palearctic
Japanese quail	*Coturnix coturnix*	Palearctic
Chicken	*Gallus gallus*	Oriental
Ringnecked pheasant	*Phasianus colchicus*	Palearctic (east)
Green pheasant	*Phasianus versicolor*	Palearctic (Japan)
Pea fowl	*Pavo cristatus*	Oriental
Turkey	*Meleagris gallopavo*	Nearctic
Pigeon	*Columba livia*	Palearctic
Spotted dove	*Streptopelia chinensis*	Palearctic
Barred dove	*Geopelia striata*	Oriental
Skylark	*Alauda arvensis*	Palearctic (west)
Yamagara	*Parus varius*	Palearctic (east)
Chinese thrush	*Garrulax canorus*	Palearctic (east)
Pekin robin	*Leiothrix lutea*	Palearctic, Oriental
Mockingbird	*Mimus polyglottus*	Nearctic, Neotropical
Shama thrush	*Copsychus malabaricua*	Oriental
Bush warbler	*Cettia diphone*	Palearctic
Indian myna	*Acridotheres tristis*	Oriental
White eye	*Zosterops japonica*	Oriental (Japan)
Strawberry finch	*Estrilda amandava*	Oriental
European sparrow	*Passer domesticus*	Palearctic (west)
Western meadowlark	*Sturnella neglecta*	Nearctic (west)
Cardinal	*Richmondena cardinalis*	Nearctic, Neotropical
Brazilian cardinal	*Paroaria cristata*	Neotropical (south)
House finch	*Carpodacus mexicanus*	Nearctic, Neotropical

neri) were introduced into several streams and reservoirs with success. More recently, a Neotropical fish known as the tacunaré (*Cichla ocellaris*) was introduced into a few reservoirs with apparent success. This is interesting because the tacunaré is a member of the family Cichlidae that occurs in the Neotropical, Ethiopian, and Oriental regions, and although adapted to tropical conditions, it was unable to cross the Pacific saltwater barrier until assisted by humans.

New Zealand. New Zealand must surely stand at the forefront of regions of similar size whose zoogeography has been modified by human agency. Indeed, so great are the human-induced faunal changes that one feels an entirely new zoogeographical region has been created.

The story of human-induced faunal change begins with the arrival of the pre-Polynesian pioneers, but it is not yet known whether they introduced the so-called Maori dog or Polynesian rat. In many cases, from the few midden remains found, it has been inferred that the pre-Polynesian settlers were hunters. They hunted, among other animals, the now-extinct Moa birds, large flightless birds that wandered over the South Island. Because it has been established that the Moas occurred well within the period of human occupance of New Zealand, these hunters have been held responsible for their extinction. They are sometimes characterized as the Moa hunter culture, but there is scant evidence that Moa populations were large enough to have permitted the degree of hunting specialization suggested by this designation.

We have seen, however, that island biotas are fragile, particularly where the islands are distant from mainlands. Thus we may assume that the advent of man in New Zealand had a marked effect on at least parts of the faunal components of the island ecosystems. If the people depended on land animals for food it is likely that some of the exploited taxa suffered serious population reverses, if not total extinction. Moreover, the later Polynesian introduction and spread of agriculture, specifically shifting cultivation, must have resulted in the profound alteration of many of New Zealand's ecosystems. This means that some, perhaps many, animal species had difficulty in retaining suitable habitats. Native New Zealand land vertebrates are mostly birds (two species of bats represent all the native mammals), so the impact must have been concentrated on this class.

The present available evidence suggests that the late Pleistocene environments of New Zealand (this was before *Homo* arrived) were limiting in the biological sense and that forested areas were reduced in area. If this interpretation of the fossil pollen record is correct, the faunal richness or poverty must be examined against this ecological backdrop. The islands may have been impoverished faunally by having passed through an ecological stress period before man arrived. And the scene that greeted man may have been one of ecosystems in a still delicate and fragile period of recovery and reconstitution in the post-Pleistocene. Perhaps one must try to understand the ecological relationships of the first human settlers with the island ecosystems within this framework.

Although desultory attempts at settlement were made in the years immediately following Captain Cook's exploration of the coasts of New Zealand, no important

European settlement took place until about 1840 and in the first years, settlement was concentrated on South Island. However, earlier there had been attempts to establish exotic animals, chiefly farm animals, to provide a food base for the settlers that would arrive later. Captain Cook himself is credited with the introduction (and liberation) of goats and pigs. In the first half of the nineteenth century, various animals were deliberately and accidentally introduced, but in the second half of that century, major human efforts were expended to establish exotic animals from an impressive share of the major taxonomic units. No major world zoogeographical region was overlooked, but most of the deliberate and accidental introductions were from the western parts of the Palearctic and Nearctic regions.

To aid these activities, so-called acclimatization societies were established; half a dozen were active at one time or another. In addition, the New Zealand colonial government did much to foster the introduction of exotic animals. Unfortunately, the government often based its efforts on concepts of ecology that we now recognize to be erroneous so the ecological results were often undesirable.

Of some fifty-three species of mammals that were the object of introduction attempts, thirty-four became established in a wild state. Of these, several became extremely troublesome in respect to their impact on the ecosystems into which they were introduced. For example, with the benefit of hindsight, it appears most incredible that especially vigorous efforts were made to establish the European rabbit (*Oryctolagus*) and the European hare (*Lepus europaeus*). Weasels and other mustelids had to be introduced to "control the rabbits." Although a wide array of vertebrates and invertebrates were introduced, we shall focus most of our discussion on the vertebrates—mammals, birds, and fish. Reptiles and amphibians will not be discussed because they have played little part in the story of human-in-duced animal introductions into New Zealand.

Representatives of seven mammalian orders from the Nearctic, Palearctic, Oriental, and Australian zoogeographical regions have been successfully introduced into New Zealand. Of these orders, members of the order artiodactyla (the even-toed ungulates) have had the greatest ecological impact.

The artiodactiles were especially attractive candidates for introduction because the English, who liked to hunt found not one land mammalian species suitable for hunting. Moreover, this order includes some of the more economically important domesticated mammals such as sheep, goats, and cattle. As previously mentioned, these three species managed to establish feral populations, but only the goat became well established. Feral sheep and cattle are rare, and local in distribution. Artiodactiles that were established for the purpose of hunting include the Himalayan thar (*Hemitragus jemlaicus*), axis deer (*Axis axis*), and sambar (*Cervus unicolor*) from the Oriental region; the chamois (*Rupicapra*), red deer "stag" (*Cervus elaphus*), and fallow deer (*Dama dama*) from the Palearctic; the Japanese deer (*Cervus nippon*) from the eastern Palearctic; and the North American elk (*Cervus canadensis*), Virginia white tail deer (*Odocoileus virginianus*), and moose (*Alces americanus*) from the Nearctic. Of this large assemblage, the red deer (Fig. 8.12) has been the most significant in terms of ecological impact.

The first red deer introductions and liberations took place in the last half of the nineteenth century, and over the years, many more such introductions occurred in both North and South Islands. The animal evidently found conditions much to its liking, for the rate of its population growth caused a great deterioration of the vegetation cover on watersheds, with the attendant further ecological damage. Thus, red deer, which are the principal game species

Figure 8.12 *Unusually fine example of a red deer* (Cervus elaphus), *New Zealand.*

over many parts of Eurasia, are classified as vermin in New Zealand, and the government has expended great efforts to eliminate or at least control their population. A bounty system has been introduced whereby hunters are paid for each deer killed, and the marketing of deer skins has been vigorously encouraged. These efforts have managed to check the size of red deer populations to some extent, but the animal is so well established that it resists most of the efforts directed against it. The large population in relation to the available food appears to be the chief reason for the small size of the adult deer and the fact that males seldom attain the great spread of horn that makes them a favorite trophy animal in their native regions. Although much study remains to be done on the ecology of red deer in New Zealand, it is clear that they are a major influence on the islands' ecosystems.

The other artiodactiles have not had so obvious an impact on the ecology, but each species does exert some influence. The general result of the introductions of this group of hooved mammals is that their populations tend to get out of hand easily and it sometimes requires drastic measures to keep them in check. This has certainly been true of the Himalayan thar and the feral goats.

There are several reasons for such population growth, but a major one is the lack of large predators. (At one time there was talk in New Zealand of overcoming this lack by introducing a species of big cat. Fortunately, reason prevailed and such an introduction was not attempted.) It is possible that the extraordinary assemblage of large mammals will eventually achieve some kind of ecological equilibrium in the islands. However, it is certain that New Zealand will always be different ecologically and biogeographically than it was around 1769.

In addition to the artiodactiles, many other mammals have been established in the wild in New Zealand. One that has given a great amount of ecological trouble is the wooley, or Australian opossum (*Trichosurus vulpecula*), a marsupial native to Australia. It was first liberated in the middle of the nineteenth century to provide a forest product of furs, and a number of later liberations also took place. Some time elapsed before it became established, and, indeed the government had to protect it from illegal trapping during the early decades. This period of protection and clandestine trapping appears almost ridiculous now in light of the fact that the animal has become one of New Zealand's most serious ecological problems. Although it required time to become established, this exotic then exploded in numbers—similar to the way the European starling did in North America. The obvious effect of this mammal on the island's ecosystems is that it has been extremely destructive to vegetation, particularly trees. Thus it may be having some impact on the ecology of the native bird species.

Turning to birds, at least 130 species have been brought to New Zealand since the time of Captain Cook, but of these only 24 have become established in a wild condition. However, four of the six major zoogeographical regions are represented

by these successful introductions, namely, the Nearctic, Palearctic, Oriental, and Australian. Most of these introductions were made to provide sport for hunters or to provide song birds familiar to the settlers who, in the early years, came chiefly from the British Isles. Although no careful ecological study has yet been published, it appears that the most harmful introduction was that of the European starling (*Sturnus vulgaris*). This species' fertility is well known, and its tendency to gather in large flocks when roosting or feeding plus its taste for many orchard crops makes it a great pest.

The geographically isolated position of New Zealand, which is a major reason for the paucity of native mammals, also accounts for the fact that there are no known native primary freshwater fish. The settlers soon set about rectifying this biogeographical "oversight." Greatest interest was shown in the salmonids, and in spite of excruciatingly slow and unreliable transport in the nineteenth century, attempts were made to introduce these and other exotic fish. After various attempts had failed, brown trout (*Salmo fario*) were successfully introduced into South Island and have since spread widely. Other successful introductions were American brook trout (*Salmo fontinalis*); rainbow trout (*Salmo gairdneri*), which were spectacularly successful; mackinaw trout (*Cristivomer namaycush*); quinnat salmon (*Oncoryhnchus* sp.); sockeye salmon (*Oncoryhnchus* sp.); carp (*Cyprinus carpio*), which, as usual, was a serious ecological mistake; tench (*Tinca tinca*); American catfish (*Pimelodus* sp.); and perch (*Perca fluviatalis*).

A large array of invertebrates have also been introduced into the islands. Some of these, such as the honey bee (*Apis mellifera*), which is now well established in the woods and forests, have been deliberate introductions. In contrast to this useful and probably ecologically sound introduction are the many insects that arrived unnoticed with cargoes. Many of these have become serious economic pests. Virtually nothing is yet known of the ecology of exotic insects in New Zealand, but it is certain that the insect fauna of the islands have been profoundly altered by man, and these alterations must have resulted in fundamental ecological changes.

Given the unbalanced nature of the native vertebrate fauna of New Zealand, it is not surprising that the group most affected by man is the birds. Earlier, we noted that the first human settlers may have been responsible for the extermination of the moas. In the period since Cook's arrival, at least three birds—a subspecies of the New Zealand laughing owl, the New Zealand quail, and the Stephen Island wren—have been exterminated. And a number of others—perhaps as many as eleven species—are on the verge of extinction. Of these, the most famous is the takahe (*Notornis mantelli*), which was at first declared extinct because it was known only from fossil and subfossil materials. During the last half of the nineteenth century, an occasional specimen that had only recently been killed was found, and there were also rare sightings recorded. Nevertheless, the bird was declared extinct. Then in 1948–1949, a concerted search was made of the isolated marshy ecosystems on South Island and the bird was found to be still living. It is now accorded the most careful protection possible.

From the various accounts of recent avian extinctions and near extinctions in New Zealand, it appears that many of the animals involved had small populations when Europeans first took notice of the bird life. In addition, the later population reductions of those forms reported as common in the earlier days of European settlement appear closely related to forest destruction—especially the destruction of southern beech (*Nothofagus*). Considering the magnitude of the ecological assault by man in this region, one is impressed by the number of avian species that have managed to survive. The most exasperating feature is that there has been so little study made of the ecology of the birds—threatened and nonthreatened. Thus, we are left to wonder what the overall ecosystemic impact of avian extinctions or population reductions has been in this nation.

INFLUENCES ON WATER, SOILS, AND MICROCLIMATES

Hawaiian Islands. The hydrography of the islands of Hawaii is, in the main, poorly developed, as one would expect of such limited land areas. The streams follow short and often precipitate courses over youthful geological structures to the always near Pacific Ocean. Little can be said about pre-European modification of the hydrography. It is known that there were areas where some terraced agriculture was developed, and it may be that some minor amount of water diversion accompanied these activities. But for the most part, the Polynesians depended on the natural rainfall to provide the water needed for their crops and located their villages where freshwater for domestic use could be easily obtained from natural sources.

Later, when plantation agriculture was established, a need grew for modification of the existing patterns of water supplies. Thus, dams and canal systems were developed by the sugar cane and pineapple growers and by civil governments which had to provide for the increasing requirements of the larger settlements. Most of the high islands have had their basic hydrographies changed during the past century.

Soil erosion during the pre-European period appears to have been of limited significance. However, in modern time, soil erosion has often presented serious problems, particularly in association with the cultivation of pineapples (Fig. 8.13). The erosion is due chiefly to the practice of ignoring the need to plant on the contour. Slopes have sometimes been plowed and planted at right angles to the contour—which all but assures accelerated soil loss. In some of the pineapple-growing districts of Oahu, there has been important soil loss from the fields and even erosion of the dirt roads to levels below that of the surrounding fields.

At least as dramatic but more widespread are soil losses caused by overgrazing. As we have seen, mismanagement of grazing animals may set off a series of events that result in major alteration and serious degrading of an ecosystem. And so it has been in several parts of Hawaii. Evidences of accelerated soil erosion are visible wherever livestock have been kept for fifty or seventy-five years; in some instances, the ecosystems are so degraded that they are all but useless. This is best illustrated on the island of Kahoolawe, where the overgrazing of sheep, cattle, and

Figure 8.13 *Erosion in a sugar plantation, Hawaii.*

goats has laid bare the top soil to winds and periodic torrential rainfall. The result has been the removal of almost all the soil—to a depth of almost five meters in some places.

With so much recent modification of the plant cover of this island region, it follows that there has been significant changes in the microclimates. Although there are a number of studies of the microclimates of the cultivated areas it is difficult to compare conditions with those that occurred under forest cover in the same region. But microclimatic changes are always concomitant with vegetation alteration, and such changes are ecologically significant even though we lack the studies necessary to document the assertion.

New Zealand. In pre-European times, there was little alteration of the hydrography of New Zealand. It may be assumed that where native agriculture was oldest and most intense, runoff was somewhat modified; and it is possible that under local circumstances, marked changes occasionally occurred. But the Europeans must have encountered an essentially non-human-modified hydrography when they first settled.

Modifications started with the extensive clearing of the forests and the widespread practice among sheepherders of setting fire to the natural and human-induced grasslands (Fig. 8.14) to encourage the growth of new feed for the stock. These practices led to the laying bare of great tracts of land, with attendant rapid runoff during periods of precipitation. It is a point of debate in New Zealand whether the frequent floods have always been a natural part of the scene. Certainly, flooding does not depend on human agency, but forest destruction such as that which occurred in New Zealand all but guarantees that floods will increase in number.

The other major aspect of hydrographic change is related to the construction of

Figure 8.14 *Tussock grassland, South Island, New Zealand.*

numerous hydroelectric plants on both North and South Islands. New Zealand has few rivers of any great length, but it does have an impressive number of streams, which carry large quantities of runoff water to the ocean. Because the little nation has almost no known petroleum and the coal is insufficient to meet the power needs of the people, it is not surprising that great importance has been attached to developing hydroelectric plants. Large sums of government revenues have been allocated for this purpose.

The many hydro plants in New Zealand are located on both islands and have recently been tied together by a cable running under Cook Straits. Most of the to-be-developed hydro power is in the western part of South Island; it is here that the largest scheme is now being developed. This involves damming a river and diverting water through a tunnel and down a slope of more than 200 meters to a power station. From there it will travel through a huge tunnel to the ocean. In terms of the New Zealand economy, the cost of this one development is large, but the economic demand seems to justify the expense. By this and other hydro sites, New Zealand is in the process of bringing under human control a large share of its once wild streams.

Accelerated soil erosion has been a major ecological and economic problem in

New Zealand for over a century. It is tied closely to a number of causes—forest removal, the burning of grasslands, the overgrazing of sheep (which, in turn, is often in response to market conditions for wool and the demands of financing institutions), and the feeding habits of introduced animals. In short, in New Zealand we can see clearly that accelerated soil erosion is often a part of a system, malfunctioning, to be sure, and that often a large number of interlocking factors are responsible for the condition.

In terms of total area affected, accelerated soil erosion is the most important aspect of human influence on the earth's crust in New Zealand. However, in more local situations, dramatic alterations resulted from mining, which was an important economic factor in the nineteenth century. Most of the mining was for alluvial gold and this was focused on South Island in the Otago district. The primary discovery was made in 1861—just in time to attract the miners who had been unsuccessful in the California gold rush of the preceding decade. As is usual with placer mining when it is conducted on a significant scale, many piles of tailings and the holes from which gold-bearing materials were removed were left behind.

The industrial plant of New Zealand is far too small to be a significant factor in the contamination of the global atmosphere.

In addition, geomorphic configurations and latitudinal position of New Zealand are such that only rarely, and then only locally, do smog conditions prevail. Nevertheless, as repeatedly noted, we must assume that man has brought into being entire new complexes of microclimates in rural and urban areas.

8.3 Madagascar

CULTURAL BACKGROUND

Few areas present a more tangled puzzle in their prehistory than the enormous island of Madagascar. Although dwarfed by the African land mass four hundred kilometers away, Madagascar contains almost 575,000 square kilometers—an area greater than France, Belgium, and the Netherlands combined.

At present there is no general agreement about who were the first humans to arrive on the island and how they got there. The various arguments fall into two categories: the first arrivals were Africans of small physical stature; the first arrivals were from insular Southeast Asia and spoke a language of the Malayo-Polynesian language family. Most scholars believe that the Southeast Asians were the first arrivals but then argue further over the route they used. One group thinks they followed a coastal route along the underbelly of Asia and the east flank of Africa. Another group thinks they took an open sea voyage across the Indian Ocean. Neither route can be substantiated at present.

The first *public* notice of the island was made by the Arabs, who were on the scene about A.D. 900. They established settlements among the native Malagasy peoples and began to import African slaves. In 1500, Portuguese explorers arrived and wrested control from the Arabs. The Portuguese, in turn, were pushed aside by the French and British. Finally, in 1896, the French gained full control. In 1960, however, the Malagasy Republic became independent, but it still retains close political and economic ties with France.

The present number of carbon 14 dates is far from sufficient to permit even an informed guess as to when man first arrived on Madagascar. If the Indonesians were in fact the first persons on the island, we may assume an arrival date sometime within the Christian era—perhaps between A.D. 300 and 500. For the purposes of discussion, we shall go along with this hypothesis—that the Indonesians were first on Madagascar, that they were followed by Arabs and Africans (perhaps alternating at times), who were, in turn, followed by Europeans.

The prehistory of the Indonesian settlers is so poorly known that trying to determine their land-use practices is extremely difficult. It does appear, however, that they introduced shifting cultivation with taro, rice, bananas, sugar cane, and, possibly, coconuts. When and by which routes the African component arrived is unknown but it is certain that Africans were responsible for the introduction of cattle, which are now a major element in the ecosystems of the island.

The Arabs undoubtedly contributed to the land-use practices of the people, but we do not yet know the details. The Portuguese influenced the cultural scene and were probably responsible for the introduction of one or both of the current staple crops—manioc and maize—although these could have come from Africa after their introduction to that continent.

Of the several European national groups that once held power in Madagascar, the French altered the culture of the island most significantly. They introduced not only a European form of government but also water sanitation, better housing, more angular street patterns, electricity, and better surface transport routes. They also made French the official language of Madagascar.

No reliable estimates of human population around 1500 exist, but it appears that in terms of the carrying capacity of the island, the population was small. In recent years, however, the population has been increasing rapidly, and now there are approximately 6.5 million persons—approximately twelve persons per square kilome-

ter. This may not appear excessive, but about 90 percent of the working population is engaged in agricultural (including pastoral) activities. Furthermore, the cattle population is estimated to be over 10 million. Thus, we have a setting for some unusual pressures on the ecosystems.

INFLUENCES ON VEGETATION

The prehistory of Madagascar is so incompletely known that it is almost impossible to determine the impact of the first invaders (whoever they were) upon the island's vegetation cover (Map 8.1a below, left). Moreover, current human interfer-

Map 8.1a

Map 8.1b

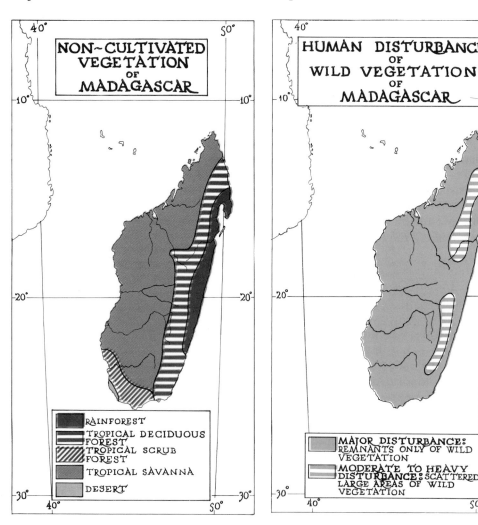

NON-CULTIVATED VEGETATION OF MADAGASCAR

- RAINFOREST
- TROPICAL DECIDUOUS FOREST
- TROPICAL SCRUB FOREST
- TROPICAL SAVANNA
- DESERT

HUMAN DISTURBANCE OF WILD VEGETATION OF MADAGASCAR

- MAJOR DISTURBANCE: REMNANTS ONLY OF WILD VEGETATION
- MODERATE TO HEAVY DISTURBANCE: SCATTERED LARGE AREAS OF WILD VEGETATION

ence with the vegetation cover is so great that it is difficult to hypothesize about what the natural vegetation was, say 2000 years ago. Map 8.1*b* on page 284 presents a general picture of the current conditions of the vegetation cover. To some extent, the present arrangement reflects environmental variables that have nothing to do with human agency. One example is the distribution of tropical rain forest along much of the eastern side of the island. Whatever man has done, his acts have been either amplified or damped by the climates of the island.

As has been repeatedly noted, shifting cultivation carries vegetation change with it. The degree, nature, and duration of such change is a function of time and human population size. Because we have no data on population size for the earlier period of human occupancy of this island, we can only draw attention to the *presence* of shifting cultivation, probably from the beginning of human occupance.

Even when we turn our attention to the period immediately after the first written records (approximately A.D. 900), we are not on much surer ground. We may assume but cannot prove that it was the Arabs who introduced coffee and were perhaps responsible for introducing cloves and the practice of growing rice in paddies.

As mentioned in earlier chapters, the Portuguese were often influential in spreading crop plants along their trade routes, and it is probably they who introduced maize, manioc, and some other important crops. The French, being interested in developing the economy of the island, were also instrumental in encouraging the commercial production of vanilla, sisal, and peanuts. (The three are of New World origin, but it is not clear who was responsible for their initial introductions into Madagascar.)

The accidental introductions of exotic plant species into Madagascar are said to number at least 1000. However, only one species has become established in habitats that were not disturbed by human agency. But, inasmuch as such habitats must now be rare, this should not be considered too important.

The major human influence on the vegetation of Madagascar is the removal of forest to make room for additional pastureland (Fig. 8.15). Not only were cattle adopted from East African herding cultures, but also the practice of employing cattle as a basic part of the status system, giving rise to herds kept on lands unable to support their numbers and thereby increasing pressure to extend the grazing land at the expense of tree cover. The result has been a wholesale destruction of the forest.

Although the extension of cattle pas-

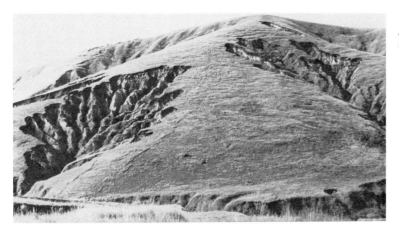

Figure 8.15 *Accelerated soil erosion after forest removal, Madagascar.*

tures appears to be the largest single cause of this destruction, the growth of human population must also be taken into account, as well as the fact that this population lives almost entirely on the land and must farm it in order to survive. Most Malagasy agriculture is of a subsistence nature; it has recently been estimated that for the overwhelming majority of citizens, the annual per capita income does not exceed $40. The implications of such poverty for the continued, growing pressure on the vegetation cover cannot be exaggerated.

So profound have the changes been in the forest cover that some authorities believe that old, mature forest can now be found only in the most remote mountain areas. The result of these changes, which are proceeding only slightly abated, is that almost all of Madagascar's ecosystems have been profoundly affected by man. One wonders if the ecological base of human existence has not been so eroded that any respectable future for the growing masses of humanity in this new republic is possible.

INFLUENCES ON ANIMAL LIFE

Because we know little about the prehistory of Madagascar, our discussion of human influences on animals must begin artificially recently. By the time records were kept, cattle had not only been introduced but had become well established as part of the general social and economic structure. It appears that their ecological impact was, for many years, localized—a reflection of the small human population size. Today, there is a general assemblage of small animals—dogs, cats, fowl—which arrived at various times. In marked contrast with New Zealand and many other islands, man introduced few animals into Madagascar. This must be explained by the cultural background of the settlers and the island's cultural history during recent centuries. For, unlike the British in New

Zealand, the Americans in Hawaii, and the Spaniards in the Caribbean, Europeans never supplanted the native Malagasy population in Madagascar. Thus, we shall consider the possible roles of man on animal extinctions or population reductions.

Madagascar is traditionally treated as part of the Ethiopian region, but we should view it as a zoological region by itself. Its native faunal elements came from both Africa and Asia and show the imbalance typical of oceanic islands. The island is a biogeographical oddity with a curious collection of plants and animals.

The most outstanding feature of the vertebrate fauna is the assemblage of lemurs (*Lemuridae*), which are primitive primates not now found anywhere else in the world. In addition, there is a large and varied avifauna, which has apparent relationships with Africa and Asia although a large component is endemic. A striking feature of the avifauna—as with the mammalian fauna—is the *lack* of elements common in Africa. This is also true of the reptiles and amphibians. As far as the modern vertebrate assemblage is concerned, the island did not have a Cenozoic land connection with Africa. There are no *primary* freshwater fishes—which lends further credence to the argument that Madagascar has long been isolated from a continental connection. Biogeographically, then, Madagascar is a large oceanic island possessing an unbalanced fauna.

A striking ecological aspect of Madagascar's fauna is that a major part of it requires tree cover in order to live, and the removal of trees has a deleterious effect on it. But, as we noted earlier, forest removal is precisely what has been occurring for the past century, especially for the last fifty years. It has progressed to the point where many forest animals, especially the lemurs and other lemur-like primates, are threatened. Unfortunately, there have been few studies of the lemurs on Madagascar, so there is almost no way we can determine how much these animals have been influenced by forest cutting and removal. Nevertheless, there is agreement among wildlife authorities that the primate fauna of the entire island is seriously threatened. Should forest cutting continue, there is little reason to expect that more than a remnant of this unique fauna will survive for many more years. The avifauna has been greatly affected by the same ecological changes. Some forest species, not found elsewhere in the world, are considered threatened. Were the situation better known, many other birds might be found to be in that category.

A FOOTNOTE TO THE ISLAND OF MADAGASCAR: THE MASCARENE ISLANDS.

Lying between 650 and 800 kilometers east of Madagascar are two of the most beautiful little islands in the world, which were first accorded human attention when a Portuguese captain found them in 1505. His name, Mascarenhas, was given to them (Mascarene Islands). In the order of size, they are named Réunion and Mauritius. Réunion (Fig. 8.16) is topped by a peak of volcanic origin that reaches slightly more than 3000 meters high; Mauritius, although also of volcanic origin, is not higher than about 800 meters. Both is-

lands, when discovered, were reported as being heavily forested and must have presented a lovely sight. Human settlement did not follow rapidly—it was not until the early eighteenth century that settlements of any appreciable size were made. Then followed a varied succession of British, Dutch, and French attempts to gain sovereignty, with the French, in the end, winning out on Réunion, and the British on Mauritius. From the slow beginnings, there later grew a human population too large for the islands to support, and the Mascarene Islands have become almost a synonym for serious human overpopulation. That this has had profound effects on the ecosystems of the islands goes without saying, but the most dramatic ecological event took place long before man was numerous there. The once common group of birds called dodos was completely destroyed through direct and indirect human acts—providing the world with the cliche "dead as a dodo."

The dodos were large, rather funny looking flightless birds which had no means of protection against the pigs, cats, goats, deer, rabbits, and monkeys introduced by the Portuguese. These mammals seriously interfered with the ground nesting activity of the dodos. In addition, the birds were often sought by meat-hungry mariners, who found them to be

Figure 8.16 *Advanced soil loss and gully erosion, Réunion.*

extremely easy prey. The result was that the dodos were gone from Réunion by about 1750 and from Mauritius shortly after 1680. They may have held out on Rodriguez, a more distant island, until the end of the eighteenth century.

Although popular attention has long fastened on the dodo when animal extinctions in this group of islands are discussed, many other bird species were similarly obliterated. It is believed that on Réunion Island, in addition to dodos, seven bird forms were exterminated and on Mauritius, in addition to dodos, eight bird species were eliminated. This becomes more significant when one learns that these extinctions represent more than half of the known Recent bird species on these islands.

INFLUENCES ON WATERS, SOILS, AND MICROCLIMATES (MADAGASCAR).

Very little study has been directed toward understanding the influences man has had on the water and microclimates of Madagascar. Thus, we shall confine our discussion to the soils.

As noted above, we do not know when cattle were first introduced into the island, although we are fairly certain that it was within the Christian era (Fig. 8.17). In any case, as herd and human populations grew, so did the area cleared for pasture. The accounts of travelers in the nineteenth century mention cleared forest tracts but convey the impression that much, if not most, of the island was still forested. (Probably large parts of those forests were second growth, because shifting cultivation was the standard agricultural system.) The sharp increase in human population with the attendant growth in cattle population probably dates from the time, late in the nineteenth century, when the French government introduced sanitation and medical facilities. However, whatever the causes, the Malagasy people have been increasing rapidly and removing tree cover rapidly.

The zoological implications of forest removal have been noted, but another and perhaps even more serious *ecological* effect of forest removal in Madagascar has been accelerated soil erosion. If Madagascar has one overriding ecological problem, it is soil erosion, induced by human misuse of the land. Over large parts of the island, accel-

Figure 8.17 *Country scene, Malagasy Republic.*

erated soil erosion has reached disaster proportions and, in spite of government efforts, this ecological scourge threatens to become much worse before it is brought under control.

Here we should note that most soil studies of Madagascar show that the island, by and large, does not have soils of high agricultural potential. The soils are tropical red soils and the island has long been known as the "red island" because of its soil color.

It may appear strange that an island so surrounded by mystery and so remote from tourist routes should be the theater for an ecological tragedy. But such is the case, and we see again that it is not only industrial man who misunderstands and destroys his ecological patrimony.

Chapter Nine

Human Influences on Marine Ecosystems

So vast are the expanses of the oceans and seas and so relentless their waves they have long symbolized nature's power and immunity from human influences. But, like the land, marine ecosystems have been influenced by man from early times.

It is not known when man first turned his attention to the oceans, but it has been suggested that the seashore was the favorite habitat for early man far back in the Pleistocene. However, there is little evidence to support such a view. Examining the habitats favored by living primates (other than man) we find that only rarely does a primate species regularly forage along the shore and that entering the water is even more rare.

Along the shorelines of many parts of the world, shell middens are common. These are garbage accumulations consisting largely of mollusk shells that were tossed aside after the meat was extracted. Some shell middens are so large that they are a distinctive geomorphic feature of the landscapes whereas others are so small that they easily escape attention. The ages of shell middens vary from place to place, but in general are of moderate antiquity—not

so old as hunting cultures, which seem often to have preceded the heavy dependence on marine mollusks. This suggests that man turned to marine sources for food only after the animal food resource of terrestrial ecosystems was no longer adequate, because fundamentally his economy was based on hunting.

Until relatively recently in the time span of hominid evolution, oceans were barriers to human dispersal just as they were to the dispersal of other organisms. Except for lands separated by narrow waterways that could be swum or crossed by raft, man was bound to the land on which he was born until about 15,000 years ago. And even then, navigation was along the shore, not offshore—at least there is no known evidence to the contrary. The ability to cross broad expanses of ocean was only recently acquired, but its acquisition enabled a new, dramatic chapter to be written in the history of human migrations.

The Mediterranean Sea was a cradle of marine navigation. It was on this sometimes calm and sometimes violent body of water that hundreds of generations developed the boat-building and sailor skills that made possible the ocean voyages of a later time. First the Egyptians, then the Phoenicians, the Greeks, and the Romans developed the basic skills which later allowed Mediterranean sailors to venture upon the ocean in search of Cathay.

But other navigation centers also developed—the Norse, around the North Sea and beyond, and the Chinese, from the Sea of Japan to at least the Philippine Islands. Perhaps the most dramatic were the Polynesian voyagers who, as we saw in Chapter 8, set off from the Asian mainland and discovered almost every island in the low latitudes of the Pacific Ocean.

The marine environment was not only a transport medium but also a source of food and raw materials. Man was no longer confined to the intertidal zone for food when he learned to use even the simplest boats. Moreover, fishing vessels and fishing gear are still being improved, for

the oceans continue to be regarded as one of the important sources of food and materials.

Virtually every nation that has a marine coast engages in some form of marine exploitation, but the technical sophistication of their efforts varies. In 1973, Japan led the world in the size and species diversity of its fish catch, the geographic dispersal of its marine fisheries, and its technological sophistication. However, a major contender for the technological lead is the U.S.S.R., and many European nations are also technologically advanced in marine fishing.

The oceans are now more important to our species than ever before. They provide a highway for the transport of bulk goods; they are used as dumping grounds; and they are a major source of animal products, a growing source of certain minerals, and, along the shore, a recreational area.

9.1 Intertidal Zone

The intertidal zone is the part of the marine environment where the substrate is periodically covered and uncovered by water as a result of tidal action. The width of the zone depends on the vertical range of tides and the slope and configuration of the shoreline. In some parts of the world, the maximum range between low and high tide is only a few centimeters, and in other areas, the range may exceed 13 meters. The intertidal zone often has many plants and animals, for it has a rich mixture of light and oxygen, as well as hiding places and other necessities of a vast host of living things. The substrate varies; it may include exposed bedrock, mud flats, sandy stretches, or loose cobble that rattle every time a wave hits.

The intertidal zone also varies in its locations. It may be an unprotected rocky headland which receives interminably the

pounding of the ocean's waves, or a system of bays and tiny coves, where the force of the waves are reduced or all but eliminated. There are also estuaries, in which the lower parts of rivers mingle with the marine waters. And along many coasts, saltwater marshes crossed by tidal channels (sloughs) are also an important part of the intertidal zone (Fig. 9.1(a) through 9.1(d)).

Primitive gathering activity, mentioned earlier, probably was the earliest human influence in this zone. If the size of the shell middens are indicative, some gathering cultures exerted extremely heavy pressure on the mollusks and other animals. Although most shell middens represent the harvests of many seasons, we must conclude that this predation may often have been of local biological significance.

The intertidal zone cannot withstand prolonged, intensive, and uninterrupted

Figure 9.1(a) *Rocky headland, Point Lobos, California.*

Figure 9.1(b) (*Below, left*): *Small bay, Massachusetts.*

Figure 9.1(c) (*Above, right*): *Estuary, North Carolina.*

Figure 9.1(d) *Salt water marsh.*

Figure 9.2(a) *Sea otter* (Enhydra lutris) *feeding on clams, California.*

Figure 9.2(b) *Dugong* (Dugong dugon).

human exploitation of its animal life, so we must assume that the gathering folk periodically eased up. The diminished resource probably forced the gleaners to look to alternate sources for food supplies. If they hadn't, the entire resource would have been destroyed. Even within the contemporary period, there is ample evidence that it is easy to overexploit this zone. For example, only a few decades ago, abalone (*Haliotis* sp.), a delicious marine mollusk, was one of the most abundant animals in rocky tide pools along much of the Pacific Coast of the United States. Today, it is rare to find one tiny abalone in the intertidal;

the commercial abalone is now being harvested by deep-sea divers well off the shore. Few coastal areas of even moderate human settlement do not have a history of a favored food animal that became scarce in the intertidal; whether the animal is a sand, rock, or mud dweller makes little difference.

However, it is not only the smaller animals that have been affected by human actions. Although some of the animals we shall discuss are not confined to the intertidal zone, it is often while they are in this zone or close to it that they are most subject to human predation.

Figure 9.2(c) *Steller's sea cow* (*now extinct*).

Figure 9.2(d) *Swimming manatee* (Trichecus).

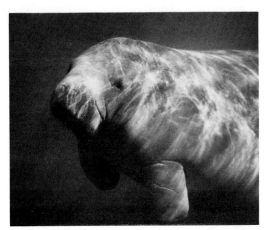

Mammals have been the most persecuted of the larger animals. These include the sea otter (*Enhydra lutris*); seals (several genera); walruses (*Odobenus rossmarus*); and all members of the marine mammal order Sirenia, including manatees (*Trichechus* sp.), dugongs (*Dugong dugon*), and Steller's sea cow (*Hydrodamalis gigas*), which is now extinct. Marine turtle populations have been greatly reduced through exploitation for food and raw materials (Fig. 9.2(a) through 9.2(d)).

Sea otters are found along parts of the west coast of North America; both a northern and southern subspecies are recog-

nized. The northern subspecies has withstood human pressures better than its southern relative. It was long believed that the southern form had been completely eliminated because of gross overhunting for its valuable fur. However, a small number were missed by the hunters; and these, having been given protection, formed the base for the population that may now exceed 100.

Members of the order Pinnipedia, which includes seals, sea lions, and walruses, have long been hunted for meat, hides, and other materials, such as the oil that can be rendered from some species. They have also been persecuted by fishermen along parts of the west coast of North America. Sea lions (*Zalophus* and other genera) in particular have drawn the gunfire of fishermen, who understandably resent the damage one of these animals can cause if it becomes entangled in a fishnet. In addition, fishermen believe that sea lions compete too successfully for the available fish, and this also results in the illegal slaughter of the animals.

When marine hunting was less industrialized than now, many species of pinnipeds were killed for the oil in their bodies, but the practice declined in most areas when pinniped numbers no longer justified the hunt. On the other hand, certain seal species have pelts that are much sought after in the fur trade, so these animals sometimes continue to be exploited by man. Perhaps the best known of this group is the northern fur seal (*Callorhinus ursinus cynocephalus*). So strenuous were the efforts of sealers at the end of the nineteenth and beginning of the twentieth centuries that the once abundant herds were reduced to about 130,000 animals. By international agreement between the United States, Russia, Canada, and Japan, hunting of this animal was placed under strict (and effective) control. The response was an increase in the population to an estimated 1.7 to 1.9 million animals; they are now harvested under strict controls (Fig. 9.3).

Another mammal, the sirenian known as Steller's sea cow, did not have such a happy fate. This huge mammal (it attained a length of 7 meters) first became known to Europeans in 1741, when Vitus Bering and Steller, a German naturalist, were shipwrecked in the Bering Straits region. They came upon moderate numbers of these animals on the shore and found them to be extremely easy to kill and their meat to be highly edible. When later sealers and sea otter hunters invaded the area, they killed these animals with little or no thought to protecting the species. By about 1830, the animals were all but gone. It appears that they were not numerous even when "found," and almost nothing was learned of their biology. It may well be that Steller's sea cow was a relict genus living far from the maximal habitats of sirenians, because the other two living genera occupy lower latitude positions.

The two living sirenians are the manatees and the dugongs. The dugongs are distributed around the Indian Ocean, along the east coast of Africa, intermittently eastward to Northern Australia, and north to the Ryuku Islands (Nansei-Shoto), a chain of small islands between Taiwan and Kyushu (Japan). The manatees are distributed along parts of the Atlantic littoral and Caribbean Islands from coastal Virginia (rarely) south to the Amazon River, but there are large distributional gaps over this range, such as in the Guiana region of South America. Manatees also occur along the West African coast from Senegal to approximately midway in Angola; they have been reported in

Figure 9.3 *Northern fur seal* (Callorhinus ursinus).

the Lake Chad drainage, but this has not been substantiated. However, they have been definitely recorded far up the Niger River, even beyond the great bend at Tombouctou.

Manatees are well adapted to fresh water but appear able to live just as well in salt water. They subsist chiefly on aquatic vegetation that they find in shallow waters. It is said that their meat has a good quality, so they probably have been long hunted—perhaps far back into the prehistoric period. When Europeans arrived in the areas where manatees occur, they, too, found the meat edible and there are various accounts of hunting these shy and inoffensive vegetarians. Thus, the animal's range is reduced, and its occurrence is rare in some places where it once was common. For example, it appears that the manatee is now less common along the African coast, but detailed data are lacking. Few studies exist on the ecology of manatees, but considerable interest has recently been shown in their herbivorous habit. For example, experiments have been conducted to determine if they could be used to reduce unwanted vegetation growing in irrigation canals in the coastal areas of Guyana, South America.

Dugongs have received more attention than manatees from biologists, so we know their numbers are declining over much of their range. Although a detailed population study has not been conducted over this large area, the reports of local investigators and fishermen's comments lead to the conclusion that the dugong may be a threatened species. The meat of the animal is well regarded in many parts of its range although not everywhere—a fact that may have contributed to its survival. Occasionally, local dugong populations are reduced even when they are not the direct object of hunters. For example, in Northern Australia, dugongs have suffered because they are caught in nets set as part of a government-sponsored shark eradica-

tion program. The dugong is offered protection in some parts of its range, but enforcement is frequently difficult or impossible.

A discussion of the larger animals in this zone whose populations have been reduced through human agency would not be complete if marine turtles were excluded. Principal among these reptiles is the green turtle (*Chelonia mydas* ssp.), which ranges widely over the tropical marine waters of the world (Fig. 9.4(a)). The flesh of these animals has been highly regarded by man since early times, but the animals were not in danger until world trade made it possible to ship them, either alive or butchered, to markets far from their native ecosystems. Heavy overexploitation made this turtle rare in many areas where it had formerly been abundant. Programs are currently underway to protect the remaining green turtles and to allow them to rebuild their populations so they can again be harvested.

Another marine turtle that must be included here is the Atlantic hawksbill (*Eretmochelys imbricata*), which is the source of most tortoise shell. The plastics industry probably helped give this animal a chance of surviving, but now there is a growing demand for true tortoise shell so its future is again in doubt. Even though this turtle is given protection in some areas, the law is extremely difficult to enforce (Fig. 9.4(b)).

Although exploitation of the biota is a significant aspect of human influence in this zone, it is often less important than the effects of physical and chemical alterations. Such effects became significant when large numbers of industrial people settled near the oceans. They viewed the oceans as convenient places to discharge the waste materials that were the byproducts of their activities. And until recently, little thought was given to the possible ecological consequences, but it is now evident that the oceans, particularly the intertidal zone, have been greatly affected. Today,

Figure 9.4(a) *Green turtle* (Chelonia mydas).

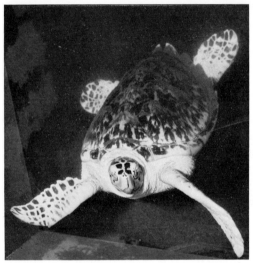

Figure 9.4(b) *Atlantic hawksbill turtle* (Eretmochelys imbricata).

few intertidal zones adjacent to human settlements are free of contamination. The contaminants range from a complex of industrial wastes to sewage, petroleum spills, and thermal changes (which, though not pollutants, nevertheless alter the environment).

No study has yet been published that presents the entire history of man's impact on the ecosystems in the intertidal zone. Therefore, we must focus our attention on recent times and events. However, it is unlikely that before recent times man was able to exert much influence on this or any other part of the marine environments by *chemical* or *physical* changes.

The rapid growth of industry, large port cities (the four largest cities in the world—Shanghai, Tokyo, London, and New York—are port cities) and industrial complexes on or near a coast have had increasingly profound influence on the biota of the intertidal zone. It is not always apparent which element of land use is responsible for most of the changes in an area, but it is certain that sewage and industrial wastes together account for most of the chemical, physical, and biotic changes.

Sewage is probably the oldest source of human contamination in this zone, but the records are too recent to inform us about, say, the "classical" cities on the Mediterranean. Probably the earliest well-documented area is that of the lower River Thames, that is, the estuary of the Thames. Water pollution of the Thames was discussed in Chapter 4, so we shall only briefly mention this estuary, which became so filthy that some fish species were unable to live in it. Most of the pollution in the early period, as previously stated, came from the practice of dumping untreated human feces into the river where it passed London. It is alleged that by the fourteenth century, the river at and below London was seriously polluted with human body wastes. This may have had little negative effect on the biota, however, for records indicate that sport fishing was still productive up to around the middle of the nineteenth century. In fact, it may be that the fecal material provided nutrients leading to increased fish populations.

The biotically lethal period of the lower Thames seems to coincide with the development of major industry, which used the Thames to discharge waste products. Then, changes in the fish catch became noticeable, and, presumably, the entire biota was subjected to serious pressures. In recent years, the government has exerted efforts to clean up the Thames estuary, with the result that fishermen are again catching fish species that have been unreported for many decades.

However, sewage remains a serious source of pollution in the intertidal zone in many parts of the world. It is not uncommon for oceans and bays to be posted as unsafe for swimming because of their sewage content. There are also coastlines, though not posted as unsafe, that are extremely unpleasant, such as some of Italy's west coast from Napoli north to the Rivera in the summer. Along that shore, the air sometimes becomes so evil smelling because of the sewage-fouled water that only the most intrepid bathers venture into the water. It is a common practice in much of the Mediterranean area to dump untreated or inadequately treated sewage into the sea, and the effects of this practice have as yet received little scientific attention. It must also be noted that use of the ocean as a place to dump untreated or partially treated sewage is a worldwide phenomenon.

Human wastes in sewage are not necessarily destructive, providing there is sufficient oxygen and that mixing with sea water is adequate. Where mixing is reduced, *eutrophication* (nutrient enrichment) occurs. This has far-reaching implications for the biota. Although a certain amount of enrichment might be beneficial and lead to an increase in primary production, at the present time the phenomenon has

been so little studied that we cannot discuss its possible merits or demerits. Sewage is a complex material that includes not only the wastes of human metabolism but also many other elements from other sources.

The sewage that used to flow from sewers and privies into the oceans was chemically very different than it is now. Mainly, the changes have come about through the addition of chemicals associated with household cleaning, particularly phosphates. A violent controversy has been raging around the world respecting the use of phosphates in cleaning agents (as noted in earlier chapters) because they enrich the waters into which they flow and this enrichment can produce undesirable ecological effects. Although rapid eutrophication appears to occur principally in lakes and rivers, enriching chemicals also find their way into the intertidal zone. There can be no question that they also play a role in the ecology of the marine ecosystems into which the effluent is introduced. In one study of San Francisco Bay, California, it was found that the number of animal *species* was reduced where the effluent was introduced into bay waters, although the total number of *individuals* remained as high as in areas that were not receiving such input. This study is mentioned here to illustrate that it is unwise to make assumptions about the effects of effluent in a marine ecosystem. That there will be some impact is certain, but the direction of the impact cannot always be predicted and it may be that a controlled degree of enrichment is desirable. At this time, we can say only that man is having measurable impact on the intertidal zone in all parts of the world where sewage is allowed to reach the sea.

Of more recent vintage and possibly of greater ecological importance is the introduction of chemicals from manufacturing and agriculture into this zone. The number of chemicals now being flushed into the oceans is large and growing. In fact, it might be impossible to list every chemical involved because new compounds are continually being developed and pressed into industrial use. It is often months or years after the first use of such compounds that they are discovered concentrated in the gut of a crustacean or in muscle tissue of a fish. This problem is now widely recognized but it is difficult to solve because there is a lack of data on the normal chemical levels of the water in a given area before a particular chemical was introduced.

Mercury (Hg) illustrates the problem rather well. Until the 1950s, it was not generally recognized that mercury posed a problem in the environments into which it was released. Although long known to be highly toxic if ingested, it was thought that the heavy metal sank to the bottom of the water and then into the substrate, where it would remain beyond reach of man or other organisms. This belief was shown to be false by the discovery of mercury poisoning in the people of Minimata, a seaside town in Japan, in 1960. The disease—since labeled the Minimata disease or *itai-itai*, which means "it hurts, it hurts"—was traced to mollusks and fish living in the bay into which the effluent of a factory using mercury was discharged. The mercury had become methylated and in this form had moved through the food web from algae to animal to man. The discovery was quickly made known to the medical world, and it became evident that methyl mercury poisoning was not limited to this site in Japan. It appears that many marine animals are capable of concentrating the mercury at levels that might prove toxic to man if he consumes them.

The present degree of mercury contamination in the intertidal and other parts of the oceans is not clear, for, as noted above, we often cannot determine the *increase,* if any, in the mercury levels in the sea and the organisms in the sea. Mercury enters the oceans naturally by weathering, erosion, and fluviatile processes and has been doing so for geological ages. It must therefore be assumed that life in the ocean is adapted to a certain level of Hg and that man is also adapted to a certain intake of this metal with his food. The question is, what constitutes too heavy an intake in man and other animals. We have greatly increased the liberation of Hg into the world's ecosystems in the last century, and use of this metal continues to increase. But at present we cannot estimate the long-term effects on the affected ecosystems. It is undoubtedly wise to monitor the inputs of mercury and even to limit them, particularly in intertidal ecosystems, for these ecosystems appear to contain the biota best able to concentrate methyl mercury to toxic levels. There is a pressing need for research on this and related topics.

Mercury is only one of many chemical elements that find their way into the intertidal zone. Another heavy metal that has been increasingly introduced is lead (Pb). Lead is used in many manufacturing processes, but perhaps the most important source of contamination is gasoline, to which lead is added to prevent knocks. Because much of this lead finds its way into marine ecosystems by way of air transport, it is probably dispersed much more widely in the marine environments than is mercury. It has been estimated that the lead content of ocean waters has increased from about 0.01–0.02 to 0.07 micrograms per kilogram of sea water since leaded gasolines have been in use.

9.2 Continental Shelf and Pelagic Zones

Most of the important marine fisheries of the world are located on the continental shelves and pelagic areas (including, for the sake of this discussion, banks). Thus, the fisheries have minor and major influences on the biota of these areas, which are subject to pollution from accidental as well as deliberate discharge of a wide variety of substances. The discharges may also cause changes in the biota. As studies continue in the next few years, we shall be better able to evaluate the biological consequences of oceanic pollution.

During most of human history, man's most significant influence on the oceans was caused by his exploitation of marine animals for food and raw materials. We do not know the origin of marine fishing or the extent that the exploitation by earlier, preliterate peoples influenced marine biota. Presumably, such influence was modest and caused few ecological upsets in marine ecosystems. The modest sizes of human populations and the limited technologies almost assure that human impact was not important until well within the period of written history.

In the fifteenth and sixteenth centuries, European nations engaged in a flurry of sea voyages that led to the spread of European power around much of the world. This "Age of Discovery" produced, among other things, a rapid advance in marine technology. Sailing vessels became more seaworthy and better designed to stay at sea for longer periods of time. A large part of this technology was derived first from improvements in the sails and gear of fishing boats and the navigational skills acquired by seagoing fisherfolk. But as ship design came to focus on cargo and military vessels, it became the turn of the fishing vessels to benefit from technical changes. European fishermen ventured out on the Atlantic in search of cod and other valuable fish species; and the once shore-based whale fishers sailed into the north Atlantic.

Whale hunting began in some parts of the world before written records. We know, for example, that there was aboriginal whaling off both the east and west coasts of North America, but there is no indication that these activities depleted whale stocks in any important way. The era of commercial whaling began (at least for Europe) in or around the Bay of Biscay during the eleventh century. This fishery was shore based, so the fishing was conducted close to shore; the whales, when killed, had to be towed to land to be flensed (butchered). Whale fat was rendered for the oil, and certain parts of the animals, particularly the tongues, were salted and sold widely in various parts of Western Europe. However, shore-based whaling gradually changed as the size and reliability of the ships increased. The site of the whale fishery shifted northward in the Atlantic until it became concentrated in waters north of the Arctic Circle. The whales that figured most significantly in this fishery were species of balleen (whalebone) whales belonging to the family Balaenidae—the right and bowhead whales (Fig. 9.5(a) and 9.5(b). These huge animals remain afloat after being killed, and yield oil, balleen, and other products, In the era before plastic and inexpensive steel rods, whalebone (balleen) was a useful raw material. Whale oil was used widely as an illuminant, a lubricant, and in a variety of industrial processes.

Near the beginning of the eighteenth century, some of the English colonists in North America began fishing for sperm whales (*Physeter catodon*). This activity in its early years was of little economic importance, but by the end of the eighteenth century, it had grown so much that Nantucket Island was recognized as the principal port for sperm whale fishers.

Sperm whales are unusual in many respects when compared with right, bowhead, and other whales. For example, adult male sperm whales are, on the average, one-third longer than adult females (Fig. 9.6). The sperm oil, which is in the heads of the males, was the chief object of the fishery. This oil can be converted into the highest quality candles and is useful in manufacturing processes that require special lubricants. Although most sperm whales are distributed in the lower latitudes in both the Atlantic and Pacific Oceans, solitary males often swim into middle latitude waters. The early whalers from Nantucket concentrated on these.

Over the years, the American whaling ship evolved into an extremely seaworthy vessel—one that could remain away from home port for several years (although reprovisioning was necessary). American whalers developed techniques for trying the whale on board ship and learned to hunt for other cetaceans as well. Sperm oil, however, remained the real prize. The War of 1812 put most of the whaling ships out of action, but afterward, the American whalers invaded the Pacific waters in force; the peak years of the fishery followed. Finally, in 1858, when kerosene became generally available, the death knell of the

Figure 9.5(a) *Right whale* (Eubalaena).

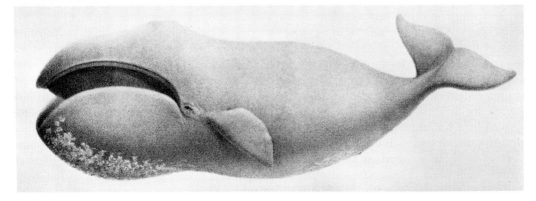

Figure 9.5(b) *Bowhead whale* (Balaena mysticetus).

Figure 9.6 *A pod of sperm whales* (Physeter catodon).

Figure 9.7 *Blue whale* (Balaenoptera musculus) *on a factory ship.*

American whale fishery was sounded. It did not cease, however, until the early part of this century.

Although Americans left the whale fishery, whaling was not over, but the hand-rowed whale boats and hand-hurled harpoons were no longer economically desirable. Norwegians developed a cannon—the Sven gun—that fired a harpoon into the whale. When it was deep in the animal, a bomb in the head of the harpoon exploded. This cannon was mounted on a steam-driven ship—and the whale never again was a match for man.

By the latter part of the nineteenth century, whaling in the North Atlantic and the North Pacific was sharply declining; what had once been rich whaling grounds contained only small whale populations. Then, explorations into Southern Hemisphere waters led to the discovery that whale stocks there were large, particularly of the blue whale (*Balaenoptera musculus*). The blue whale is the largest of the whales; in fact, it is the largest animal known to have lived on this planet (Fig. 9.7). Before the days of the harpoon gun, this whale was almost never hunted, because its great size and fast swimming made it all but im-

possible to catch. Afterward, however, it became the basis of the modern whale fishery. The modernization included the development of huge, floating whale factories capable of processing all that is commercially valuable in a whale, which is hoisted aboard via a tunnel in the stern, and the development of special killer vessels to carry the gun. Each factory ship has a fleet of killer vessels attached to it. In the most recent period, aircraft and electronic equipment have also been used in the annual whale hunt in Antarctic waters.

The results are what might have been expected. Virtually every whale species of commercial value that was not completely protected is now greatly reduced in number. Ironically, the blue whale, which earlier in this century was the mainstay of the whale fishery, is now so reduced in

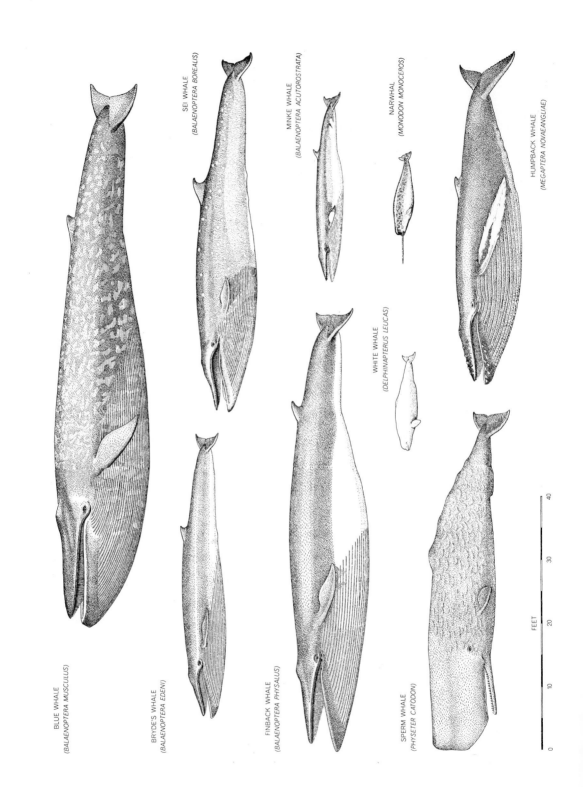

Figure 9.8 *Principal species of baleen whales. (From "The Last of the Great Whales" by Scott McVay. Copyright © August 1966 by Scientific American, Inc. All rights reserved.)*

304

number that there is concern that it may not recover and will pass into extinction within the next few decades. Even more ironic, the annual whale quotas that are established for nations still engaged in the large-scale fishing of whales (Japan, U.S.S.R.) are in terms of blue whale units (b.w.u.). It has finally become illegal to kill a blue whale, but the b.w.u. system continues (Fig. 9.8).

This system obscures the fact that the other whale species of present commercial importance—the fin-backed whale (*Balaenoptera physalus*), humpback whale (*Megaptera modosa*), and sei whale (*Balaenoptera borealis*)—are now receiving most of the impact of the commercial whalers. It is not surprising that there is every indication, mostly through sharply reduced catch sizes, that these animals are or soon will be threatened species. There is no excuse for this. Throughout this book, we have tried to take a reasoned view of man's alteration of the earth and have even suggested that some of the changes are desirable. But no such case can be made for the commercial whale fishery of this century. It cannot even be excused by saying that whalers had no previous experience with the depletion of whale stocks, for in the nineteenth century it was seen that whalers, even with relatively crude tools could reduce whale

stocks so much that it was not commercially rewarding to hunt the remainders. Such depletion was the reason that whalers of an earlier day ventured into Arctic waters and that American whalers moved from the Atlantic to the Pacific. It was only at the "last moment" that the gray whale (*Eschrictius glaucus*) escaped extinction by the passage and enforcement of protective laws. Thus, there was more than ample experience. The history of whaling must rank as one of the larger of our biological crimes.

What will be the ecological results of the near extermination of the whales? Because the ecology of whales, particularly of those species of commercial importance, is little known, we will have to wait, perhaps for decades, for answers to this question. It may be that the removal of the whales (or their marked reduction in numbers) will not produce identifiable reactions in marine ecosystems. Except for the sperm whale, the commercial whales feed low in the trophic levels of the oceans. They subsist largely on a variety of small crustaceans known to the whaling trade as *krill,* a complex mixture of small organisms among which euphasiids (small shrimplike animals) often predominate (Fig. 9.9). It may be that removal of whale stocks will result in the increase of krill organisms, which in

Figure 9.9 *Krill from a whale's stomach.*

turn, may lead to an increase in the population of other animal species that feed on the same organisms. Thus, the whales may not be missed by the oceans. However, this is only conjecture because we do not understand the ecological positions of the krill-feeding whales in the oceans. Sperm whales appear to feed chiefly on various species of squids (marine cephalopods), but again, we do not know what happens when this whale is removed or greatly reduced in number in the marine areas it has occupied. The prudent view is that we ought not to *assume* that only moderate damage will take place.

Fishing.   Most of the world's important commercial finfish fisheries are located on the continental shelves or in waters above banks, which may be some distance from continental coasts. There are also pelagic fisheries, the most notable being for members of the tuna group, but the pelagic fisheries contribute only a modest part of the total catch from marine waters. Man has been exploiting the continental shelves for fish for many centuries—long before numerical records were first kept. In fact, even today, a significant part of the catch is not measured because it does not enter the type of marketing channels where it can be counted. Most of the fisheries data we have are on fisheries in those areas where government agencies try to keep a more or less accurate count of catch size. Unfortunately, such records often do not go back far in time. Also, because of varied types of fishing gear and other changes, the data are of limited value in helping us learn about the impact on exploited populations and on the ecological relationships within the ecosystems subjected to the marine harvests. Certain fish, such as codfish in the Atlantic, have been the object of centuries-long exploitation. Recently, the size of the annual catch of some of the more sought-after species have been declining. Although this may be the result of overfishing, there may be

other reasons, such as water pollution or ecological changes that have little or nothing to do with human activities.

However, we are justified in hypothesizing that any long-term, fairly heavy drawdown on the biomass of a species affects most or all of the ecosystems in which the animal lives. This is almost certain, because the animals divert energy and nutrients to themselves and have a multitude of influences on the complex interactions of the environments they occupy. Thus, the yearly removal of large parts of the biomass of fish must result in myriad changes. However, we can do little more than speculate about the nature of the changes.

One probable change is the alteration of the gene pool of the exploited species. Although this has received little experimental attention, the possibility has not gone unremarked. For example, Dr. Ernst Mayr, in his book *Animal Species and Evolution,* suggested that alterations have probably occurred in the gene-pool structure of salmon in the northeast Pacific region as a result of intensive fishing pressure. And what is possibly true for salmon must also hold for all other heavily exploited fish taxa. Thus, in a curious way, man is "domesticating" a great host of fish species through his inadvertent manipulation of their gene pools.

As commercial fishing techniques become ever more sophisticated and the total pressure on the fish resource in the oceans grows, we must assume that other changes, aside from that of population reduction, will occur. We may discover a multitude of changes only after it is too late to undo them.

Pollution.   So vast are the expanses and depths of the oceans that until recently it appeared absurd to suggest that man could pollute any marine waters except those in the intertidal zone. Recent events, however, have shown that the oceans are being polluted through human agency, but the scope and ecological importance of the pollution are still undetermined. The pollutants reach the offshore waters by several routes: deliberate or accidental discharge of oil and other materials from ships; discharge of pollutants into streams entering the ocean; discharge of pollutants along the shore; and air transport of minute droplets (aerosols) of chemicals such as pesticides. The routes of most of the contaminants are not yet known, because many include both mechanical transport (currents and wind) and biological transport (through food webs).

The insecticide DDT, for example, has been moved by various means and has reached extremely remote areas. It has been recovered from the fatty tissues of the cahow or Bermuda petrel (*Pterodroma cahow*), a bird that spends much of its time in the middle of the Atlantic, where it feeds on fish. This bird, which was once abundant, according to reports from the seventeenth century, became rare chiefly because of the introduction of rats (*Rattus*) to the islands where it nested. It became so rare that it was thought to be extinct, but in recent years a few were found still breeding in the Bermudas. It was hoped that the species would recover if it was given careful protection, but breeding success was low. Examination of the tissues of the dead cahows showed that they contained an average of 6.44 parts per million (ppm) of DDT. Because the birds feed exclusively on the high seas, the only apparent source seems to be contaminated fish living far from land. It must be pointed out that the presence of DDT in the tissues is not proof that the insecticide caused the low breeding success, but it *does* show that DDT is present in animals living great distances from land. However, we do not know how the DDT got into the fish—it may have been through a combination of physical and biological transport.

It is true that the ecological dangers of oil spills are greatest close to shore. However, oil spills also occur far from land, and the slicks can persist and travel by wind and current over considerable distances. They may even reach the shore area.

It has been common practice for ships to throw all refuse overboard while at sea. This might appear unimportant, but some maritime routes are heavily traveled, so the garbage heaved overboard may be highly visible. The ecological importance of this practice is unknown.

The oceans have also been used as dumping grounds for the garbage and debris of land-based man. A familiar sight in many ports is a garbage scow being towed out to sea to discharge its cargo. Sometimes, however, this garbage washes up on some adjacent shore in rows of foul-smelling offal. Garbage is a complex material, and no one can judge how a particular load will break down in the oceans and how the various residues will affect marine life.

But garbage may actually be safe when compared with other things deliberately dumped into the oceans, such as sealed metal packages containing radioactive materials or poison gas. This practice is highly questionable at best; there is a rapidly growing awareness that the oceans are not suitable for the disposal of such wastes or, for that matter, of *most* wastes.

In our discussion of the intertidal zone, we examined the problem of the inflow of treated and untreated human sewage. The chief purpose of treating sewage is to reduce or eliminate the biological aspect that might cause diseases such as typhoid. But even after treatment, an almost unbelievable collection of chemical substances remains whose ultimate fate in oceans cannot even be guessed at. A major city located close to marine waters, such as Los Angeles or New York, must discharge vast quantities of chemicals into the marine waters nearby. It is common practice to flush down domestic drains all kinds of chemicals—from medicines to household chemicals. To this is added an even greater array of chemicals introduced from manufacturing sources. What happens to them? More importantly, do they get into the food webs of organisms beyond the intertidal zone? A partial answer to these questions is given by methyl mercury found in swordfish (*Xiphias gladius*) tissues. The concentration of methyl mercury is so high in some swordfish, a pelagic fish, that it has alarmed health officials and caused the withdrawal of the fish from markets. Although we still do not know exactly how the mercury finds its way into the swordfish, one likely source is the wastes discharged by manufacturing plants. These wastes find their way to marine waters and then through the food webs.

Because we have only recently come to recognize that man is having an ecological impact on marine ecosystems, we do not yet know the details of this impact. However, we do know that we can no longer accept the view that marine waters are too vast to be affected by human actions. In fact, the evidence strongly suggests that far from being limitless dumping grounds for mankind's wastes, the oceans and seas must be given care and protection to keep them from going the way of many terrestrial ecosystems.

Suggested Readings

Chapter One

Clapham, W. B., *Communities and Ecosystems.* London: Macmillan. 1970.

Darlington, Philip J. Jr., *Zoogeography: The Geographical Distribution of Animals.* New York: Wiley. 1957.

Davis, Wayne H. (ed.), *Readings in Human Population Ecology.* Englewood Cliffs, N.J.: Prentice-Hall. 1971.

Ehrlich, Paul R., and Anne H. Ehrlich, *Population Resources Environment: Issues in Human Ecology,* 2d ed. San Francisco: Freeman. 1972.

Farnworth, Edward G., and Frank B. Golley (eds.), *Fragile Ecosystems.* New York: Springer-Verlag. 1974.

Farvar, M. Taghi, and John P. Milton (eds.), *The Careless Technology: Ecology and International Development.* Garden City: The Natural History Press. 1972.

Kormondy, Edward J., *Concepts of Ecology.* Englewood Cliffs: Prentice-Hall. 1969.

Krebs, Charles J., *Ecology.* New York: Harper & Row. 1972.

Marsh, George P., *Man and Nature.* New York: Scribers. 1864.

Odum, Eugene P., *Fundamentals of Ecology,* 3d ed. Philadelphia: Saunders. 1971.

Polunin, Nicholas, *Introduction to Plant Geography.* New York: McGraw-Hill. 1960.

Russell-Hunter, W. D., *Aquatic Productivity.* London: Macmillan. 1970.

Smith, Robert Leo, *Ecology and Field Biology.* New York: Harper & Row. 1966.

Spencer, J. E., and William L. Thomas, Jr., *Cultural Geography: An Evolutional Introduction to Our Humanized Earth.* New York: Wiley. 1969.

Strahler, Arthur N., and Alan H. Strahler, *Environmental Geoscience: Interaction between Natural Systems and Man.* Santa Barbara, Calif.: Hamilton. 1973.

Thomas, William L., Jr. (ed.), *Man's Role in Changing the Face of the Earth.* Chicago: University of Chicago Press. 1956.

Ucko, Peter J., and G. W. Dimbley (eds.), *The Domestication and Exploitation of Plants and Animals.* London: Duckworth. 1969.

Watts, David, *Principles of Biogeography.* New York: McGraw-Hill. 1971.

Whittaker, Robert H., *Communities and Ecosystems.* London: MacMillan. 1970.

Chapter Two

Attwell, Roelf I. G., "Some Effects of Lake Kariba on the Ecology of a Floodplain of the Mid-Zambezi Valley of Rhodesia," *Biological Conservation,* vol. 2, no. 3, pp. 189–196, 1970.

Bartlett, Harley H., *Fire in Relation to Primitive Agriculture and Grazing in the Tropics.* Annotated Bibliography, vol. 11. Ann Arbor, Mich.: University of Michigan Press. 1957.

Bishop, Walter W., and J. Desmond Clark (eds.), *Background to Evolution in Africa.* Chicago: University of Chicago Press. 1967.

Brokensha, David (ed.), *Ecology and Economic Development in Africa.* Berkeley, Calif.: University of California Press, Inst. of International Studies, Research Series No. 9. 1965.

Butzer, Karl W., *Environment and Archeology: An Ecological Approach to Prehistory,* 2d ed. Chicago: Aldine Atherton. 1971.

Clark, J. Desmond, *The Prehistory of Southern Africa.* London: Penguin. 1959.

Clark, J. Desmond, *The Prehistory of Africa.* New York: Praeger. 1970.

Cloudsley-Thompson, J. L., *Animal Twilight: Man and Game in Eastern Africa.* London: Foulis & Co. 1967.

Cloudsley-Thompson, J. L., *The Zoology of Tropical Africa.* London: Weidenfeld and Nicolson. 1969.

Darling, F. Fraser, *Wild Life in an African Territory.* London: Oxford University Press. 1960.

Dorst, Jean, *A Field Guide to the Larger Mammals of Africa.* New York: Houghton Mifflin. 1970.

Evans-Pritchard, E. E., *The Nuer.* Oxford: Clarendon Press. 1940.

Garlick, J. P., and R. W. J. Keay (eds.), *Human Ecology in the Tropics.* New York: Pergamon Press. 1970.

Hance, William A., *The Geography of Modern Africa.* New York: Columbia. 1964.

Hance, William A., *African Economic Development,* Rev. ed. New York: Praeger. 1967.

Harper, Francis, *Extinct and Vanishing Mammals of the Old World.* New York: American Committee for International Wild Life Protection. No. 12. 1945.

Hickling, C. F., *Tropical Inland Fisheries.* London: Longmans. 1961.

Jacks, G. V., and R. O. Whyte, *The Rape of the Earth: A World Survey of Soil Erosion.* London: Faber. 1939.

Hopkins, Brian, *Forest and Savanna.* London: Heinemann. 1964.

Lee, Richard B., and Irven DeVore (eds.), *Man the Hunter.* Chicago: Aldine. 1968.

Light, Richard Upjohn, *Focus on Africa.* New York: American Geographical Society. 1941.

Lowe-McConnell, R. H. (ed.), *Man-Made Lakes.* New York: Academic. 1966.

Maberly, C. T. Astley, *The Game Animals of Southern Africa.* Johannesburg: Nelson. 1963.

Matheson, J. K., and E. W. Bovill, *East African Agriculture: A Short Survey of the Agriculture of Kenya, Uganda, Tanganyika, and Zanzibar and of Its Principal Products.* London: Oxford University Press. 1950.

Moss, R. P. (ed.), *The Soil Resources of Tropical Africa.* Cambridge: Cambridge. 1968.

Murdock, George Peter, *Africa: Its People and Their Culture History.* New York: McGraw-Hill. 1959.

Nye, P. H., and D. J. Greenland, *The Soil under Shifting Cultivation.* Harpenden: Commonwealth Bureau of Soils, Technical Comm. No. 51. 1960.

Oliver, Roland, and J. D. Fage, *A Short History of Africa.* 2d ed. Baltimore: Penguin. 1966.

Owen, D. F., *Animal Ecology in Tropical Africa.* Edinburgh & London: Oliver & Boyd. 1966.

Phillips, John, *Agriculture and Ecology in Africa: A Study of Actual and Potential Development South of the Sahara.* London: Faber. 1959.

Sidney, J., "The Past and Present Distribution of Some African Ungulates," *Transactions Zoological Society, London,* vol. 30, pp. 1–430, 1965.

Stebbing, E. P., *The Forests of West Africa and the Sahara: A Study of Modern Conditions.* Edinburgh: Chambers. 1937.

Tall Timbers Fire Ecology Conference, *Fire in Africa.* Tallahassee: Proceedings, Annual Tall Timbers Fire Ecology Conference, no. 11. 1972.

Trapnell, C. G., and J. N. Clothier, *The Soils, Vegetation and Agricultural Systems of North-Western Rhodesia.* Lusaka: government printer. 1957.

UNESCO, *A Review of the Natural Resources of the African Continent.* Paris: UNESCO. 1963.

Wills, J. Brian (ed.), *Agriculture and Land Use in Ghana.* London: Oxford University Press. 1962.

Chapter Three

Andrus, J. Russel, *Burmese Economic Life.* Stanford, Calif.: Stanford. 1947.

Brandis, Dietrich, *Indian Trees: An Account of the Trees, Shrubs, Woody Climbers, Bamboos, and Palms Indigenous or Commonly Cultivated in the British Indian Empire.* London: Constable. 1906.

Buck, John Lossing, *Chinese Farm Economy: A Study of 2866 Farms in Seventeen Localities and Seven Provinces in China.* Chicago: University of Chicago Press. 1937.

Butzer, Karl W., *Environment and Archeology: An Ecological Approach to Prehistory.* Chicago: Aldine Atherton. 1971.

Chang, K. C., *The Archeology of Ancient China.* New Haven, Conn.: Yale. 1968.

Coventry, B. O., "Denudation of the Punjab Hills," *Indian Forest Records,* vol. 14, part 2, pp. 49–78, 1929.

de Young, John, *Village Life in Modern Thailand.* Berkeley, Calif.: University of California Press. 1955.

Dobby, E. H. G., *Southeast Asia.* London: University of London Press, Ltd. 1950.

Dube, S. C., *Indian Village.* Ithaca, N.Y.: Cornell. 1955.

Eberhard, W., *A History of China.* Berkeley, Calif.: University of California Press. 1960.

Ellerman, J. R., and T. C. S. Morrison-Scott, *Checklist of Palaearctic and Indian Mammals. 1758–1946.* London: British Museum (Natural History). 1951.

Fairservis, Walter A., *The Roots of Ancient India.* New York: MacMillan. 1971.

Fei, Hsio-Tung, *Peasant Life in China.* Fair Lawn, N.J.: Oxford University Press. 1946.

Fisher, James, Noel Simon, and Jack Vincent, *Wildlife in Danger.* New York: Viking. 1969.

Fisher, W. B., *The Middle East.* 5th ed. London: Methuen. 1963.

Fürer-Haimendorf, C. von, *The Aboriginal Tribes of Hyderabad.* Vol. 1. *The Chenchus.* London: Macmillan. 1943.

Geertz, C., *Agricultural Involution: The Processes of Ecological Change in Indonesia.* Berkeley, Calif.: University of California Press. 1966.

Gourou, Pierre, *Land Utilization in French Indochina.* Washington, D.C.: Institute of Public Relations. 1945.

Gourou, Pierre, *The Tropical World.* London: Longmans. 1961.

Li, Chi, *The Beginnings of Chinese Civilization.* Seattle, Wash.: University of Washington Press. 1957.

Lydolph, Paul E., *Geography of the U.S.S.R.,* 2d ed. New York: Wiley. 1970.

Michael, H. N. (ed.), *The Archaeology and Geomorphology of Northern Asia.* Toronto, Canada: University of Toronto Press. 1964.

Peltzer, K. J., *Pioneer Settlement in the Asiatic Tropics.* New York: American Geographical Society, special pub. no. 29. 1945.

Pfeffer, Pierre, *Asia: A Natural History.* New York: Random House. 1968.

Richards, R. W., *The Tropical Rainforest: An Ecological Study.* New York: Cambridge University Press. 1952.

Richardson, S. D., *Forestry in Communist China.* Baltimore: Johns Hopkins. 1966.

Sauer, Carl O., *Agricultural Origins and Dispersals.* New York: American Geographical Society. 1952.

Schaller, G. B., and N. M. Simon, "The Endangered Large Mammals of Asia," *I U C N Publications,* n.s., no. 18, vol. 2, pp. 11–23, 1970.

Spate, O. H. K., and A. T. A. Learmonth, *India and Pakistan.* London: Methuen. 1967.

Spencer, J. E., *Shifting Cultivation in Southeastern Asia.* University of California Pubs. in Geography, vol. 19. 1966.

Spencer, J. E., and William L. Thomas, *Asia East by South,* 2d ed. New York: Wiley. 1971.

Stamp, L. Dudley, *A History of Land Use in Arid Regions.* Paris: UNESCO. 1961.

Suslov, S. P., *Physical Geography of Asiatic Russia.* San Francisco: Freeman. 1961.

Tate, G. H. H., *Mammals of Eastern Asia.* New York: Macmillan. 1947.

Tempany, H., and D. H. Grist, *An Introduction to Tropical Agriculture.* New York: Longmans. 1958.

Tuan, Yi-Fu, *China.* Chicago: Aldine. 1969.

UNESCO, *Study of Tropical Vegetation: Proceedings of the Kandy Symposium, 1956.* Paris: UNESCO. 1958.

UNESCO, *Salinity Problems of the Arid Zones.* Paris: UNESCO. 1961.

Wang, C. W., *The Forests of China, with a Survey of Grassland and Desert Vegetation.* Cambridge, Mass.: Harvard. 1961.

Wheeler, M., *The Indus Civilization,* 3d ed. London: Cambridge. 1968.

Wittfogel, H., *Oriental Despotism and Hydraulic Society.* New Haven, Conn.: Yale. 1957.

Chapter Four

Butzer, Karl W., *Environment and Archeology: An Ecological Approach to Prehistory,* 2d ed. Chicago: Aldine Atherton. 1971.

Childe, V. G., *The Prehistory of European Society.* Harmondsworth: Pelican. 1958.

Clapham, John H., and Eileen Power (eds.), *The Cambridge Economic History.* Vol. 1. *The Agrarian Life of the Middle Ages.* London: Cambridge. 1941.

Clark, J. G. D., "Farmers and Forests in Neolithic Europe," *Antiquity,* vol. 19, pp. 57–71, 1945.

Clark, J. G. D., "Forest Clearance and Prehistoric Clearing," *Economic History Review,* vol. 17, pp. 45–51, 1947.

Darby, H. C., "Domesday Woodland," *Economic History Review,* 2d series, vol. 3, pp. 21–43, 1950.

Darby, H. C., "The Clearing of the English Woodlands," *Geography,* vol. 36, pp. 71–83, 1951.

Darling, F. Fraser (ed.), *West Highland Survey: An Essay in Human Ecology.* London: Oxford. 1955.

East, W. Gordon, *A Historical Geography of Europe,* 4th ed. New York: Dutton. 1961.

Ellerman, J. R., and T. C. S. Morrison-Scott, *Checklist of Palaearctic and Indian Mammals. 1758 to 1946.* London: British Museum (Natural History). 1951.

Elton, Charles S., *The Ecology of Invasions by Animals and Plants.* London: Methuen. 1958.

Hoffman, George W. (ed.), *A Geography of*

Europe Including Asiatic U.S.S.R., 2d ed. New York: Ronald. 1961.

Houston, J. M., *The Western Mediterranean World.* New York: Praeger. 1967.

Iversen, J., *The Influence of Prehistoric Man on Vegetation.* Danmarks Geologiske Undersøgelse, series 4, vol. 3, no. 6, 1949.

Kurten, B., *Pleistocene Mammals of Europe.* Chicago: Aldine. 1968.

McVean, D. N., and J. D. Lockie, *Ecology and Land Use in Upland Scotland.* Edinburgh: Edinburgh University Press. 1969.

Orme, A. R., *Ireland.* Chicago: Aldine. 1970.

Peterson, Roger T., Guy Montfort, and P. A. D. Hollom, *A Field Guide to the Birds of Britain and Europe,* 2d ed. Boston: Houghton Mifflin. 1966.

Polunin, O., *Flowers of Europe: A Field Guide.* London: Oxford. 1969.

Ritchie, James, *The Influence of Man on Animal Life in Scotland.* London: Cambridge. 1920.

Semple, Ellen C., *The Geography of the Mediterranean Region. Its Relation to Ancient History.* New York: Holt. 1931.

Smith, C. T., *An Historical Geography of Western Europe before 1800.* London: Longmans. 1967.

Thomas, William L., Jr. (ed.), *Man's Role in Changing the Face of the Earth.* Chicago: University of Chicago Press. 1956.

Veen, Johan van, *Dredge, Drain, Reclaim: The Art of a Nation.* The Hague: M. Nijoff. 1955.

Zeuner, F. E., *A History of Domesticated Animals.* London: Methuen. 1963.

Chapter Five

Berndt, Ronald M., and Catherine H., *Aboriginal Man in Australia.* Sydney: Angus & Robertson. 1965.

Blake, S. T., "The Plant Communities of Western Queensland and Their Relationships, with Special Reference to the Grazing Industry," *Proceedings of the Royal Society of Queensland,* vol. 49, no. 16, pp. 156–204, 1937.

Butzer, Karl W., *Environment and Archeology: An Ecological Approach to Prehistory,* 2d ed. Chicago: Aldine Atherton. 1971.

Cleland, J. B., "Our Natives and the Vegetation of Southern Australia," *Mankind,* vol. 5, pp. 149–162, 1957.

Crocker, R. L., and J. G. Wood, "Some Historical Influences on the Development of the South Australian Vegetation Communities and Their Bearing on Concepts and Classification in Ecology," *Transactions Royal Society South Australia,* vol. 71, pp. 91–136, 1947.

Dare, H. H., *Water Conservation in Australia.* Sydney: Simmons. 1939.

Ealy, E. M. H., "Ecology of the Euro, *Macropus robustus* (Gould), in North Western Australia. 1: The Environment and Changes in Euro and Sheep Populations," *CSIRO Wildlife Research,* vol. 12, pp. 9–25, 1967.

Ebert, Charles H. V., "Irrigation and Salt Problems in Renmark, South Australia," *The Geographical Review,* vol. 61, no. 3, pp. 355–369, 1971.

Elkin, A. P., *The Australian Aborigines: How to Understand Them,* 4th ed. Sydney: Angus & Robertson. 1964.

Elton, Charles S., *The Ecology of Invasions by Animals and Plants.* London: Methuen. 1958.

Fenner, Charles, *South Australia. A Geographical Study.* Melbourne: Whitcombe & Tombs. 1931.

Finlayson, H. H., *The Red Centre.* Sydney: Angus & Robertson. 1935.

Forrest, J., *Explorations in Australia.* London: Sampson Low. 1875.

Freeman, Otis (ed.), *Geography of the Pacific.* New York: Wiley. 1951.

Gilbert, J. M., "Forest Succession in the Florentine Valley, Tasmania," *Papers and Proceedings of the Royal Society of Tasmania,* vol. 93, pp. 129–151, 1959.

Giles, E., *Australia Twice Traversed.* London: Sampson Low. 1889.

Gill, E. D., "The Australian Aborigines and the Giant Extinct Marsupials," Australian Natural History, vol. 14, pp. 263–266.

Gould, R. A., "Notes on Hunting, Butchering and Sharing of Game among the Ngatatjara and Their Neighbors in the West Australia Desert," *Kroeber Anthropological Society Papers,* vol. 36, pp. 41–65, 1967.

Holmes, J. M., *The Erosion-Pastoral Problem of the Western Division of New South Wales.* Sydney: University of Sydney Publications in Geography, no. 2. 1938.

Holmes, J. M., *Soil Erosion in Australia and New Zealand.* Sydney: Angus & Robertson. 1946.

Holmes, J. M., *The Murray Valley: A Geographical Reconnaissance of the Murray Valley and a New Design for its Regional Organization.* Sydney: Angus & Robertson. 1948.

Hurley, P. J., *Red Cedar: The Story of the North Coast.* Sydney: Dymock's Book Arcade. 1948.

Ingersoll, Jean M., "The Australian Rabbit," *American Scientist,* vol. 2, no. 2, pp. 265–273, 1964.

Jacobs, M. R., "A History of the Use and Abuse of the Wooded Lands of Australia," *Australian Journal of Science,* vol. 19, pp. 132–139, 1957.

Jones, R., "The Geographical Background of the Arrival of Man in Australia," *Archaeology and Physical Anthropology in Oceania,* vol. 3, pp. 186–215, 1968.

Keast, A., R. L. Crocker, and C. S. Christian (eds.), *Biogeography and Ecology in Australia.* The Hague: Junk. 1959.

McKnight, Tom L., "Barrier Fencing for Vermin Control in Australia," *The Geographical Review,* vol. 59, no. 3, pp. 330–347, 1969.

Marshall, A. J., *The Great Extermination: A Guide to Anglo-Australian Cupidity, Wickedness and Waste.* London: Heinemann. 1966.

Merrilees, D., "Man the Destroyer: Late Quaternary Changes in the Australian

Marsupial Fauna," *Journal Royal Society of Western Australia,* vol. 51, pp. 1–24, 1968.

Mulvaney, D. J., *The Prehistory of Australia.* London: Thames and Hudson. 1969.

Peron, M. F., *A Voyage of Discovery to the Southern Hemisphere.* London: Phillips. 1809.

Rees, Henry, *Australasia: Australia, New Zealand and the Pacific Islands,* 3d ed. London: McDonald & Evans. 1968.

Rolls, Eric C., *They All Ran Wild: The Story of Pests on the Land in Australia.* Sydney: Angus & Robertson. 1969.

Roughley, T. C., *Fish and Fisheries of Australia.* Sydney: Angus & Robertson. 1966.

Wadham, S., *Australian Farming, 1788–1965.* Melbourne: F. W. Cheshire. 1967.

Wadham, S., R. K. Wilson, and J. Wood, *Land Utilization in Australia,* 4th ed. Melbourne: Melbourne University Press. 1964.

Chapter Six

Allen, Glover M., *Extinct and Vanishing Mammals of the Western Hemisphere with Marine Species of All the Oceans.* Lancaster, Penn.: American Committee for International Wild Life Protection. 1942.

Bernarde, M. A., *Our Precarious Habitat* (rev. ed.). New York: Norton. 1973.

Blair, F. W., A. P. Blair, P. Bradkord, F. R. Cagle, and G. A. Moore, *Vertebrates of the United States,* 2d ed. New York: McGraw-Hill. 1968.

Brady, N. C., *Agriculture and the Quality of Our Environment.* Washington, D.C.: American Association for the Advancement of Science. Pub. 85. 1967.

Burch, W. R., Jr., N. H. Cheek, Jr., and L. A. Taylor (eds.), *Social Behavior, Natural Resources, and the Environment.* New York: Harper & Row. 1972.

Carson, Rachel, *Silent Spring.* Boston: Houghton Mifflin. 1962.

Christman, R. F., B. W. Mar, E. B. Welch, R. J. Charlson, and D. A. Carlson, *The Natural Environment: Wastes and Control.* Pacific Palisades, Calif.: Goodyear. 1973.

Cole, H. H. (ed.), *Introduction to Livestock Production, Including Dairy and Poultry,* 2d ed. San Francisco: Freeman. 1966.

Dale, E. E., *The Range Cattle Industry.* Norman, Okla.: University of Oklahoma Press. 1930.

Darling, F. Fraser, and John P. Milton (eds.), *Future Environments of North America.* Garden City, N.Y.: The Natural History Press. 1966.

Dasmann, R. F., *The Destruction of California.* New York: Macmillan. 1965.

Driver, H. E., *The Indians of North America.* Chicago: University of Chicago Press. 1961.

Ehrlich, P. R., and A. H. Ehrlich, *Population Resources Environment,* 2d ed. San Francisco: Freeman. 1972.

Elton, Charles S., *The Ecology of Invasions by Animals and Plants.* London: Methuen. 1958.

Frederick, D., W. L. Howenstine, and J. Sochen, *Destroy to Create.* Hinsdale, Ill.: Dryden Press. 1972.

Garvey, G., *Energy, Ecology, Economy.* New York: Norton. 1972.

Giddings, J. C., and M. B. Monroe, *Our Chemical Environment.* San Francisco: Canfield Press. 1972.

Goldman, M. I., *Controlling Pollution: The Economics of a Clean America.* Englewood Cliffs, N.J.: Prentice-Hall. 1967.

Graham, F., Jr., *Since Silent Spring.* Boston: Houghton Mifflin. 1970.

Guthrie, J. D., *Great Forest Fires of America.* Washington, D.C.: U.S. Dept. of Agriculture. 1936.

Hall, E. R., and K. R. Kelson, *The Mammals of North America.* 2 vols. New York: Ronald. 1959.

Huth, Hans, *Nature and the American: Three Centuries of Changing Attitudes.* Berkeley, Calif.: University of California Press. 1957.

Jennings, J. D., and E. Norbeck (eds.), *Prehistoric Man in the New World.* Chicago: University of Chicago Press. 1964.

Kuchler, A. W., *Potential Natural Vegetation of the Conterminous United States.* New York: American Geographical Society, special pub. 36. 1964.

Leopold, Aldo, *A Sand County Almanac.* New York: Oxford. 1949.

Leopold, A. Starker, "Deer in Relation to Plant Succession," *Transactions North American Wildlife Conference,* vol. 15, pp. 571–580, 1950.

McHarg, Ian, *Design with Nature.* Garden City, N.Y.: Natural History Press. 1969.

Martin, P. S., and H. E. Wright, Jr. (eds.), *Pleistocene Extinctions: The Search for a Cause.* New Haven, Conn.: Yale. 1967.

Nash, Roderick, *Wilderness and the American Mind.* New Haven, Conn.: Yale. 1967.

National Academy of Sciences, *Eutrophication: Causes, Consequences, Correctives.* Washington, D.C.: National Academy of Sciences. 1970.

Ridker, R. G., *Economic Costs of Air Pollution.* New York: Praeger. 1967.

Roe, F. G., *The North American Buffalo: A Critical Study of the Species in Its Wild State.* Toronto, Canada: University of Toronto Press. 1951.

Rudd, R., *Pesticides and the Living Landscape.* Madison, Wis.: University of Wisconsin Press. 1964.

Sauer, Carl O., "Grassland Climax, Fire, and Man," *Journal of Range Management,* vol. 3, pp. 16–21, 1950.

Schorger, A. W., *The Passenger Pigeon, its Natural History and Extinction.* Madison, Wis.: University of Wisconsin Press. 1955.

Shelford, Victor E., *Animal Communities in Temperate America,* 2d ed. Chicago: University of Chicago Press. 1937.

Shelford, Victor E., *The Ecology of North America.* Urbana, Ill.: University of Illinois Press. 1963.

Thomas, William L., Jr. (ed.), *Man's Role in Changing the Face of the Earth.* Chicago: University of Chicago Press. 1956.

Thomas, William L., Jr. (ed.), *Man, Time and Space in Southern California. Annals of the Association of American Geographers, Supplement,* vol. 49, no. 3, part 2, 1959.

Thomlinson, R., *Demographic Problems, Controversy over Population Control.* Belmont, Calif.: Dickenson. 1967.

Turk, A., J. Turk, and J. T. Wittes, *Ecology Pollution Environment.* Philadelphia: Saunders. 1972.

Udall, Stewart L., *The Quiet Crisis.* New York: Holt. 1963.

U.S. Department of the Interior, *Rare and Endangered Fish and Wildlife of the United States* (rev. ed.), Washington, D.C.: Bureau of Sport Fisheries and Wildlife. 1968.

Wagner, R. H., *Environment and Man.* New York: Norton. 1971.

Willey, G. R., *An Introduction to American Archaeology. Vol. 1: North and Middle America.* Englewood Cliffs, N.J.: Prentice-Hall. 1966.

Williamson, S. J., *Fundamentals of Air Pollution.* Reading, Mass.: Addison-Wesley. 1973.

Chapter Seven

Beard, J. S., "The Classification of Tropical America Vegetation Types," *Ecology,* vol. 36, no. 1, pp. 89–100, 1955.

Bennett, Charles F., *Human Influences on the Zoogeography of Panama.* Berkeley, Calif.: University of California Press. Ibero-Americana: 51. 1968.

Borah, Woodrow, and Sherburne F. Cook, *The Aboriginal Population of Central Mexico on the Eve of the Spanish Conquest.* Berkeley, Calif.: University of California Press. Ibero-Americana: 45. 1963.

Budowski, G., "Tropical Savannas a Sequence of Forest Felling and Repeated Burning," *Turrialba,* 1 & 2, pp. 23–33, 1956.

Coe, Michael D., and Kent V. Flannery, *Early Cultures and Human Ecology in South Coastal Guatemala.* Washington, D.C.: The Smithsonian Institution. 1967.

Cook, S. F., *Soil Erosion and Population in Central Mexico.* Berkeley, Calif.: University of California Press. Ibero-Americana: 34. 1949.

Daugherty, Howard E., "The Impact of Man on the Zoogeography of El Salvador," *Biological Conservation,* vol. 4, no. 4, pp. 273–278, 1972.

Denevan, William M., "The Upland Forests of Nicaragua: A Study in Cultural Plant Geography," *University of California Publications in Geography,* vol. 12, no. 4, pp. 251–320, 1961.

Driver, H. E., *Indians of North America.* Chicago: University of Chicago Press. 1961.

Gordon, B. L., *Human Geography and Ecology in the Sinú Country of Colombia.* Berkeley, Calif.:

University of California Press. Ibero-Americana: 39. 1957.

Henderson, David A., "Land, Man and Time," in Russel C. Ewing (ed.), *Six Faces of Mexico,* pp. 103–160. Tucson, Ariz.: University of Arizona Press. 1966.

Idyll, C. P., "The Anchovy Crisis," *Scientific American,* vol. 228, no. 6, pp. 22–29, 1973.

James, P. E., *Latin America,* 4th ed. New York: Odyssey. 1942.

Johannessen, C. L., *Savannas of Interior Honduras.* Berkeley, Calif.: University of California Press. Ibero-Americana: 46. 1963.

Leopold, A. Starker, *Wildlife of Mexico.* Berkeley, Calif.: University of California Press. 1959.

Meggers, B. J., E. S. Ayenshu, and W. D. Duckworth (eds.), *Tropical Forest Ecosystems in Africa and South America: A Comparative Review.* Washington, D.C.: The Smithsonian Institution. 1973.

Parsons, James J., "African Pasture Grasses to the American Tropics," *Journal of Range Management,* vol. 25, no. 1, pp. 12–17, 1972.

Roseveare, G. M., *The Grasslands of Latin America.* Aberystwyth, Great Britain: Imperial Bureau of Pastures and Field Crops. Bulletin 36. 1948.

Sauer, Carl O., "Man in the Ecology of Tropical America," *Proceedings 9th Pacific Science Congress,* vol. 20, pp. 104–110, 1958.

Sauer, Carl O., *The Early Spanish Main.*

Berkeley, Calif.: University of California Press. 1966.

Schmieder, O., "The Pampa—a Natural or Culturally Induced Grassland?" *University of California Publications in Geography,* vol. 2, pp. 255–270, 1927.

Simpson, L. B., *Many Mexicos,* 3d ed. Berkeley, Calif.: University of California Press, 1952.

Smith, C. T., "Depopulation of the Central Andes in the 16th Century," *Current Anthropology,* vol. 11, no. 4/5, pp. 453–464, 1970.

Sternberg, H. O'Reilly, "Man and Environmental Change in South America," in *Biogeography and Ecology of South America,* pp. 413–445. The Hague: Junk. 1968.

Steward, J., and L. C. Faron, *Native Peoples of South America.* New York: McGraw-Hill. 1959.

Verdoorn, F. (ed.), *Plants and Plant Science in Latin America.* Waltham, Mass.: Chronica Botanica. 1945.

West, R. C., and J. P. Augelli, *Middle America: Its Lands and Peoples.* Englewood Cliffs, N.J.: Prentice-Hall. 1966.

West, R. C. (ed.), *Natural Environment and Early Cultures.* Vol. 1 in R. Wauchope (ed.), *Handbook of Middle American Indians.* Austin, Tex.: University of Texas Press. 1964.

Wilbert, J. (ed.), *The Evolution of Horticultural Systems in Native South America: Causes and Consequences. A Symposium.* Caracas, Venezuela: Editorial Sucre. 1961.

Chapter Eight

Atkinson, I. A. E., "Relations between Feral Goats and Vegetation in New Zealand," *Proceedings New Zealand Ecological Society,* vol. 11, pp. 39–44, 1964.

Barrau, J. (ed.), *Plants and the Migration of Pacific Peoples: A Symposium.* Honolulu: Bishop Museum Press. 1963.

Bates, M., and D. P. Abbott, *Coral Island: Portrait of an Atoll.* New York: Scribners. 1958.

Carlquist, S., *Island Life: A Natural History of the Islands of the World.* New York: Garden City. 1965.

Cruxent, J. M., and I. Rouse, "Early Man in the West Indies," *Scientific American,* pp. 42–52, November, 1969.

Cumberland, K. B., "A Century's Change: From Natural to Cultural Vegetation in New Zealand," *Geographical Review,* vol. 31, no. 4, pp. 529–554, 1941.

Cumberland, K. B., "Moas and Men: New Zealand about A.D. 1250," *Geographical Review,* vol. 52, no. 2, pp. 151–173, 1962.

Cumberland, K. B., *Southwest Pacific: A Geography of Australia, New Zealand and Their*

Pacific Island Neighbourhoods, 4th ed. Christchurch: Whitcombe & Tombs. 1968.

Dumont, Rene, *Evolution des Campagnes Malagaches.* Tananarive: Imp. Off. 1959.

Fosberg, F. R. (ed.), *Man's Place in the Island Ecosystem: A Symposium.* Honolulu: Bishop Museum Press. 1963.

Greenway, J. C., *Extinct and Vanishing Birds of the World.* New York: American Committee for International Wildlife Protection. 1958.

Harris, D. R., "Invasion of Oceanic Islands by Alien Plants," *Institute of British Geographers. Transactions and Papers,* no. 31, pp. 67–82, 1962.

Harris, D. R., *Plants, Animals, and Man in the Outer Leeward Islands, West Indies: An Ecological Study of Antigua, Barbuda and Anguilla.* University of California Publications in Geography. Vol. 18. 1965.

Merrill, G., "The Historical Record of Man as an Ecological Dominant in the Lesser Antilles," *Canadian Geographer,* no. 11, pp. 17–22, 1958.

Sauer, C. O., *The Early Spanish Main.* Berkeley, Calif.: University of California Press. 1966.

Schmitt, R. C., *Demographic Statistics of Hawaii: 1778–1965.* Honolulu: University of Hawaii Press. 1968.

Suggs, R. C., *The Island Civilizations of Polynesia.* New York: New American Library. 1960.

Thomson, G. M., *The Naturalization of Animals and Plants in New Zealand.* London: Cambridge. 1922.

Tomichi, P. Q., *Mammals in Hawaii.* Honolulu: Bishop Museum Press. 1969.

Waibel, L., "Place Names as an Aid to the Reconstruction of the Original Vegetation of Cuba," *Geographical Review,* vol. 33, no. 3, pp. 376–396, 1943.

Watters, R. F., *Land and Society in New Zealand: Essays in Historical Geography.* Wellington: A. H. & A. W. Reed. 1965.

Wodzicki, K. A., *Introduced Mammals of New Zealand.* Wellington: Department of Scientific and Industrial Research. 1950.

Chapter Nine

Bardach, J., *Harvest of the Sea.* New York: Harper & Row. 1968.

Borgstrom, Georg, *Hungry Planet.* Toronto, Canada: Collier-Macmillan. 1967.

Ekman, Sven, *Zoogeography of the Sea.* London: Sidgwick & Jackson. 1967.

Fairbridge, R. W., *The Encyclopedia of Oceanography.* New York: Reinhold. 1966.

Henderson, D. A., *Men and Whales at Scammon's Lagoon.* Los Angeles, Calif.: Dawson's Book Shop. 1972.

Mackintosh, N. A., *The Stocks of Whales.* London: Fishing News Books. 1965.

Matthews, W. H., F. E. Smith, and E. D. Goldberg, *Man's Impact on Terrestrial and Oceanic Ecosystems.* Cambridge, Mass.: MIT Press. 1971.

Parsons, James J., *The Green Turtle and Man.* Gainesville, Fla.: University of Florida Press. 1962.

Russell-Hunter, W. D., *Aquatic Productivity.* New York: Macmillan. 1970.

Slijper, E. J., *Whales.* London: Hutchinson. 1962.

Weyl, P. K., *Oceanography: An Introduction to the Marine Environment.* New York: Wiley. 1970.

Glossary

Albedo. The reflectivity of a surface. A white surface has a very high albedo but that of a black surface is very low.

Alluvium. Silt, sand, and gravel deposited by running water.

Arthropod. The jointed-legged animals, namely, insects, spiders, and their relatives.

Arboreal. An organism that lives on or in trees or requires trees for a significant part of its life needs.

Autochthonous. Native to a given place or region in that it evolved there.

Autotrophs. Organisms, chiefly green plants, capable of self-nourishment.

Bioclimate. The portion of the atmosphere with which an organism is in contact.

Biological community. A group of plants and animals living and interacting together.

Biome. The largest terrestrial ecosystem convenient to recognize.

Biotope. In general, the smallest ecosystem one can usefully recognize.

Chitemene. A system of slash and burn (shifting) cultivation practiced in Zambia.

Community. See *Biological community*.

Crepuscular. Animals active chiefly at twilight or just prior to sunrise (usually the former).

Deciduous. When used in reference to vegetation means the shedding of leaves and sometimes the bark in a regular seasonal rhythm.

Decomposers. Fungi and bacteria which reduce, that is, decompose, organic matter in ecosystems.

Defoliant. A chemical compound applied to trees to cause them to shed their leaves.

Disease vector. Any organism capable of physically carrying a pathogenic organism from one organism to another. Examples of disease vectors are *Anopheles* mosquitoes transmitting malaria and *Glossina* flies transmitting African sleeping sickness.

Ecological diversity. The number of ecological situations (habitats) occurring in a given area.

Ecology. The scientific study of the interactions between living organisms and their biotic and physical environments.

Ecological succession. The orderly succession of plant and animal assemblages in a given place through time. Succession typically moves from simple, unstable ecosystems to stable, complex (climax) ecosystems.

Ecosystem. A unit of biological organization made up of all the organisms in a given area interacting with the physical environment so

319

that a flow of energy leads to a characteristic trophic structure and material cycles within the system.

Ecotone. Zone of ecological stress where two or more ecosystems come together.

Elluviate. The physical transport of fine soil particles from upper to lower levels in the soil.

Epiphytes. Plants that employ other plants for physical support but not for nutrients, *e.g.,* orchids.

Endemic. Any organism having a limited geographic distribution.

Eutrophication. Accelerated nutrient enrichment (aging) of water.

Evapotranspiration. Water given up through evaporation and by passing through the leaf stomata (transpiration).

Exotic. A plant or animal introduced into a given area through human agency.

Feral animal. An animal that was once kept in a domesticated state but which has escaped and maintains itself in a wild condition.

Food chain. Path followed by a nutrient from the lowest to highest trophic level in an ecosystem.

Food web. All the food chains in an ecosystem.

Hectare. A land measure equal to 10,000 square meters, or approximately 2.47 acres.

Heterotrophs. Organisms requiring other organisms for food.

Homeostasis. A steady state or equilibrium condition.

Hydrography. Surface and subsurface configurations of freshwater flow and impoundment.

Hydrosere. A complete ecological succession in an aquatic environment.

Isothermal. Nonvarying temperature conditions.

Kilometer. Linear distance of 1,000 meters, equal to approximately .62 miles.

Kogón. Any area dominated by the coarse grass *Imperata.*

Laterite. A red, bricklike material that was once a soil.

Lingua franca. A language that serves as a common language in an area where there is great linguistic diversity.

Loess. Wind-deposited dust; sometimes weathers into fine *loess soil.*

Macroclimate. The average atmospheric conditions in a large region over many years.

Megafauna. The total collection of the larger vertebrates in any region.

Mesolithic. A period in the cultural evolution of Old World societies when there was marked experimentation with plant and animal domestication leading to the *Neolithic.*

Microclimate. See *Bioclimate.*

Montane. Mountain or highland environment.

Neolithic. A period in the cultural evolution of Old World societies when there was a shift from food gathering to food production, that is, to agriculture and animal husbandry.

Paleolithic. A period in the cultural evolution of Old World societies extending from the origins of the genus *Homo* (approximately two million years ago) to the beginning of the *Mesolithic.*

Pathogen. A disease-causing organism.

Permafrost. Arctic and Antarctic condition in which the subsoil is permanently frozen.

Photoperiod. Daylight length.

Photosynthesis. The process by which light acts on water and carbon dioxide in the presence of chlorophyll to synthesize carbohydrates required for plant nutrition.

Phytogeography. Plant geography; the geographical distribution of plants.

Phytophysiognomic. The physical appearance of vegetation.

Pioneer species. Plant or animal species capable of becoming established in early seral stages of an ecosystem.

Pleistocene. The most recent major ice age. Generally the Pleistocene is considered to have begun approximately two million years ago and to have ended eight to ten thousand years ago.

Polyglot. High language diversity in a given area.

Primary production. *Gross primary production* is the total production through photosynthesis in a given ecosystem. *Net primary production* is the part of the production that remains stored in plants after the plant's needs are met.

Salinization. Process whereby salts in water or soil are increased above the normal levels.

Seral stage. An identifiable part of an ecological succession; complete series is called a *sere*.

Slash and burn cultivation (Also, *Shifting cultivation*). Agricultural systems that involve cutting trees and/or shrubs, allowing the debris to dry, burning the debris, and then planting the cleared area.

Taxon, pl. **Taxa.** Any recognized unit of biological classification, *e.g.,* subspecies, species, genus, family, order, class, phylum.

Terrestrial. Living on land as opposed to living in water environments.

Tilth. Agricultural term referring to the ease or difficulty with which soil can be worked by plows or other agricultural tools.

Trophic structure. Levels of food and energy consumption within an ecosystem.

Xerophytes. Plants adapted to conditions of scarce soil moisture.

Veld. Grasslands in southern Africa.

Zoonosis. A disease that affects humans and other vertebrate animals.

Illustration Credits

CHAPTER ONE Figure 1.1: Hans Mann/ Monkmeyer. Figure 1.2: Karl Weidmann/ National Audubon Society. Figure 1.3: Karl Weidmann/National Audubon Society. Figure 1.4: Almasy. Figure 1.5: (a) Emil Schulthess/Black Star; (b) Jacques Six; (c) Transafrica; (d) Leonard Lee Rue III/Monkmeyer. Figure 1.6: Naud/Afrique Photo. Figure 1.7: Grant Heilman. Figure 1.8: Editorial Photocolor Archives. Figure 1.9: (a) George Gerster/Rapho-Photo Researchers; (b) Alan Pitcairn/Grant Heilman; (c) Howard Sochurek/Woodfin Camp; (d) G. R. Roberts. Figure 1.10: Bill Anderson/Monkmeyer. Figure 1.11: Grant Heilman. Figure 1.12: U.S. Forest Service. Figure 1.13: Pro Pix/Monkmeyer. Figure 1.14: Reprinted with permission of Macmillan Publishing Co., Inc. from *Communities and Ecosystems* by R. H. Whittaker. Copyright © 1970 by Robert H. Whittaker. Figure 1.15: Courtesy University of Wisconsin Press. From Robert F. Inger, Arthur D. Hasler, F. Herbert Bormann, and W. Frank Blair, eds., *Man in the Living Environment,* The University of Wisconsin Press, Madison; © 1972 by the Board of Regents of the University of Wisconsin. Figure 1.16: Grant Heilman. Figure 1.17: Reprinted with permission of Macmillan Publishing Co., Inc. from *Natural Ecosystems* by W. B. Clapham, Jr. Copyright © 1973 by W. B. Clapham, Jr. Figure 1.19: Reprinted by permission from A. N. Strahler, *Physical Geography* 3rd ed. Copyright © 1969 by John Wiley & Sons, Inc. Figure 1.20: (a) Loren McIntyre/Woodfin Camp; (b) Almasy. Figure 1.21: (a) F.A.O.; (b) United Nations; (c) Klaus D. Francke /Peter Arnold; (d) Grant Heilman. Figure 1.22: Grant Heilman. Figure 1.23: (a) Federico Patellani/Camera Press-Transworld Feature Syndicate; (b) J. Crichton/Camera Press-Transworld Feature Syndicate; (c) Almasy; (d) Camera Press-Transworld Feature Syndicate. Figure 1.24: Grant Heilman. Figure 1.25: (a) Pro Pix/Monkmeyer; (b) Charlie Ott/National Audubon Society. Figure 1.26: Copyright Information Canada Photothèque.

CHAPTER TWO Figure 2.1: Constance Stuart. Figure 2.2: J. I. Bishop/Monkmeyer. Figure 2.3: Pickerelle/I.D.A.-United Nations. Figures 2.4 and 2.5: Naud/Afrique Photo. Figure 2.6: George Gerster/Rapho-Photo Researchers. Figure 2.7: Courtesy of the American Museum of Natural History. Figure 2.8: (a) Norman Myers/Bruce Coleman; (b) Argus Africa News Service/Photo Trends. Figure

2.9: Animals, Animals © 1974. Figure 2.10: Mark Boulton/National Audubon Society. Figures 2.11, 2.12, 2.13 and 2.14: Transafrica. Figure 2.15: George Gerster/Rapho-Photo Researchers.

CHAPTER THREE Figure 3.1: Joseph E. Spencer. Figure 3.2: F.A.O. Figure 3.3: Almasy. Figure 3.4: Wide World Photos. Figure 3.5: Marc Riboud/Magnum. Figure 3.6: Vautier/Asie Photo. Figure 3.7: United Nations. Figures 3.8 and 3.9: Eastfoto. Figures 3.10 and 3.11: Paolo Koch/Rapho-Photo Researchers. Figure 3.12: Sovfoto. Figure 3.13: Silberstein/Monkmeyer. Figure 3.14: Almasy. Figure 3.15: F.A.O. Figure 3.16: David Channer/Camera Press-Transworld Feature Syndicate. Figure 3.17: Lynn McLaren/Rapho-Photo Researchers. Figure 3.18: Paolo Koch/Rapho-Photo Researchers. Figure 3.19: F.A.O.

CHAPTER FOUR Figure 4.1: (a) Charles Bennett; (b) Walter Chandoha; (c) J. Allan Cash/Rapho-Photo Researchers. Figure 4.2: Almasy. Figure 4.3: Charles Bennett. Figure 4.4: Elizabeth H. Burpee/D.P.I. Figure 4.5: Editorial Photocolor Archives. Figure 4.6: (a) Courtesy Italian Government Travel Office; (b) Helena Kolda/Monkmeyer. Figure 4.7: Charles Bennett. Figure 4.8: (a) Alvin E. Staffan/National Audubon Society; (b) Hall Harrison/Grant Heilman; (c) Gordon S. Smith/National Audubon Society. Figure 4.9: Robert Perron. Figure 4.10: Courtesy Consulate General of Netherlands. Figure 4.11: Editorial Photocolor Archives. Figure 4.12: Charles Rotkin/Photography for Industry.

CHAPTER FIVE Figure 5.1: Department of Lands, Queensland, Australia. Figures 5.2, 5.3 and 5.4: G. R. Roberts. Figure 5.5: Ray Skobe/DeWys. Figure 5.6: Australian Information Service. Figure 5.7: R. Van Nostrand/National Audubon Society. Figure 5.8: Courtesy Professor Tom McKnight, Department of Geography, UCLA. Figure 5.9: John R. Brownlie/Bruce Coleman. Figures 5.10 and 5.11: Douglass Baglin. Figure 5.12: G. R. Roberts.

CHAPTER SIX Figure 6.1: Dr. E. R. Degginger. Figures 6.2 and 6.3: U.S. Forest Ser-

vice. Figure 6.4: Culver Pictures. Figure 6.5: Grant Heilman. Figure 6.6: (a) Grant Heilman; (b) U.S. Forest Service. Figures 6.7, 6.8 and 6.9: Grant Heilman. Figure 6.10: South Dakota Department of Game, Fish and Parks, Pierre, S.D. Figure 6.11: (a) Courtesy Royal Ontario Museum, Toronto, Canada; (b) Courtesy of the American Museum of Natural History. Figure 6.12: Richard Weymouth Brooks. Figure 6.13: (a) U.S.D.A.; (b) George Gerster/Rapho-Photo Researchers. Figure 6.14: Courtesy U.S. Department of the Interior, Bureau of Reclamation. Figure 6.15: Max & Kit Hunn/National Audubon Society. Figure 6.16: Grant Heilman. Figure 6.17: U.S. Forest Service. Figure 6.18: U.S.D.A., Soil Conservation Service. Figure 6.19: (a) Grant Heilman; (b) The Bettmann Archive; (c) *Mining Engineering*, March 1974; (d) Bucky Reeves/National Audubon Society. Figure 6.20: George Gerster/Rapho-Photo Researchers. Figure 6.21: *Mining Engineering*, July 1973. Figure 6.22: (Top left) Tim Kantor/Rapho-Photo Researchers; (top right) Georg Gerster/Rapho-Photo Researchers; (bottom) Bruce McAllister/EPA-Documerica. Figure 6.23: Courtesy Statewide Air Pollution Research Center, Riverside, California. Figure 6.24: Dennis Cowals/EPA-Documerica.

CHAPTER SEVEN Figure 7.1: Robert Perron. Figure 7.2: United Nations. Figure 7.3: American Geographical Society. Figure 7.4: Vautier/Decool. Figure 7.5: United Nations. Figure 7.6: Courtesy Professor James Parsons, Department of Geography, University of California, Berkeley. Figure 7.7: Rodolfo Grieco/Inter-Prensa. Figures 7.8 and 7.9: Herbert Lanks/Black Star. Figure 7.10: Charles Perry Weimer. Figures 7.11 and 7.12: Georg Gerster/Rapho-Photo Researchers. Figure 7.13: Nat Norman/Rapho-Photo Researchers. Figure 7.14: Dan Budnick/Woodfin Camp. Figure 7.15: Courtesy of Ministry of Information and Culture, Georgetown, Guyana. Figure 7.16: Peter Anderson, Time-Life Picture Agency. Figure 7.17: Vautier/Decool.

CHAPTER EIGHT Figure 8.1: Carl Frank. Figure 8.2: Fujihara/Monkmeyer. Figure 8.3: Hilda Bijur/Monkmeyer. Figure 8.4: Editorial Photocolor Archives Figure 8.5: Courtesy of The Whaling Museum, New Bedford, Massachusetts. Figure 8.6: Keith Gunnar/National Audubon Society. Figure 8.7: Courtesy of the New Zealand Embassy. Figure 8.8: G. R. Roberts. Figure 8.9: Courtesy of the New Zealand Embassy. Figure 8.10: G. R. Roberts. Figure 8.11: R. R. Lippincott/Camera Hawaii. Figure 8.12: Courtesy of the New Zealand Embassy. Figure 8.13: Werner Stoy/Camera Hawaii. Figure 8.14: G. R. Roberts. Figure 8.15: F.A.O. Figures 8.16 and 8.17: Patrick Morin/Monkmeyer.

CHAPTER NINE Figure 9.1: (a) C. G. Maxwell/National Audubon Society; (b) Daniel D. Sullivan; (c) Robert Perron: (d) Gordon S. Smith/National Audubon Society. Figure 9.2: (a) Karl W. Kenyon/National Audubon Society; (b) Robert Lackenbach/Black Star; (c) Courtesy of the American Museum of Natural History; (d) Patricia Caulfield/Animals, Animals © 1974. Figure 9.3: Fred Breummer. Figure 9.4: (a) Miami Seaquarium; (b) A. W. Ambler/National Audubon Society. Figure 9.5: From *Marine Mammals of the Northwest Coast of America* by Charles M. Scammon. Figure 9.6: Courtesy of The Whales Research Institute, Tokyo. Figure 9.7: Sekai Bunka Photo. Figure 9.9: Courtesy of The Whales Research Institute, Tokyo.

Index